Teaching Statistics in School Mathematics-Challenges for Teaching and Teacher Education

New ICMI Study Series

VOLUME 14

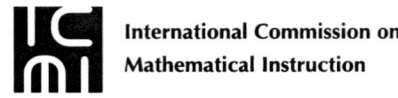

International Commission on
Mathematical Instruction

Published under the auspices of the International Commission on
Mathematical Instruction under the general editorship of

Bill Barton, President   Jaime Carvalho e Silva, Secretary-General

For further volumes:
http://www.springer.com/series/6351

Information on the ICMI Study program and on the resulting publications can be obtained
at the ICMI website http://www.mathunion.org/ICMI/ or by contacting the ICMI Secretary-
General, whose email address is available on that website.

Carmen Batanero • Gail Burrill • Chris Reading
Editors

Teaching Statistics in School Mathematics-Challenges for Teaching and Teacher Education

A Joint ICMI/IASE Study:
The 18th ICMI Study

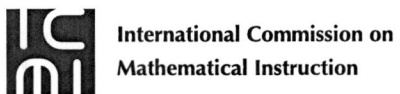

**International Commission on
Mathematical Instruction**

 Springer

Editors
Dr. Carmen Batanero
Departamento de Didáctica de la
Matemática
Universidad de Granada
Facultad de Ciencias de la Educación
Campus de Cartuja
18071 Granada, Spain
batanero@ugr.es

Gail Burrill
Division of Science and Mathematics
Education
Michigan State University
240 Erickson Hall
East Lansing, MI 48824
USA
burrill@msu.edu

Dr. Chris Reading
SiMERR National Centre
University of New England
Armidale, NSW 2351
Australia
creading@une.edu.au

ISSN 1387-6872
ISBN 978-94-007-1130-3   e-ISBN 978-94-007-1131-0
DOI 10.1007/978-94-007-1131-0
Springer Dordrecht Heidelberg London New York

Library of Congress Control Number: 2011934262

Cover design: eStudio Calamar S.L.

Printed on acid-free paper

Springer is part of Springer Science+Business Media (www.springer.com)

Contents

Preface

1 Introduction

The teaching of statistics in secondary school has a long tradition in countries like France, Spain and the United Kingdom. However, because statistics is becoming increasingly important in modern society, the relevance of developing statistical thinking in students across all levels of education has grown. Consequently, the new curricula published in the past years in many countries like Brazil, Costa Rica, South Africa, Spain, the United Arab Emirates and the United States of America include statistics from the first year of primary school level (6 year-old children).

Changes in what is expected in the teaching of statistics do not just concern the amount but also the quality of the content. Until recently, statistics in many school curricula was reduced to tasks in which learners were given small organised data sets and were asked to produce specific graphs, compute simple statistics (e.g., the mean or median) or answer simple direct questions. This formula-based approach to statistics resulted in students who were ill-prepared for tertiary level statistics and adults who were statistically illiterate.

The current recommendations, even for primary school levels, suggest a data-orientated approach to the teaching of statistics where students are expected to: design investigations; formulate research questions; collect data using observations, surveys, and experiments; describe and compare data sets; and propose and justify conclusions and predictions based on data. Learners are expected to deal with data in significant contexts and to take a critical stance on the analysis and interpretation of data and especially the abuse of data and statistics. The importance of developing statistical thinking and reasoning and not just statistical knowledge in students is being emphasised in many curricula.

Concurrent with these changes, the International Statistical Institute (ISI) started to pay more attention to teaching statistics in schools in the mid seventies, when the socio-economic conditions in developed countries, frequent use of quantitative information in newspapers and more widespread use of personal computers led to increasing demands on statistics education for the general citizen. The International Conferences on Teaching Statistics (ICOTS) were started in 1982 by the ISI and later continued by the International Association for Statistical Education (IASE) to bring together statistics teachers at all levels, and from all disciplines and countries, every four years.

Changing the teaching of statistics in schools will depend on the extent to which teachers can be convinced that statistics is one of the most useful themes for their students and the extent to which these teachers are adequately prepared to teach statistics at school level. Although interest in the education and professional development of mathematics teachers has increased in the past 20 years and there is now a body of research results on this issue, current literature seems to indicate that we are not in the same place in the specific case of statistics. This is shown by an analysis of research literature, for example papers published in the *Journal of Mathematics Teacher Education*, as well as in survey papers and handbooks quoted throughout this book that pay little attention to the teaching of statistics.

This book is a consequence of and presents the results from the Joint ICMI/IASE Study, *Teaching Statistics in School Mathematics-Challenges for Teaching and Teacher Education*, organised by the International Commission on Mathematical Instruction (ICMI; www.mathunion.org/ICMI/) in collaboration with the International Association for Statistical Education (IASE; www.stat.auckland. ac.nz/~iase/) and intended to address the lack of attention to teaching statistics by promoting research specifically focussed on the education and professional development of teachers to teach statistics.

2 Study Background

Since the mid-1980s, ICMI has found it important to involve itself directly in the identification and investigation of issues or topics of particular significance to the theory or practice of contemporary mathematics education and to invest effort in mounting specific ICMI studies on these themes.

At the same time, in the past three decades a statistics education research community has developed, linking people from various backgrounds (statisticians involved in teaching statistics in service courses at the university; statisticians working in statistical offices; mathematics educators; researchers in statistics education; educators; and psychologists), leading to the creation of the International Association for Statistical Education (IASE), in 1991. Conversations between ICMI and the IASE made clear there was a common interest in organising a Joint Study related to current problems in the teaching of statistics within school mathematics. This interest arose from the fact that, in spite of recommendations to increase the presence of statistics teaching at the school level, students in these levels do not acquire a statistical literacy adequate to function in an information-based society and to progress in the study of statistics at higher levels such as university or professional training.

The invitation from ICMI to collaborate on a Joint Study was accepted by the IASE. Subsequently, IASE suggested that this Joint Study merge with the 2008 IASE Round Table Conference, intended as part of a series of conferences started by the ISI and held by the IASE every four years. As a consequence of this agreement, the Joint ICMI/IASE Study Conference was held at the Instituto Tecnológico y de Estudios Superiores, Monterrey (ITESM), Monterrey Campus, Mexico in July 2008. This book is the final outcome from this Conference.

3   Joint ICMI/IASE Study Conference

Many people have been involved, first in the Joint ICMI/IASE Study conference and then in the production of this book. The work started with the appointment of an International Programme Committee in 2005 whose members worked collaboratively to prepare the Discussion Document. This document described the aims, topics and related research questions for the Joint Study, included a Call for Papers and was released in October 2006. The document was published in main mathematics and statistics education journals and also disseminated through statistics and mathematics education conferences and associations.

A specification of the Joint Study was its inter-disciplinary character, and therefore, the Programme Committee invited participation from mathematicians, mathematics educators, statisticians, (including official statisticians working at statistical agencies), and statistics educators, as well as psychologists and university lecturers of other disciplines where statistics is used as a tool. The Committee was specifically interested in inviting people with different levels of experience, including people who were well known in the area, new researchers who were just forming their views and teacher educators who were training the future mathematics teachers who would be delivering statistics at school levels.

Preliminary papers were received by October 2007 and reviewed by external referees over the next few months. Statistics and mathematics educators from all across the world contributed to the selection and improvement of the papers in the refereeing process. The papers selected by the International Programme Committee after the refereeing process were rewritten and received between March and April, 2008. The papers accepted covered a variety of topics and came from around the world, including both developed and developing countries.

The conference theme: *Teaching Statistics in School Mathematics: Challenges for Teaching and Teacher Education* had appeal for both mathematicians and statisticians, proving that the time was ripe for collaboration between the ICMI and the IASE. All together 109 participants representing 33 countries from all parts of the world participated in the conference. The Joint Study Conference was structured around six different topics, each organised by two members of the International Programme Committee. The six topics, briefly described below, served as an initial focus for potential papers and to organise the working groups in the conference.

1. *The current situation of teaching statistics in* schools, organised by Dani Ben-Zvi and Chris Reading. The interest in this group was a reflection of the status of statistics in the curricula of different countries; comparing the statistical content included in national curricula and tests and how the teaching of statistics at the school level specifically compared to teaching other topics in the school mathematics curriculum. The working group was also interested in analysing the differences between statistical literacy and reasoning and the teaching of statistics through project work.

2. *Teachers' attitudes, knowledge, conceptions and beliefs in relation to statistics education*, organised by Carmen Batanero and Gail Burrill. This group discussed teachers' beliefs and attitudes towards statistics and the effect of these beliefs and attitudes on the way statistics is taught. A second interest was in the analysis of the mathematical and pedagogical knowledge teachers need to teach statistics and on research instruments and strategies useful for determining the knowledge of statistics and of teaching statistics that teachers possess.

3. *Analysing current practices in teacher education regarding the teaching of statistics*, organised by Doreen Connor and Lionel Pereira-Mendoza. The aim of this group was to compare current training of teachers to teach statistics in different countries and to analyse the role of technology, current materials and teaching practice in developing teachers' competence to teach statistics.

4. *Empowering teachers to teach statistics: A look into the future*, organised by Joachim Engel and Maxine Pfannkuch. While Topic 3 analysed current practices in training teachers, in Topic 4 the focus was on innovative proposals or materials to change the current practice and improve the preparation of teachers.

5. *Training teachers in developing countries*, organised by Jun Li and Victor Polaki. A common concern of both the ICMI and the IASE has been related to the provision of research and teaching opportunities in statistics in developing or transitional countries. In the Joint Study Conference a working group was organised to reflect on the specific problems these countries have in training their teachers and in developing statistics education in their schools.

6. *Building collaboration between mathematics and statistics educators in teacher education,* organised by Joan Garfield and Maria Gabriella Ottaviani. Given the current interest from national statistical offices and statistics associations in developing statistical literacy for all citizens, this working group analysed examples from these institutions of collaborations in developing teaching materials or offering support to statistics teachers. Other examples of collaboration included collaborative projects between different university departments, university and schools or even between different countries.

The conference papers were distributed to participants before the conference and were published in the conference proceedings, edited by Carmen Batanero, Gail Burrill, Chris Reading and Allan Rossman, and published by ICMI and IASE. These proceedings are available from the IASE publication webpage at www.stat.auckland.ac.nz/~iase/publications. Each paper was assigned a reactor who read the paper before the conference and discussed the paper in special discussion sessions organised within the conference. Results from these discussions and conclusions from the working groups were presented in a final plenary session and also served as a basis to structure this book.

Other plenary sessions included an Opening lecture on the theme *Preparing teachers to meet the challenges of statistics education,* by Joao Pedro da Ponte, Portugal, and three panel sessions. The first, *Fundamental ideas in statistics and how they affect the training of teachers* was coordinated by Gail Burrill, United States of America. Since different curricula around the world include different statistical content at school level, the presenters offered their reflections and analyses about what basic statistical ideas and types of reasoning would be needed to educate statistically literate citizens. Speakers were Martha Aliaga, United States of America; Rolf Biehler, Germany; and Ernesto Sánchez, Mexico.

The second panel session, *The interplay of probability and statistics in teaching and in training the teachers* was organised by Maria Gabriella Ottaviani, Italy. Although the main focus of the conference was statistics, this panel reflected on possible relationships between statistics and probability in the curriculum and how the different views of probability (classical, frequentist, subjective) affect the teaching of statistics. Speakers were Manfred Borovcnik, Austria; Jean Claude Girard, France; and Delia North, South Africa.

Technology today is changing not only the way we work in mathematics and statistics but also the way we teach these topics. Dave Pratt, United Kingdom organised a discussion around *Technology in the teaching of statistics: Potentials and challenges in preparing the teachers.* Speakers were Dani Ben-Zvi, Israel; Doreen Connor, United Kingdom; and Anthony Harradine, Australia.

The conference was held at the Monterrey Campus of the Instituto Tecnológico y de Estudios Superiores de Monterrey (ITESM), (www.mty.itesm.mx/), a well-established Mexican educational institution that was founded in 1943 with campuses distributed throughout the country and other Latin American countries. This institution and in particular the Mathematics and Statistics Department supported the conference, offering its facilities and organising different social activities that provided opportunities for participants to interact informally. Other institutions supporting the conference were the American Statistical Association, the Mexican Statistical Association and the Centro de Investigaciones y Estudios Avanzados (CINVESTAV). The Programme Committee is indebted to these institutions as well as to the local organising committee: Blanca Ruiz (Chair), Armando Albert and Tomás Sánchez, all lecturers of ITESM, and Ernesto Sánchez, CINVESTAV, México.

The Joint Study Conference was held the week before ICME-11 (the International Congress on Mathematical Education). The following weekend an *Encuentro Latinoamericano de Educación Estadística* (ELEE, Latin American Statistics Education Meeting) was organised. The aim of this meeting was to gather together Latin-American statistics educators and teachers taking part in either the Joint ICMI/IASE Study Conference or ICME 11 with the purpose of exchanging experiences, expanding their statistics education knowledge, widening their network of contacts and establishing projects for future collaboration. The ELEE meeting was attended by about 80 participants and included panel discussions, presentations, workshops, posters and attendance at the closing sessions of the Joint ICMI/IASE conference.

4 Structure of This Book

To produce a monograph that covers the state of art for the Joint ICMI/IASE Study, the editors fixed a tentative content and a tentative structure for the book in the Call for Papers. This structure took into account the papers presented and the discussions held at the conference and tried to assure coherence and completeness in the monograph. The Call for Papers was distributed to participants in the conference, who were encouraged to form teams to prepare a common chapter in the book when the papers presented in the conference dealt with complementary themes.

The book is organised into four main parts, each consisting of several chapters. Part I: *Global Perspective* derives from the conference work in Topics 1 and 5 and some papers presented in Topic 3. This part offers examples of how statistics is conceived in the mathematics school curriculum around the world, including developed and transitional countries, and how mathematics teachers who are responsible for the teaching of statistics are currently trained. It consists of short chapters organised around two themes. The first one, the statistics school curricula around the world, deals with curricular issues in Brazil, United States of America, Uganda and South Africa. The second theme (or section) discusses the particular experiences of training teachers in Germany Honduras-Costa Rica, Iran, the United States of America, and the Philippines.

Parts II, III and IV consist of chapters each of which considers a different theme. The chapters take into account previous research presented in the conference and discuss the topic in a general way. Specific research or experiences are included in some cases, but particular examples are not the central focus of these chapters. Instead they complement the theme of the part and serve to enlighten general discussion on that theme. Chapters in Part II, *Fundamentals for Teaching Statistics,* include discussion of the following topics of importance in the teaching of statistics, three of which (the fundamental statistical ideas, the role of probability in the statistics curriculum, and the challenges set by technology) were debated at the conference plenary panels. Other topics that arose in the working group discussions as relevant for the teaching of statistics (modelling in probability and statistics, differences/complementarities between statistics and mathematics, the role of assessment in teaching and learning, and teaching statistics through investigative projects) are also discussed in this part.

Research on teachers' knowledge and professional development in statistics, that is, chapters in Part III, *Teachers' Beliefs, Attitudes and Knowledge,* result from the conference work in Topic 2. This part also includes a collective effort to present a state-of-the-art summary of the research on this topic and implications for training teachers to teach statistics, as well as suggestions about how to advance research in this area. After reflection on the components of teachers' attitudes, beliefs and classroom practices and how they are interrelated and affect teaching and learning of statistics, the part contains a series of chapters each focussing on teachers' knowledge or learning about a different statistical topic. Three chapters are focussed on models for teachers' statistical knowledge and how to measure this type of knowledge.

Chapters in Part IV, *Challenges and Experiences in Teacher Training,* derive from the conference working groups 3 and 4 and analyse questions and activities of relevance in training the teachers. This part starts with an expansion of the opening lecture (Preparing teachers to meet the challenges of statistics education) and then discusses challenges and possibilities that real data, case analysis, statistical investigations, technology and distance training offer to educate the teachers. Ways to develop students' and teachers' statistical thinking and literacy are also discussed. This part finishes with an overview of relevant examples of collaboration from statistical offices and associations to improve the preparation of mathematics teachers to teach statistics in different countries. In addition, the book includes an overview and introduction to the different parts, written by the editors.

5   Final Notes

The book is directed to both mathematics and statistics educators, including in-service teachers, students preparing to be teachers, teacher educators, people involved in curricular development in statistics as well as researchers in statistics and mathematics education and can be of interest to any in that audience. A primary goal of the book is to help teacher educators and educational authorities to clearly perceive the current need for all the students to be statistically literate and able to reason statistically, the differences and complementarities between statistical and mathematical thinking and literacy, and consequently the relevance of adequately preparing mathematics teachers to teach statistics.

Parts III and IV contain very useful information about the knowledge required by teachers, their current difficulties related to teaching statistics and possible strategies for educating the teachers. These parts can be of interest to teachers themselves, as an important part of research summarised in these chapters both in learning difficulties or teaching strategies is applicable to students. In the same way, the basic ideas for teaching statistics described in Part II are common in both the training of students and teachers.

This book is designed to be useful to researchers in mathematics education and statistics education with the hope that it will foster further research in the problems related to educating teachers to teach statistics at different school levels, from primary to secondary school. Finally, we hope this book will prove helpful towards improving the teaching of statistics at school level and increasing the statistical literacy and the statistical thinking of both teachers and students.

While it was a large task, the editors found that the experience of editing this book and working with such a varied group of international authors has been a privilege for us. We recognise that we could not have completed this book without the collaboration and cooperation of many people. Consequently, we are most grateful for the dedication, expertise, and professionalism of authors and referees, for the advice and feedback from ICMI and IASE officers, and particularly we are most grateful for the work of the International Programme Committee, both in the planning of the Joint ICMI/IASE Study Conference and in the initial stages the of production of the book.

Carmen Batanero   Universidad de Granada, Spain
Gail Burrill   Michigan State University, USA
Chris Reading   University of New England, Australia

International Programme Committee

- Carmen Batanero, Chair, Didáctica de la Matemática, Universidad de Granada, Spain;
- Bernard Hodgson, Ex-officio, representing ICMI, Département de Mathématiques et de Statistique, Université Laval, Québec, Canada;
- Allan Rossman, Ex-officio, representing the International Association of Statistical Education, Department of Statistics, California Polytechnic State University, United States of America;
- Armando Albert, Mathematics Department, Instituto Tecnológico de Estudios Superiores de Monterrey (Instituto Tecnológico y de Estudios Superiores de Monterrey), Mexico;
- Dani Ben-Zvi, Faculty of Education, University of Haifa, Israel;
- Gail Burrill, Division of Science and Mathematics Education, Michigan State University, United States of America;
- Doreen Connor, Royal Statistical Society Centre for Statistical Education, Nottingham Trent University, United Kingdom;
- Joachim Engel, Department of Mathematics and Computer Science, University of Education, Ludwigsburg, Germany;
- Joan Garfield, Educational Psychology, University of Minnesota, United States of America;
- Jun Li, Department of Mathematics, East China Normal University, Shanghai, China;
- Maria Gabriella Ottaviani, Dipartimento di Statistica, Probabilita' e Statistiche Applicate, University of Rome "La Sapienza", Italy;
- Lionel Pereira Mendoza, Mathematics Education, National Institute of Education, Singapore;
- Maxine Pfannkuch, Department of Statistics, The University of Auckland, New Zealand;
- Mokaeane Victor Polaki, Department of Science Education, Faculty of Education, National University of Lesotho, Lesotho;
- Chris Reading, The National Centre of Science, Information and Communication Technology, and Education for Rural and Regional Australia, Faculty of Education, Health and Professional Studies, University of New England, Australia.

Part I
Global Perspective

Gail Burrill

Chapters presented in this part are organised into two sections, each composed of short chapters that present examples of how the teaching of statistics is conceived in different curriculum around the world (Chaps. 1–4) and how teachers are trained to teach statistics at school level in different countries (Chaps. 5–9).

Chapters 1–4 derived from presentations in Topic 1 of the Joint Study Conference: *The current situation of teaching statistics in schools.* Presentations and discussions in this topic showed that school curricula in general are detailed, communicated and enacted in various ways among the countries of the world. Some countries have a well-defined national curriculum followed by the vast majority of school systems. Others have a curriculum on paper, but implementation is not universal. A few have no nationally mandated curriculum. The curricula for statistics share this diversity, although nearly universally, statistics is incorporated into the mathematics curriculum. Chapters 1–4 give a brief window into this diversity, with discussions of the statistics programmes in Brazil, the United States, Uganda, and South Africa. Brazil utilises statistics as a way to focus on social and political facets of society, South Africa on preparing students to be consumers of data, while in Uganda, statistics seems to be envisioned as a mathematical body of knowledge to be learned. In South Africa, national assessments drive the inclusion of statistics in the implemented curriculum, but this is not true of all countries. The use of technology in the study of statistics also differs; Brazil recommends the use of technology to minimise the tedious nature of data processing and maximise data analysis and to simulate random experiments that can help students develop an intuitive meaning of probability while in Uganda technology is not allowed at the elementary level and typically not available at other levels.

Campos, Cazorla, and Kataoka describe the statistics curriculum and methodological guidelines for implementing the objectives prescribed by the Ministry of Education

G. Burrill (✉)
Michigan State University, 240 Erickson, East Lansing, MI 48824, USA
e-mail: burrill@msu.edu

in Brazil. The methodological guidelines position statistics as a tool for understanding the social context in which students live and suggest that statistics be considered as essential for the formation of a critical attitude on current social, political, cultural and scientific issues in the study of interdisciplinary or cross-cutting themes.

In contrast, at the time of the study, the United States had no national curriculum, with each state responsible for its own mathematics and statistics standards. Newton, Dietiker, and Horvath report on an analysis of the standards from 41 states about the role of statistical reasoning and the statistical process. The findings suggested that procedures are overemphasised in the curricular expectations as defined by the standards (particularly in the lower grades) with little expectation that the curriculum encourage statistical reasoning.

Uganda has a compulsory curriculum, but according to Opolot-Okurut and Opyene-Elu the statistical content in the curriculum is mostly formula-based, and only 10% of the curriculum at the elementary level is on statistics. The focus is primarily on simple exploratory data skills; other content is optional, which, in practice, means that teachers concentrate on mathematics not on statistics. And often texts are not available, which means teachers have few resources to use in teaching statistics.

Wessels describes the revised approach to statistics in South Africa. The goals align with the goals of preparing learners for the social and economic needs they will face as adults in the twenty-first century and as consumers of interpretations of data.

Chapters 5–9 discuss the training of teachers to teach statistics and is a consequence of specific examples of courses for teacher preparation and professional development presented in the Conference Topic 3, *Analysing current practices in teacher education regarding the teaching of statistics*, while more general topics related to the training of teachers are included in Part IV. As countries increasingly are recognising the need to shift their curricula to include statistics and probability, those responsible for teaching this content are primarily teachers trained to teach mathematics. Providing appropriate preparation and professional development for teachers to teach statistics is done through programmes offered by government institutions, professional statistics associations, academic and teacher education institutions, private organisations, and in some countries, through collaborative efforts among these entities. The papers in Chaps. 5–9 include a discussion of a university-based programme in the United States, university outreach initiatives in Germany, collaborative efforts among institutions in Iran and in the Philippines, and a comparison of the training to teach statistics in Costa Rica and Panama.

Froelich describes a new curriculum in statistical content that requires the cooperation of mathematics, mathematics education, and statistics faculty for future secondary mathematics teachers at a major state university in the United State. The curriculum, however, does not focus on how to teach statistics.

According to Martignon, curricula across all states in Germany now include mandatory competencies in data analysis and statistical reasoning from elementary school to grade 12 with the focus in most states on statistical literacy. To prepare teachers to carry out these new mandates, some universities have introduced regular seminars for future and experienced teachers on educational problems in stochastics, and some states include statistical questions on data analysis and visualisation in the central final examinations for future teachers.

The Iranian Ministry of Education designed a new course in statistics for all students in the second or third year of high school that emphasised statistical reasoning and the use of technology as a tool for analysis according to Persian and Rejali. Several professional organisations offered programmes to prepare teachers to teach this course, with much of the work initiated by the Iranian Statistical Society. Along with the Isfahan Mathematics House and the Mathematics Teachers' Society of Isfahan, they started an annual team-based statistics competition among high school students and with the help of the Iranian Statistics Research and Training Centre (ISRTC) developed an electronic site in Farsi for the popularisation of statistics.

Reston and Bersales describe examples in the Philippines of how individuals, universities, government and private organisations work together to achieve reforms. To better prepare the teachers to implement a revised school curriculum that includes some statistics and probability, the Department of Education organised a programme for elementary mathematics teachers delivered by five teacher education institutions. Two government organisations, the Philippine Statistical System and the Commission on Higher Education, collaborated with the Philippine Statistical Association to organise several reform efforts including the development and implementation of a nation-wide course in probability and statistics for teachers and the preparation of texts and reference material for teachers.

Sorto contrasts the preparation of teachers to teach statistics in two South American countries, Panama and Costa Rica reporting on opportunities to study statistics during teacher preparation programmes and in structured professional development activities in each country.

In summary, papers included in Chaps. 1–4 and 5–9 represent the variation in statistics curricula and teacher training found around the world. The examples can provide a base for comparison with situations in other countries and highlight the need to recognise the relevance of improving the statistics education of students and teachers in every country.

Chapter 1
Statistics School Curricula in Brazil

Tânia M.M. Campos, Irene M. Cazorla, and Verônica Y. Kataoka

Abstract In Brazilian basic education, National Curricular Parameters recommend the inclusion of probability and statistics as part of mathematics. Despite the innovative character and methodological guidelines focused on the formation of a scientific spirit and on civic preparation, teaching probability and statistics faces difficulties because of lack of training for teachers, didactic materials, and availability of software, among others. Therefore, statistics educators in Brazil have hard but promising work ahead.

1 Introduction

Basic education in Brazil consists of elementary and secondary education, totalling 12 years. Elementary education is comprised of two phases: the first consists of cycles 1 (6–8 year-olds) and 2 (9–10 year-olds), and the second consists of cycles 3 (11–12 year-olds) and 4 (13–14 year-olds), totalling nine grades. Secondary education consists of three grades (15–17 year-olds).

To establish content themes and their development, according to the specificities of each school level, the Ministry of Education (Ministério da Educação) prepared a document in 1997 known as the National Curricular Parameters (NCP). These parameters were developed first for application in cycles 1–2 of elementary education (Ministério da Educação, 1997), then for cycles 3–4 of elementary education (Ministério da Educação, 1998), and finally for secondary education

T.M.M. Campos (✉) and V.Y. Kataoka
Universidade Bandeirante de São Paulo, Av. Bras Leme, 3029, São Paulo,
SP, 02022-011, Brazil
e-mail: taniammcampos@hotmail.com; veronicayumi@terra.com.br

I.M. Cazorla
Universidade Estadual de Santa Cruz, Campus Soane Nazaré de Andrade, km 16, Ilhéus,
BA, 45662-900, Brazil
e-mail: icazorla@uol.com.br

C. Batanero, G. Burrill, and C. Reading (eds.), *Teaching Statistics in School Mathematics-Challenges for Teaching and Teacher Education: A Joint ICMI/IASE Study*, DOI 10.1007/978-94-007-1131-0_1, © Springer Science+Business Media B.V. 2011

(Ministério da Educação, 2002, 2006). The NCP were designed with the aim of "establishing quality goals to assist students to face the world today as participatory, reflective and independent citizens, knowledgeable of their rights and duties" (Ministério da Educação, 1997, p. 4).

This chapter presents an analysis of the mathematics in the NCP related to probability and statistics, using a methodology for content analysis (Bardin, 2006).

2   Objectives and Contents

Probability and statistics are inserted in the NCP recommendations in mathematics. In elementary education, they are part of one of the four content blocks, the "Information Handling" block. In secondary education, they are part of one of the three blocks, the "Data Analysis" block. This status shows the recognition of the importance of developing statistical reasoning in the intellectual and civic formation of students.

- The objectives of cycle 1 for elementary education are: (a) to develop procedures to collect, organise, communicate, and interpret data through tables, charts, and representations that are frequently used in daily lives; and (b) to understand that most events of everyday life are random in nature by exploring concepts of chance and uncertainty that arise intuitively in situations where the student performs experiments and observes events (Ministério da Educação, 1997).
- The objectives of cycle 2 for elementary education are: (a) to appreciate the use of statistical language as a means of communication and to facilitate ways to solve and communicate strategies and results; and (b) in the field of probability, to identify characteristics of predictable or random events from problem situations (Ministério da Educação, 1997).
- The objectives of cycle 3 for elementary education are: (a) to encourage the formulation of hypotheses from systematic observations of quantitative and qualitative aspects of reality, establishing interrelationships between those aspects (variables) by making use of mathematical knowledge, and to select, organise, and produce relevant information in order to interpret and assess those relations critically; and (b) to promote understanding of patterns and trends in data by drawing inferences from the frequencies and measures of central tendency of a population sample (Ministério da Educação, 1998).
- The objectives of cycle 4 for elementary education are: (a) to build the sample space of equally likely events by using the multiplicative principle or simulations to estimate the probability of the success of an event; and (b) to go beyond the reading of information and think more critically about the meaning in the information. Thus, the proposed topic should go beyond mere description and representation of data to include investigation of the data and decision-making based on that investigation (Ministério da Educação, 1998).
- The objectives of secondary education are: (a) to enable students to master the language of probability; (b) to raise some equiprobability hypotheses; and (c) to

associate statistics with observed results and frequencies of corresponding events and make use of such statistical frequencies to estimate the probability of a given event (Ministério da Educação, 2002, 2006).

The content suggested for the proposed objectives can be grouped under four categories, ranging in intensity and complexity as students progress through the levels of education.

- Collecting, organising, and representing data: simple and two-way tables, absolute and relative frequency; construction of bar charts, pie charts, line graphs, histograms, and frequency polygons.
- Interpreting data: for cycles 1 and 2 of elementary education, data interpretation involves essentially the reading of tables and graphs. In cycles 3 and 4 and in secondary education, in addition to reading data, students are expected to produce and interpret a number of statistical measures, including measures of central tendency (mean, median, and mode) and measures of variability (mean deviation, variance, and standard deviation).
- Drawing and assessing inferences: for cycle 4 elementary and secondary education only, drawing inferences from data analysis; using measures of central tendency and frequencies to estimate trends and probabilities.
- Understanding and applying probability and chance: notions of chance and uncertainty; probability of a single event; for cycle 4 elementary and secondary education only, building a tree diagram and using combinatorial analysis to calculate probability and conditional probability.

3 Methodological Guidelines

The formulation of hypotheses does not seem to be explicit in the objectives or content. However, the role of hypotheses becomes more evident in the methodological guidelines, where statistics is considered to be an essential tool for the formation of a critical attitude about current social, political, cultural, and scientific issues in the study of interdisciplinary or cross-cutting themes.

The guidelines also recognise the role of statistics in understanding the social context in which students live and therefore as a tool for their civic education. In relation to information reported in the media, the NCP emphasise statistics as a language to describe reality, recognising the relativity of statistical measures and how they can be handled in accordance with specific interests.

Another suggestion in the guidelines is that systematic observation of phenomena in several fields of knowledge may help students develop an investigative spirit where statistics is seen as part of the scientific method. Students should use simulation to study the regularities of phenomena, with empirical evidence needed to test hypotheses and inferences, even informally.

Finally, the NCP recommend the use of a calculator and a computer, especially spreadsheets, to minimise the tedious nature of data processing and maximise data analysis, as well as to simulate random experiments that can help students develop an intuitive meaning of probability, observing, for example, the relative frequency of an event over a long run of repetitions.

4 Final Considerations

In conclusion, probability and statistics education in Brazil prioritises the analysis and interpretation of data where it is seen as a language to describe reality and does not emphasise the formalism of concepts and formulas. However, although the guidelines mention terms such as population and sample and use mean and frequencies as estimates of population values and probabilities, the NCP have no discussion about sampling and the variability of sample means and make no references to quantiles (except of the median) or to box plot.

The analysis conducted in this chapter may help in discussions of the NCP guidelines during the process of teacher training, assist researchers interested in the process of teaching and learning statistics, or be useful in making comparisons with the curricula from other countries.

Finally, we should mention that implementation of the NCP guidelines in schools still faces major challenges, including: (a) initial and continued preparation of teachers; (b) didactic books, which have conceptual mistakes and present the content in a fragmented way; (c) the scarcity of didactic materials; (d) research results that are not yet available to schools; and (e) the lack of free software as well as other obstacles. Taken all together, in Brazil statistics educators still have hard but promising work ahead.

References

Bardin, L. (2006). *Análise de conteúdo (Content analysis)* (3rd ed.). Lisbon: Edições 70.
Ministério da Educação. (1997). *Parâmetros curriculares nacionais: Matemática (National curricular parameters: Mathematics)*. Brasilia, Brazil: Author.
Ministério da Educação. (1998). *Parâmetros curriculares nacionais: Matemática (National curricular parameters: Mathematics)*. Brasilia, Brazil: Author.
Ministério da Educação. (2002). *PCN ensino médio: Orientações educacionais complementares aos parâmetros curriculares nacionais: Ciências da natureza, matemática e suas tecnologias (PCN secondary education: Complementary educational guidelines to the national curricular parameters: Nature sciences, mathematics, and their technologies)*. Brasilia, Brazil: Author.
Ministério da Educação. (2006). *Orientações curriculares nacionais para o ensino médio: Ciências da natureza, matemática e suas tecnologias (National curricular guidelines for secondary school: Nature sciences, mathematics, and their technologies)*. Brasilia, Brazil: Author.

Chapter 2
Statistics Education in the United States: Statistical Reasoning and the Statistical Process

Jill Newton, Leslie Dietiker, and Aladar Horvath

Abstract Two important components of statistical literacy are statistical reasoning and the statistical process. This chapter summarises a study that analysed 41 mathematics state standards documents as they existed in 2006 to surmise the extent to which learning these components is expected of students in the United States. Most prominent among the findings were the overrepresentation of isolated statistical procedures and the corresponding scarcity of expectations addressing statistical reasoning and the statistical process.

1 Introduction

Statistical literacy has been conceptualised in many ways (e.g., Utts, 2003; Ben-Zvi & Garfield, 2004; Franklin et al., 2005); however, often highlighted as important for statistical literacy are: (a) statistical reasoning, and (b) the statistical process. Utts (2003) states that "there is less need to emphasise calculations, and more need to focus on understanding how statistical studies are conducted and interpreted" (p. 74). Similarly, Burrill and Camden (2005) propose that "students seem to be mastering statistical procedures and vocabulary but are not able to use statistical reasoning in a meaningful way" and that "an over-emphasis in school syllabi on answering questions rather than posing them, and making decisions based only on data displays produces an approach based on absoluteness of data that stifles the development of statistical thinking" (p. 4).

J. Newton (✉)
Purdue University, 100 N. University St., BRNG 4136, West Lafayette, IN 47907, USA
e-mail: janewton@purdue.edu

L. Dietiker and A. Horvath
A718 Wells Hall, Michigan State University, East Lansing, MI 48824, USA
e-mail: dietike4@msu.edu; horvat54@msu.edu

C. Batanero, G. Burrill, and C. Reading (eds.), *Teaching Statistics in School Mathematics-Challenges for Teaching and Teacher Education: A Joint ICMI/IASE Study*,
DOI 10.1007/978-94-007-1131-0_2, © Springer Science+Business Media B.V. 2011

The goal in the analysis described in this chapter was to report on expectations that students in the United States will use statistical reasoning and carry out the statistical process. The investigation was complicated by the fact that, unlike the majority of countries in the world, the United States does not have a national mathematics or statistics curriculum. Rather, each state has its own mathematics and statistics standards.

The analysis, then, became an examination of a set of state standards. The analysis was framed using the *Guidelines for Assessment and Instruction in Statistics Education (GAISE) Report* published in 2005 by the American Statistical Association (ASA). The *GAISE Report* proposes four process components of the statistical investigative process: (1) formulate questions, (2) collect data, (3) analyse data, and (4) interpret results (Franklin et al., 2005). These components are consistent with the data analysis standards proposed by the National Council of Teachers of Mathematics (NCTM, 2000): (1) formulate questions that can be addressed with data and collect, organise, and display relevant data to answer them; (2) select and use appropriate statistical methods to analyse data; and (3) develop and evaluate inferences and predictions that are based on data. As part of a larger study, this chapter summarises the analysis of the state standards using the four process components in the *GAISE Report* to address the following questions: (1) To what extent do the K-8 US state mathematics standards promote statistical reasoning? (2) To what extent do the K-8 US state mathematics standards expect students to carry out the statistical process?

2   Method

All of the statistics grade level expectations (GLEs) from 41 state standards documents were collected, and each GLE was coded into the appropriate process component (Franklin et al., 2005):

1. Formulate questions: (a) clarify the problem at hand; (b) formulate one (or more) questions that can be answered with data;
2. Collect data: (a) design a plan to collect appropriate data; (b) employ the plan to collect the data;
3. Analyse data: (a) select appropriate graphical and numerical methods; (b) use these methods to analyse the data; and
4. Interpret results: (a) interpret the analysis; (b) relate the interpretation to the original question.

Many GLEs were coded as applicable for more than one process component. For example, third graders in South Dakota are expected to "gather data and use information to complete a scaled and labelled graph". This GLE was coded as both Collect Data and Analyse Data. Expectations for statistical reasoning within each process component and expectations related to conducting the statistical process were noted.

3   Results and Discussion

General findings. In the 41 state standards documents, as they existed in 2006, 1,711 GLEs address at least one of the four process components (approximately 42 statistics GLEs per state). Across all states, the number of GLEs increases steadily from Kindergarten (98 GLEs) until grade 7 (244 GLEs), and then decreases slightly in grade 8. Table 2.1 summarises the number of GLEs coded into each process component. Results show that students are much more often expected to analyse data and interpret results than to formulate questions and collect data.

Statistical reasoning. When students were expected to go beyond statistical procedures to evaluate or reflect on these procedures, the GLE was coded as statistical reasoning. For example, sixth graders in Florida are expected to "find the range, mean, median, and mode of a set of data". GLEs of this type do not seem to require statistical reasoning. In contrast, eighth graders in Michigan are expected to "recognise practices of collecting and displaying data that may bias the presentation or analysis". Some GLEs expected both doing procedures and reasoning statistically. For example, seventh graders in Washington are expected to "formulate a question and collect data from a population, describing how the questions, collection method, and sample population affect the results". Forty of the 41 states analysed include at least one GLE that promotes statistical reasoning. Table 2.2 summarises the relative frequencies of GLEs that promote statistical reasoning across the process components.

Only 28% of the GLEs across the 41 states promote statistical reasoning. In addition, expectations for statistical reasoning were much more prevalent in GLEs addressing data collection and analysis than in question formulation and interpretation of results. The frequency of GLEs that promote statistical reasoning increases from four GLEs in Kindergarten to 113 GLEs in Grade 8, indicating that young students are expected to do little statistical reasoning.

Table 2.1   Number of grade level expectations (GLEs) by process component

	Formulate questions	Collect data	Analyse data	Interpret results	Overall
Number of GLEs	12	423	968	867	1,711

Table 2.2   Frequency of statistical reasoning grade level expectations (GLEs)

GLE	Formulate questions	Collect data	Analyse data	Interpret results	Overall
Promotes statistical reasoning	13	119	325	66	475
Includes process component	112	423	968	867	1,711
Percent of process component GLEs requiring statistical reasoning	12%	28%	34%	8%	28%

Statistical process. Only 41 of the 1,711 GLEs (approximately 2%) include the expectation that students plan and carry out the statistical process. For example, third graders in the Department of Defense schools are expected to "develop and implement a plan to collect and organise data to address a given question". However, some GLEs include several process components that may indicate the state's expectation that students move beyond isolated process components. For example, third graders in Oklahoma are expected to "pose questions, collect, record, and interpret data to help answer questions". This GLE expects students to carry out the statistical process from beginning to end (i.e., includes all four process components). Less than 30% of the GLEs include more than one process component in a single GLE, and only 7% of the total include three or four process components. However, nearly half of the states either address study design explicitly or combine all four process components into one GLE suggesting that students should carry out the entire statistical process. In addition, 15 GLEs address the iterative nature of the statistical process. For example, sixth graders in Tennessee are expected to "make conjectures to formulate new questions for future studies".

4   Conclusion

The study set out to determine whether K-8 state standards in the United States of America promote statistical reasoning and expect students to conduct the statistical process. In both cases, the answer is that most states do but to a very limited extent. The procedures associated with the statistical process are undoubtedly an important part of statistical knowledge; however, this analysis indicates that there is an overemphasis on these procedures (particularly in the lower grades) and a lack of expectation that the curriculum should go beyond these procedures to encourage statistical reasoning.

Several important implications for teacher education programmes emerge from this analysis. First, a holistic approach to the statistical process is needed in order for teachers to understand the importance of spending time assisting students with question formulation and data collection (this analysis found these process components to be underrepresented in the state standards). Second, teachers will need to be prepared to facilitate discussions with students around the expectations that promote statistical reasoning. That is, in many states (to varying degrees), statistics education has moved beyond calculating means and constructing graphs, and it is important that teachers know how to implement these new expectations. Finally, it seems important that teachers begin to see state expectations as a minimum requirement. That is, teachers working in states that expect students only to "do" the process components and that lack attention to statistical reasoning and/ or the statistical process should be encouraged to enhance their instruction to include these critical components of statistical literacy.

References

Ben-Zvi, D., & Garfield, J. B. (Eds.). (2004). *The challenge of developing statistical literacy, reasoning and thinking*. Dordrecht, The Netherlands: Kluwer.

Burrill, G., & Camden, M. (Eds.). (2005). *Curricular Development in Statistics Education: International Association for Statistical Education 2004 Roundtable*. Voorburg, The Netherlands: International Statistical Institute. Online: www.stat.auckland.ac.nz/~iase/publications

Franklin, C., Kader, G., Mewborn, D., Moreno, J., Peck, R., Perry, M., & Scheaffer, R. (2005). *Guidelines for assessment and instruction in statistics education (GAISE) report: A pre-K-12 curriculum framework*. Alexandria, VA: American Statistical Association. Online: www.amstat.org/Education/gaise/

National Council of Teachers of Mathematics. (2000). *Principles and standards for school mathematics*. Reston, VA: Author.

Utts, J. (2003). What educated citizens should know about statistics and probability. *American Statistician, 57*(2), 74–79.

Chapter 3
Statistics School Curricula for Uganda

Charles Opolot-Okurut and Patrick Opyene Eluk

Abstract This chapter describes the statistics curricula for schools in Uganda, outlining the intended curricula for the primary and secondary schools and explaining the statistics content for each year of study in the primary schools. The secondary school curriculum has ordinary and advanced level components. In the two courses offered at the ordinary level and at the advanced level, statistics is incorporated to different widths and breadths. The instructional materials and resources that are available and used for the teaching of statistics, especially the textbooks and technology, are described as well as the assessment format and practice for statistics within mathematics at the different levels.

1 Introduction

"Planning without accurate figures is a common feature of the economy of most developing countries" (Oyelese, 1982, p. 189) like Uganda, which the use of correct statistical data could remedy. Statistical data partly derive from statistical activities in schools. This chapter describes the statistics school curricula for Uganda including the contextual background of the education system, the statistics content at the primary and secondary school levels, statistics teaching resources, and the assessment format used.

C. Opolot-Okurut (✉)
Makerere University, P.O. Box 16675, Kampala, Uganda
e-mail: copolotokurut@yahoo.co.uk

P.O. Eluk
Islamic University in Uganda, P.O Box 2555, Mbale, Uganda
e-mail: patrickelukopyene@yahoo.com

C. Batanero, G. Burrill, and C. Reading (eds.), *Teaching Statistics in School Mathematics-Challenges for Teaching and Teacher Education: A Joint ICMI/IASE Study*, DOI 10.1007/978-94-007-1131-0_3, © Springer Science+Business Media B.V. 2011

1.1   Contextual Background of the Education System in Uganda

Uganda's education system follows a national curriculum for all subjects and school levels. The education system has pre-primary school, primary school, secondary school, and tertiary levels. The primary level is of 7 years duration, while the secondary level has two sublevels: an ordinary level lasting 4 years and an advanced level lasting 2 years. Mathematics and statistics are a combined course that is compulsory at both primary and ordinary levels taught by the same teacher. Teachers of statistics are initially trained to teach mathematics in preservice teacher programmes (Opolot-Okurut, Opyene Eluk, & Mwanamoiza, 2008).

2   Statistics Curriculum at Primary School Level

At the primary level, statistics topics include graphs and interpretation of information (Uganda National Examinations Board [UNEB], 1991; National Curriculum Development Centre [NCDC], 1999) intended to introduce pupils to basic statistical concepts. The statistics content covered in lower primary classes (primary one to three) includes graphical representation of simple data. In the upper primary classes (primary four to seven) the content includes scales on horizontal and vertical axes, drawing and interpreting bar graphs; tabular and graphical representation of information; simple statistical averages; and probability of simple events (NCDC, 2000). Unfortunately, teachers concentrate on teaching mathematics as only 10% of the topics in the curriculum are on statistics.

3   Statistics Curriculum at Secondary School Level

3.1   Statistics Curriculum at Ordinary Level

Courses 456-mathematics and 475-additional mathematics are offered at this level to extend the statistics introduced in the primary school. The work covered involves using existing data from textbooks:

Four-five-six mathematics: In this course, statistics topics fall under "Miscellaneous Applications" (UNEB, 2005). The content includes graphical and tabular organisation of data, summarising and interpreting data, and basic concepts of probability. According to NCDC (2008) discrete data organisation, analysis and interpretation are done in senior one; averages for ungrouped distribution are in senior two; data collection and organisation of grouped data, averages for grouped data, and experimental and theoretical probability are in senior three. In senior four, statistical concepts earlier covered are revisited, though some students fail to acquire the basic foundation in statistics.

Four-seven-five additional mathematics: In a few schools, gifted students are offered additional mathematics as an optional subject. This course contains more statistics topics than 456-mathematics (UNEB, 2005). The content includes appropriate organisation of data, data analysis and interpretation, moving averages, and index numbers. Laws of probability, discrete and continuous variables, expectation, normal and binomial distributions, confidence intervals, sampling and surveys, correlation and regression (UNEB, 2005) are also covered. Senior four students cover this course in most schools.

3.2 Statistics at Advanced Level

Two courses, principal (P425) and subsidiary (S475) mathematics, are offered at advanced level, both of which cover some statistics (UNEB, 2008). In principal mathematics, statistics falls under applied mathematics. The statistics content includes introduction to statistics, organisation of data, measures of central tendency and dispersion, index numbers, probability sample spaces, functions and distribution functions, and confidence intervals (UNEB, 2008). Statistics covers about a quarter of the mathematics curriculum. The content of subsidiary mathematics is similar to the additional mathematics for ordinary level (UNEB, 2008).

4 Statistics Teaching and Learning Resources

The teaching of statistics in Uganda is similar to that of other African countries such as Kenya (Odhiambo, 2002). Teachers who teach statistics rarely have an opportunity to develop sound knowledge of the principles underlying good statistics teaching. The topics treated and approaches used in classrooms derive from textbooks, which teachers regard as the main authority for statistics content and the *de facto* curriculum. Unfortunately, textbooks are often unavailable or inaccessible to students. Furthermore, most locally produced primary and ordinary level textbooks introduce students to statistics through "non-genuine data". This scenario contrasts with the statistics teaching involving projects and problem-solving (Moore & Roberts, 1989; Cobb, 1992). Most statistics teaching (based on personal experience) involves using mechanical formulae-based approaches without applications of real-life data, statistical reasoning, and problem solving. Meanwhile, advanced level textbooks are mostly imported and expensive. Teaching statistics in schools rarely involves using technology, and use of calculators by primary pupils is unaccepted.

5   Statistics Assessment Items

The items for Primary Leaving Examination (PLE), Uganda Certificate of Education (UCE), and Uganda Advanced Certificate of Education (UACE) are like the questions in textbooks. The 42-item test for PLE assessment includes statistics items within a single compulsory mathematics question paper from the primary syllabus. The items emphasise more mathematics than statistics, which contrasts with Egypt, where statistics is a major part of the mathematics curriculum (Assar, 2002). Statistics assessment items for UCE examinations are from the 456-mathematics syllabus. Meanwhile, the statistics assessment items for the 475-additional mathematics question paper are from the 475-additional mathematics syllabus.

Statistics items assessed in the UACE S475-mathematics paper are similar to the UCE 475-additional mathematics. Worth noting is that the number of students taking subsidiary mathematics is on the decline, as many students prefer the P425-Principal Mathematics because passing P425 offers a better opportunity for entrance to a variety of university courses.

The assessment items for the P425-Principal Mathematics statistics examinations derive from the relevant curriculum syllabus and include introduction to statistics and index numbers, but confidence intervals receive little attention (UNEB, 2007). Overall, statistics assessment in Ugandan schools, as was the case in South Africa before the introduction of Curriculum-2005 (Wessels, 2008), is formula-based. The compulsory sections of the question paper contain some statistics questions, which force students to attempt some statistics. However, such assessments produce students who are ill prepared for the statistics required in higher education and for application in adult life.

6   Conclusion

The observations described in this chapter can be useful to educators in Uganda and other countries that are engaged in improving the teaching and learning of statistics. Statistics teaching faces common challenges such as teacher-centred and examination-oriented instruction, little feedback on students' assessment, and the use of the textbook as a primary source of problems. To address these challenges teacher training must focus on preparing teachers to teach statistics effectively. Statistics teaching should embrace the use of technology, cooperative learning, and student-centred instruction. More varied and motivating forms of statistics instruction and assessment are needed to inspire learners to embrace statistics, and more research is needed to inform statistics educators about strategies to make this happen.

References

Assar, R. M. E. (2002). An experimental approach for teaching statistics in the Egyptian schools. In B. Phillips (Ed.), *Proceedings of the Sixth International Conference on Teaching Statistics.* Cape Town, South Africa: International Association for Statistics Education. Online: www. stat.auckland.ac.nz/~iase/publications

Cobb, G. W. (1992). Teaching statistics. In A. S. Lynn (Ed.), *Heeding the call for change: Suggestions for curriculum action* (MAA notes, pp. 3–43). Washington, DC: Mathematical Association of America.

Moore, T. L., & Roberts, R. A. (1989). Statistics at liberal arts colleges. *The American Statistician, 43,* 80–85.

National Curriculum Development Centre. (1999). *Uganda primary school curriculum: Syllabus for primary schools* (Vol. 1, pp. 219–275). Kampala: Author.

National Curriculum Development Centre. (2000). *Teachers' guide to Uganda primary school curriculum* (Vol. 1, pp. 241–352). Kampala: Author.

National Curriculum Development Centre. (2008). *Mathematics teaching syllabus: Uganda certificate of education.* Kampala: Author.

Odhiambo, J. W. (2002). Teaching statistics in Kenya. In B. Phillips (Ed.), *Proceedings of the Sixth International Conference on Teaching Statistics.* Cape Town, South Africa: International Association for Statistics Education. Online: www.stat.auckland.ac.nz/~iase/publications

Opolot-Okurut, C., Opyene Eluk, P., & Mwanamoiza, M. V. (2008). The current teaching of statistics in schools in Uganda. In C. Batanero, G. Burrill, C. Reading, & A. Rossman (Eds.), *Joint ICMI/IASE Study: Teaching Statistics in School Mathematics. Challenges for Teaching and Teacher Education. Proceedings of the ICMI Study 18 and 2008 IASE Round Table Conference.* Monterrey, Mexico: International Commission on Mathematical Instruction and International Association for Statistical Education. Online: www.stat.auckland.ac.nz/~iase/ publications

Oyelese, J. O. (1982). Some problems of the teaching of statistics in developing countries: The Nigeria example. In V. Barnett (Ed.), *Teaching statistics in schools throughout the world* (pp. 189–198). Voorburg, The Netherlands: International Statistical Institute.

Uganda National Examinations Board. (1991). *Uganda primary leaving certificate: Regulations and syllabuses 1991–1995.* Kampala: Author.

Uganda National Examinations Board. (2005). *Uganda certificate of education: Regulations and syllabuses 2005–2010.* Kampala: Author.

Uganda National Examinations Board. (2007). *Uganda advanced certificate of education principal mathematics and subsidiary mathematics past papers; 1988–2007.* Kampala: Author.

Uganda National Examinations Board. (2008). *Uganda advanced certificate of education: Regulations and syllabuses 2009–2013.* Kampala: Author.

Wessels, H. (2008). Statistics in the South African school curriculum: Content, assessment and teacher training. In C. Batanero, G. Burrill, C. Reading, & A. Rossman (Eds.), *Joint ICMI/ IASE Study: Teaching Statistics in School Mathematics. Challenges for Teaching and Teacher Education. Proceedings of the ICMI Study 18 and 2008 IASE Round Table Conference.* Monterrey, Mexico: International Commission on Mathematical Instruction and International Association for Statistical Education. Online: www.stat.auckland.ac.nz/~iase/publications

Chapter 4
Statistics in the South African School Curriculum

Helena Wessels

Abstract In this chapter, the status and content of statistics in South African primary and secondary school curricula are discussed. In the post-1994 school mathematics curriculum, the scope of statistics, or data handling and probability, has been broadened to promote statistical thinking through all phases of the statistical process. Because of a lack of content knowledge and knowledge of how learners develop and understand statistical concepts, however, teachers are not yet fully prepared to implement the intended curriculum. An external assessment at the end of grade 12 influences to a great extent teaching approaches and how content is covered.

1 Introduction

Outcomes-based education forms the foundation of the post-apartheid school curriculum in South Africa, promoting a learner-centred, activity-based approach. The new curriculum, revised in 2002, is divided into two parts: the General Education and Training (GET) band and the Further Education and Training (FET) band (Department of Education [DoE], 2002, 2003a). The GET band consists of three different phases: the Foundation (grades R-3, where R indicates the grade before grade 1), Intermediate (grades 4–6) and Senior Phases (grades 7–9), while the FET band spans grades 10–12. All learners in the FET band have to take either mathematics or mathematical literacy. In both the GET and FET bands, the focus of data handling and probability is on the development of skills to collect, organise and critically analyse and interpret data; in the FET band, these skills are used to

H. Wessels (✉)
Department of Curriculum Studies, Faculty of Education, Stellenbosch University,
Private Bag X1, Matieland, 7602 Stellenbosch, South Africa
e-mail: hwessels@sun.ac.za

C. Batanero, G. Burrill, and C. Reading (eds.), *Teaching Statistics in School Mathematics-Challenges for Teaching and Teacher Education: A Joint ICMI/IASE Study*, DOI 10.1007/978-94-007-1131-0_4, © Springer Science+Business Media B.V. 2011

establish statistical and probability models to solve related problems. In mathematical literacy, the development of skills to collect, organise, analyse and interpret data is also prominent, focussing on the role of learners as consumers of interpretations of data (DoE, 2003b).

In the past, data handling was taught traditionally, aiming at the drawing of simple graphs of already organised data sets, the description of single data sets, answering of direct questions about graphs and tables, and calculations of the mean, median and mode of small artificial data sets without real understanding from learners or teachers about the meaning or appropriateness of these measures. This traditional approach taught learners fragmented skills that did not prepare them to interpret data critically, communicate their opinions and conclusions, or to be statistically literate; nor did it prepare them for further studies in statistics (North & Zewotir, 2006).

The revised curriculum includes the collection, representation and critical analysis of data to draw conclusions, predict and determine chance variation. These goals are in line with the mission of the Department of Education to prepare learners for the social and economic needs they will face as adults in the twenty-first century (DoE, 2009). One of the main ideas on all levels of the curriculum is the solving of problems on social, environmental and political issues, human rights and inclusivity by using the statistical process.

Many mathematics teachers in South Africa are not yet statistically literate themselves and lack the confidence to teach statistics; more professional development initiatives are needed to prepare them to implement the curriculum successfully (Wessels, 2009). The Sixth International Conference on Teaching Statistics (ICOTS-6) held in Cape Town in 2002 resulted in considerable advantages for statistics education in South Africa. The teacher training programme that ran parallel to the conference sparked a series of professional development initiatives for teachers in South Africa that are still continuing (North & Zewotir, 2006).

2   Statistics Content in the South African Mathematics Curriculum: GET Band

In the *Foundation Phase*, the development of skills starts with sorting objects and data according to different features, while developing an awareness that the selection of attributes used for sorting the data influences the representation of the data and the subsequent conclusions and predictions. Representation should adhere to one-to-one correspondence between an item and its representation.

Central in the *Intermediate Phase* is the gaining of skills to collect and summarise data for interpretation and prediction. The influence of specific questions on the understanding of the situation and the role of representations in the clarification or concealment of different features of the data limiting interpretation and prediction should be emphasised. The constraining role of data-gathering contexts in the interpretation of and subsequent prediction from data should be made explicit through examples; for example, interviewing only a group of teenagers on music preferences for background music in a shopping mall as compared to interviewing people of all ages. Teachers should foster learners' abilities to critically consider data collection methods, the suitability of different representations to summarise a data set and interpretations and predictions made from data, using discrete data and whole numbers. Calculation of the probability of an event happening is not required, but awareness should be developed that different situations yield different probabilities and that many situations have a finite number of different possible outcomes.

Learners investigate and solve data handling problems in the *Senior Phase* by using techniques learned in previous phases. Learners should deal with discrete and continuous data, do projects and use measures of central tendency, range and standard deviation. Chance is studied through single and compound events, simple experiments showing the difference between the probability and relative frequency, and expressions of chance and probability from real life; for example, "The HIV test for babies is 74% reliable".

3   Statistics Content in the South African Mathematics Curriculum: FET Band

In the discussion of data handling for the *FET band* a distinction is made between the content for mathematics and for mathematics literacy for grade 12. The content covered in grades 10 and 11 led to the required content for grade 12 and will not be discussed separately. For mathematics it is necessary to consider the assessment requirements for the external examination as a determining factor for the content in grades 10–12. The learning outcomes and assessment standards for grade 12 have been divided into Core and Optional Assessment Standards. Core Assessment Standards are examined by means of two compulsory papers, while Optional Assessment Standards are examined by means of an optional third paper (DoE, 2008).

Data handling comprises 16.7% of the second compulsory paper and includes graphical representation of data; measures of central tendency and dispersion; box plots, scatter plots and ogives; calculation of variance and standard deviation; and fitting lines to data. Topics included in the optional paper comprise dependent and independent events, Venn diagrams, bias, error in measurement, uses and misuses of statistics, effective communication of conclusions and predictions, symmetric and skewed data, importance of sample size, regression functions and the correlation coefficient for bivariate numerical data.

It was anticipated that the Optional Assessment Standards would become compulsory after 2010 to provide for the required training of teachers, but the Department of Education has since decided that these standards will remain optional (DoE, 2009). This assessment policy has caused many teachers to drop the optional grades 10–12 Assessment Standards from the curriculum. Learners who want to write the optional paper in grade 12 have to attend extra classes after school. Concerns are that learners who have not been taught statistics in grades 10–12 will not be statistically literate when they leave school and will be ill-prepared for the interpretation of statistical information presented in the media and for statistics at the tertiary level.

The main focus of data handling in mathematical literacy is informal inferential reasoning, more specifically the critical comparison and interpretation of two data sets to draw conclusions and make predictions, taking into account sources of error and bias. The use of numerical and graphical summaries of data should be used to describe trends and the use and misuse of statistics in society. Probability values are used to make predictions about outcomes of games and other real-life situations. Mathematical literacy is examined through two papers: the first focuses on basic knowledge and routine applications of data handling and probability, while the second paper requires more reasoning about and interpretation and application of given information.

4 Conclusion

Statistics education in South Africa is still in its infancy, and much needs to be done to prepare mathematics teachers to teach the broadened statistics curriculum to promote statistical literacy. Through cooperation between the Department of Education, university mathematics education departments and Statistics South Africa, preservice and inservice teacher training courses are presented to attain this objective (North & Scheiber, 2008; Wessels, 2009).

References

Department of Education. (2002). *Revised National Curriculum Statement grades R-9, mathematics*. Pretoria, South Africa: Author.

Department of Education. (2003a). *National Curriculum Statement grades 10–12, mathematics*. Pretoria, South Africa: Author.

Department of Education. (2003b). *National Curriculum Statement grades 10–12, mathematical literacy*. Pretoria, South Africa: Author.

Department of Education. (2008). *National Curriculum Statement grades 10–12, mathematics: Subject assessment guidelines*. Pretoria, South Africa: Author.

Department of Education. (2009). *National Curriculum Statement Learning Programme Guidelines. Mathematics. Mathematical literacy. Examination guidelines grade 12*. Pretoria, South Africa: Author.

North, D., & Scheiber, J. (2008). Introducing statistics at school level in South Africa: The crucial role played by the national Statistics Office in training in-service teachers. In C. Batanero, G. Burrill, C. Reading, & A. Rossman (Eds.), *Joint ICMI/IASE Study: Teaching Statistics in School Mathematics. Challenges for Teaching and Teacher Education. Proceedings of the ICMI Study 18 and 2008 IASE Round Table Conference*. Monterrey, Mexico: International

Commission on Mathematical Instruction and International Association for Statistical Education. Online: www.stat.auckland.ac.nz/~iase/publications

North, D., & Zewotir, T. (2006). Introducing statistics at school level in South Africa. In A. Rossman, & B. Chance (Eds.), *Proceedings of the Seventh International Conference on Teaching Statistics*. Salvador, Brazil: International Statistical Institute and International Association for Statistical Education. Online: www.stat.auckland.ac.nz/~iase/publications

Wessels, H. M. (2009). *Effective teaching and learning of grade 8 and 9 statistics*. Unpublished post doctoral research report, North West University, Potchefstroom, South Africa.

Chapter 5
Developing a Statistics Curriculum for Future Secondary Mathematics Teachers

Amy G. Froelich

Abstract To support the teaching of statistics in secondary schools as recommended by the National Council of Teachers of Mathematics 2000 *Principles and Standards for School Mathematics* and the American Statistical Association 2005 *Guidelines for Assessment and Instruction in Statistics Education*, the faculty at Iowa State University designed a new curriculum in statistical content for future secondary mathematics teachers. Based on recommendations from national mathematics committees, this new curriculum engages future secondary mathematics teachers with data collection and analysis, inferential statistics, and probability, and highlights connections and differences between mathematics and statistics.

1 Introduction

In the last 20 years, the teaching of statistics as a part of the mathematics school curriculum has become more prevalent across the United States. In 1989, the National Council of Teachers of Mathematics as a part of their document *Curriculum and Evaluation Standards for School Mathematics* (NCTM, 1989) called for data analysis and probability to be taught in the nation's schools. This call was repeated and expanded in 2000 in the NCTM document *Principles and Standards for School Mathematics* (NCTM, 2000). The American Statistical Association's *Guidelines for Assessment and Instruction in Statistics Education* (*GAISE*) *Report* (Franklin et al., 2005) for the Pre-K–12 classroom calls for teaching statistics in the schools using a problem-solving framework along with a focus on the nature and sources of variability.

A.G. Froelich (✉)
Department of Statistics, Iowa State University, 3109 Snedecor Hall, Ames,
IA 50011-1210, USA
e-mail: amyf@iastate.edu

C. Batanero, G. Burrill, and C. Reading (eds.), *Teaching Statistics in School Mathematics-Challenges for Teaching and Teacher Education: A Joint ICMI/IASE Study*, DOI 10.1007/978-94-007-1131-0_5, © Springer Science+Business Media B.V. 2011

2   Curriculum Recommendations for Future Mathematics Teachers

In recognition of the changes in the status of statistics in the school mathematics curriculum, two reports on the curriculum for future mathematics teachers were issued in the last decade. Both the Conference Board of Mathematical Sciences book, *The Mathematical Education of Teachers* (the MET Report) (CBMS, 2001), and The Mathematical Association of America report, *Undergraduate Programs and Courses in the Mathematical Sciences: CUPM Curriculum Guide 2004* (the Committee on the Undergraduate Program in Mathematics (CUPM) Curriculum Guide) (CUPM, 2004), reinforced the need for statistical training in the preparation of future mathematics teachers. The CUPM Curriculum Guide states that "(t)he emphasis on data analysis in the 2000 NCTM standards … make(s) a study of statistics necessary for those preparing for secondary school teaching in mathematics" (CUPM, 2004, p. 47). This report further indicates the importance of having all mathematics majors (including future teachers) "study statistics or probability with an approach that is data-driven" (CUPM, 2004, p. 47). The MET Report goes further, by recommending that future teachers gain experience in five areas: exploring data, planning a study, anticipating patterns, statistical inference, and probability (CBMS, 2001).

At the same time, statistics educators focused on student-level outcomes and guidelines for teaching the college introductory statistics course. In 1992, the Guidelines of the American Statistical Association/Mathematical Association of America Joint Committee on Undergraduate Statistics called for the introductory statistics course to (1) emphasise statistical thinking, (2) include more data and concepts, less theory and fewer recipes, and (3) foster active learning (Cobb, 1992). Building upon this work, the *GAISE College Report* (Garfield et al., 2005) provides six recommendations for teaching the introductory statistics course and 23 student learning outcomes for all introductory statistics courses. The guidelines in Cobb (1992) were endorsed in the CUPM Curriculum Guide (CUPM, 2004) as a part of their recommendation for a data-driven approach to statistics and probability study for mathematics majors.

3   A Statistics Curriculum for Secondary Mathematics Teachers

In light of these reports, the faculty from the Departments of Statistics and Mathematics at Iowa State University decided to implement a new curriculum in statistical content for future secondary mathematics teachers. The new curriculum includes two required courses (introductory statistics and probability) and three recommended courses (applied regression modelling, design and analysis of experiments, and mathematical statistics).

Following the recommendations of the CUPM Curriculum Guide, the first required course is an introductory statistics course following the guidelines from the Cobb (1992) report and the *GAISE College Report* (Garfield et al., 2005). Course content includes descriptive statistics, data collection through random samples and random experiments, and an introduction to statistical inference, all taught with an emphasis on conceptual understanding. Completion of this course gives future teachers content knowledge in three areas (exploring data, planning a study, and statistical inference) from the MET Report (CBMS, 2001).

The second required course is a course in probability including content in standard probability distributions. At Iowa State, this course has been redesigned to place more focus on data analysis and investigations of concepts. The differences between theoretical probabilities and probabilities estimated through simulation and experimentation are emphasised through this approach (Froelich, 2009). From this course, future teachers gain content knowledge in the other two areas from the MET Report (CBMS, 2001) (anticipating patterns and probability) not covered by the introductory course.

While the two required courses are a good basis for the statistical content training of future secondary mathematics teachers, the content in these courses only mirrors the content future teachers will be responsible for teaching in the classroom. To gain a deeper understanding of statistics and its connection to mathematics, additional courses are necessary. However, as with many other education degree programmes, there is very little room to add courses to the degree programme for these future teachers and still allow for graduation from the university in a four-year time frame. Thus, the remainder of the curriculum (applied regression models, design and analysis of experiments, and mathematical statistics) is recommended, but not required, for all future secondary mathematics teachers.

The first two recommended courses give students a deeper exposure to statistical methods. The first course in applied linear regression models gives students experience in simple and multiple linear regression and an introduction to the analysis of variance. The second course in the design and analysis of experiments exposes students to different experimental designs (one-factor, two-factor, blocking, etc.) and the analysis of data from these experiments. In both courses, the focus is on data analysis and the appropriate interpretation of the results in context.

Finally, the connections between statistics and mathematics in this curriculum (CBMS, 2001) are emphasised through the mathematical statistics course. Unlike the traditional course, this course at Iowa State is focused on the development of statistical concepts through both simulation and mathematical proof. The course content includes a study of the distributions of common sample statistics, the properties of estimators, and the connections between statistical theory and practice. Throughout the course, emphasis is placed on data analysis and on the connections between mathematics and statistics.

4   Future Work

Taken together, the two required and three recommended courses give future mathematics teachers a firm foundation in the statistical content in the school mathematics curriculum. All five courses are taught with an emphasis on conceptual understanding and data analysis through the use of classroom, laboratory, and homework activities, statistical software and java applets, real data sets, and student course projects. While the pedagogy used for these classes will help future mathematics teachers develop their pedagogical content knowledge in statistics, the current curriculum is missing a concentrated study of how to teach statistics. Future plans are to develop two one-credit courses covering pedagogical content knowledge in statistics in conjunction with the two required courses in introductory statistics and probability.

5   Conclusions

The curriculum in statistical content for future secondary mathematics teachers at Iowa State was designed based on recommendations from national committee reports. The courses in this curriculum are general enough to be easily adapted to other colleges and universities in the United States. However, implementation requires the cooperation of mathematics, mathematics education, and statistics faculty (Froelich, Kliemann, & Thompson, 2008). This cooperation is the first step to providing future secondary mathematics teachers with appropriate training in statistical content in order to prepare more effective teachers of statistics in the schools.

References

Cobb, G. (1992). Teaching statistics. In L. A. Steen (Ed.), *Heeding the call for change: Suggestions for curricular action* (MAA notes 22, pp. 3–43). Washington, DC: Mathematical Association of America.

Committee on the Undergraduate Program in Mathematics. (2004). *Undergraduate programs and courses in the mathematical sciences: CUPM curriculum guide 2004*. Washington, DC: The Mathematical Association of America. Online: www.maa.org/CUPM/

Conference Board of the Mathematical Sciences. (2001). *The mathematical education of teachers*. Washington, DC: American Mathematical Society and The Mathematical Association of America. Online: www.cbmsweb.org/MET_Document/

Franklin, C., Kader, G., Mewborn, D., Moreno, J., Peck, R., Perry, M., & Scheaffer, R. (2005). *Guidelines for assessment and instruction in statistics education (GAISE) report: A pre-K-12 curriculum framework*. Alexandria, VA: American Statistical Association. Online: www. amstat.org/education/gaise/

Froelich, A. G. (2009). *Using R to teach undergraduate probability and statistics courses*. Unpublished manuscript.

Froelich, A. G., Kliemann, W., & Thompson, H. (2008). Changing the statistics curriculum for future and current high school mathematics teachers: A case study. In C. Batanero, G. Burrill, C. Reading, & A. Rossman (Eds.), *Joint ICMI/IASE Study: Teaching Statistics in School Mathematics. Challenges for Teaching and Teacher Education. Proceedings of the ICMI Study 18 and 2008 IASE Round Table Conference.* Monterrey, Mexico: International Commission on Mathematical Instruction and International Association for Statistical Education. Online: www.stat.auckland.ac.nz/~iase/publications

Garfield, J., Aliaga, M., Cobb, G., Cuff, C., Gould, R., Lock, R., et al. (2005). *Guidelines for assessment and instruction in statistics education (GAISE) college report.* Alexandria, VA: American Statistical Association. Online: www.amstat.org/education/gaise/

National Council of Teachers of Mathematics. (1989). *Curriculum and evaluation standards for school mathematics.* Reston, VA: Author.

National Council of Teachers of Mathematics. (2000). *Principles and standards for school mathematics.* Reston, VA: Author.

Chapter 6
Future Teachers' Training in Statistics: The Situation in Germany

Laura Martignon

Abstract This chapter describes some relevant features of the training in statistics and probability future mathematics teachers receive in Germany. It also discusses aspects of the stochastic courses taught in school as well as some of the relevant textbooks on the subject.

1 Introduction

National school systems differ greatly as to the amount of time allotted to different areas of mathematics in school. Germany's educational system, similar to the educational systems of many other countries (see, for instance, Coutinho, 2008; Innabi, 2008), had traditionally been reluctant to allot a significant portion of school hours to data analysis and statistics. During the last decade this has been drastically changing, and national curricula across all states of the federal republic (*Länder*) see the achievement of competencies in data analysis and statistical reasoning as mandatory from elementary school to grade 12. A crucial impetus for this change has been the weak achievements of German school students in international student assessments such as Programme for International Student Assessment (PISA) on tasks belonging to the area of "uncertainty". In fact, 25% of all questions in PISA 2003 dealt with probabilistic or statistical tasks, and the lack of competency of German students in solving such tasks was a drawback to their overall performance.

Since recent reforms in 2005 that follow the dispositions of the educational standards (*Bildungsstandards*, i.e., the adaptations of the National Council of Teachers of Mathematics Standards (NCTM, 2000) to the German educational context), school curricula in all states of the federal republic now recommend

L. Martignon (✉)
Institute of Mathematics and Computing, University of Education, Reuteallee 46,
Ludwigsburg D-71634, Germany
e-mail: martignon@ph-ludwigsburg.de

C. Batanero, G. Burrill, and C. Reading (eds.), *Teaching Statistics in School Mathematics-Challenges for Teaching and Teacher Education: A Joint ICMI/IASE Study*,
DOI 10.1007/978-94-007-1131-0_6, © Springer Science+Business Media B.V. 2011

introducing statistical concepts and methods beginning in primary school and fostering competencies in statistical reasoning through secondary school in all school types. The emphasis on data analysis and data visualisation is strong in *all* states, but only some programmes in a portion of the federal states, for instance Sachsen-Anhalt, also recommend fostering elementary probabilistic reasoning during the first school years. A crucial factor that has enhanced the interest of school teachers in school statistics has been the inclusion of questions testing statistical competencies in the final central examinations taken by school students at the end of secondary school. For many years, questions of this type, if any, were not mandatory but could be selected from an additional pool. Today questions on data analysis and data visualisation belong not only in the final examination but also in the biannual central Comparison Exams (*Vergleichsarbeiten*) for all school types, which have become mandatory since the inception of the new educational standards in 2005. This, perhaps more than any other factor, has motivated teachers to take special extra courses (*Lehrerfortbildungen*), offered regularly both in schools and universities by instructors mostly from universities, in order to receive training in statistical competencies.

2   Training Future Teachers in Statistical Thinking During Their University Studies

Perhaps even more so than in other mathematical fields, the gap between disciplinary statistics and school statistics has to be taken seriously into account when preparing future school teachers of statistics who have to be aware that they will be providing future citizens with "statistical literacy". In other words, future citizens need to be endowed with tools for interpreting statistical information in the media, for dealing with relative and absolute risks, and for understanding the effect of base rates on the predictive accuracy of medical tests. This requires assessments of the validity of features characterising financial investments like risk, liquidity, and time horizon. It means also understanding data visualisation, as commonly used in newspapers and brochures, and understanding the meaning of correlations as well as of conditional probabilities. Such goals had been absent from traditional German university courses in stochastics that combined descriptive statistics, inferential statistics, and probability, typically at a very formal theoretical level, with applications, if any, more in abstract problems rather than in day-to-day decision making. These courses had been – and, by and large, still are – taught to *all* students, regardless of whether they are interested in pursuing a mathematical career or in becoming future teachers (*Lehramtstudenten*).

Mathematics educators in Germany are now being heard in their request that stochastic education be treated with the same intensity as other mathematical areas, like geometry education or algebra education. For instance in the state of Bavaria, mathematics educators are planning to include statistical questions in all central final examinations for future teachers. In several Bavarian cities (like Erlangen, Regensburg, München, & Würzburg) courses in Educational Aspects of Stochastic

(*Didaktik der Stochastik*) have recently been introduced in the regular programmes of future teachers. In other states the introduction of such courses has been less uniform, although many universities have reacted with enthusiasm to the spirit of innovation regarding the preparation of future teachers (of both primary and secondary schools) in statistics. Some German universities with strong groups of mathematics educators have introduced regular special seminars for future teachers on educational problems in stochastics.

3   The Special Situation of the Universities of Education in Baden-Württemberg

A strong emphasis on statistics training of future teachers is being placed in the six Universities of Education (*Pädagogische Hochschulen*) in Baden-Württemberg. This type of university systematically combines instruction in content knowledge with instruction in pedagogical content knowledge. For the special case of mathematics instruction this means that a course, say, on arithmetic is taught in parallel with a course on the peculiarities of fostering pupils' understanding of numbers.

A close look at the situation of practising elementary school teachers shows that up to now they have seldom had any statistical training at all. The universities of education are working to improve this situation. One of the forms in which teachers are recently acquiring both content knowledge and pedagogical content knowledge in elementary statistics and in basic probability theory is by direct contact with staff at the universities. This contact happens in different modalities: on the one hand, students of these universities perform their practical training (Praktikum) in their schools, and on the other hand, teachers regularly participate in so-called additional crash courses in specific educational subjects (*Lehrerfortbildungen*).

The experience at the Universities of Education in Baden-Württemberg is an example of a successful conjoint action of university staff and teachers. Crash courses on elementary statistics and probabilities for future teachers of primary schools are made possible by the openness of teachers. Teachers have, by and large, been extremely helpful, giving interviews on their reactions to the new statistical and probabilistic topics in the mathematics curricula and on their personal progress in implementing this new content. Their frequent comment is that they would need more time in class for implementing these topics in statistics and probability: it is important to note, they say, that in spite of all the novelties in the curricula and all recommendations for early inclusion of statistical (and, to an even lesser degree, probabilistic) practices, the number of hours devoted to mathematics in school has remained unaltered. Teachers' main concern has therefore been that they have to "steal" hours from the usual mathematical subject matters without impoverishing children's knowledge and training in those subject matters. This issue is crucial because the biennial evaluation tests, centrally organised by the Ministry of Education, still mainly focus on arithmetic and geometry. The first evaluation tests at the end of 2007 placed no emphasis on statistical ideas. This will change, as has recently been announced by the representatives of the Ministry who periodically visit schools, although it is difficult to imagine statistics becoming preponderant, as it should become.

4   Textbooks and Software

The introduction of the German version of the software Fathom (Biehler, Hofmann, Maxara, & Prömmel, 2006) has fostered real-life applications of statistics in Germany. Biehler and his team in Paderborn are now preparing a German version of the software Tinkerplots (Biehler, 2007). Another important contribution has been the publication of the book by Büchter and Henn (2007) on Stochastics in the Springer Series for future teachers (*Leharamtstudenten*). Eichler and Vogel have written a book on elements of data analysis and probability dedicated to the first years of secondary school that will certainly have an impact on statistical education in Germany (Eichler & Vogel, 2009). The community of mathematics educators who specialise in Stochastics in the German Society of Mathematics Education (*Gesellschaft für die Didaktik der Mathematik, GDM*) has played a crucial role in promoting a more careful treatment of educational issues in stochastics courses for future teachers. In this regard the periodical *Stochastik in der Schule*, founded in 1979 and published by members of this community, should be mentioned because it regularly provides teachers and students with peer-refereed articles and excellent reviews of books and materials devoted to educational aspects of statistics and probability in school.

References

Batanero, C., Burrill, G., Reading, C., & Rossman, A. (Eds.). (2008). *Joint ICMI/IASE Study: Teaching Statistics in School Mathematics. Challenges for Teaching and Teacher Education. Proceedings of the ICMI Study 18 and 2008 IASE Round Table Conference*. Monterrey, Mexico: International Commission on Mathematical Instruction and International Association for Statistical Education. Online: www.stat.auckland.ac.nz/~iase/publications

Biehler, R. (2007). Tinkerplots: Eine Software zur Förderung der Datenkompetenz in Primar- und früher Sekundarstufe (A software supporting data analysis in primary and early secondary school). *Stochastik in der Schule, 27*(3), 34–42.

Biehler, R., Hoffman, T., Maxara, C., & Prömmel, E. (2006). *Fathom 2: Eine Einführung (Fathom 2: An introduction)*. Heidelberg: Springer.

Büchter, A., & Henn, H.-W. (2007). *Elementare Stochastik. Eine Einführung in die Mathematik der Daten und des Zufalls*. (Elementary Stochastics: An introduction to the Mathematics of data and randomness). Berlin/Heidelberg: Springer. (First Edition 2005).

Coutinho, S. (2008). Teaching statistics in elementary and high school and teacher training. In C. Batanero, G. Burrill, C. Reading, & A. Rossman (2008).

Eichler, A., & Vogel, M. (2009). *Leitidee: Daten und Zufall (Core topic: Data and randomness)*. Braunschweig: Vieweg.

Innabi, H. (2008). Teacher training program on teaching arithmetic mean by using the visual approach: A case study. In C. Batanero, G. Burrill, C. Reading, & A. Rossman (2008).

National Council of Teachers of Mathematics. (2000). *Principles and standards for school mathematics*. Reston, VA: Author.

Chapter 7
An Experience on Training Mathematics Teachers for Teaching Statistics in Iran

Ahmad Parsian and Ali Rejali

Abstract In order to have a more statistically literate society, the Iranian Statistical Society, in cooperation with the Iranian Association of Mathematics Teachers' Societies, convinced the Ministry of Education to include one statistics course in the national high school curriculum. This chapter discusses some activities of the Isfahan Mathematics House related to preparing mathematics teachers to teach statistics in high school. It may provide a model for other countries in their efforts to improve statistics education and promote statistical literacy among their citizens.

1 Introduction

A large proportion of the general population in Iran does not have a developed sense of quantitative and statistical reasoning, as seems typical in much of the world. The teachers and school system do not encourage talented students to continue their studies in the field of statistics. There is no awareness of the usefulness of statistical reasoning and methods. With lack of encouragement and because of their ignorance of the importance of statistics in all walks of life, very few talented students choose statistics as their field of study at the university, and very often statistical projects are carried out by nonspecialists.

Attempts by the Iranian Statistical Society (IRSS) to popularise statistical concepts in Iranian society were not successful, so other organisations were asked to help. A few statisticians and institutes responded. Some of the activities are listed in the paper presented at the Joint ICMI/IASE Study Conference (Parsian & Rejali, 2008).

A. Parsian
Department of Mathematics, Statistics and Computer Sciences University of Tehran, Enghelab Avenue, Tehran, Iran
e-mail: ahmad_p@khayam.ut.ac.ir

A. Rejali (✉)
School of Mathematical Sciences, Isfahan University of Technology, Isfahan 84156, Iran
e-mail: a_rejali@cc.iut.ac.ir

C. Batanero, G. Burrill, and C. Reading (eds.), *Teaching Statistics in School Mathematics-Challenges for Teaching and Teacher Education: A Joint ICMI/IASE Study*, DOI 10.1007/978-94-007-1131-0_7, © Springer Science+Business Media B.V. 2011

Some statistical concepts have been taught in Iranian schools for the last 30 years, but they were either taught as a part of mathematics courses with emphasis only on theoretical aspects or in a statistical methods course in other curricula, such as social sciences (Parsian & Rejali, 1998). Prior to year 2000, the Ministry of Education designed a new course in statistics (Rejali, 1997) that is taught to all students in the second or third year of high school, but the problem of lack of specialists in statistics in the school system remains. Mathematics teachers still teach this course without being prepared to do so.

To make the goals of statistics education clear to mathematics teachers who are teaching statistics, Isfahan Mathematics House (IMH) in cooperation with IRSS started to prepare mathematics teachers to teach statistics. It was agreed that the teachers who want to teach statistics should be familiar with statistical concepts and methods and have some experiences with statistical problems (Larsen, 2006; Jordan, 2007). They should know the difference between statistical thinking and statistical methods (Garfield, 2002; Sanchez & Blancarte, 2008), and the difference between mathematical reasoning and statistical reasoning (Meletiou, 2003; Gattuso & Ottaviani, this book). They should recognise that teaching statistics without the involvement of students in various projects does not help students to learn the art of statistical thinking (Melton, 2004; Kahn, 2005). They have to raise awareness of the importance of the subject (Gattuso, 2006), and they should believe in the importance of using statistical software to do real statistical analysis (Connor & Davies, 2008).

Having agreed on the above constructs, the organisation of lectures and workshops throughout the country was started, and expository journals accessible to mathematics teachers published papers on statistical reasoning and statistics education. In these articles, public lectures and workshops, the usefulness of statistics was illustrated by explaining real-life examples and discussing the abuse of statistics and the difference between statistical thinking and statistical methods. IMH, in cooperation with the IRSS and the Mathematics Teachers' Society of Isfahan (MTSI), started an annual team-based statistics competition among high school students of Isfahan in 2006. Moreover, IMH, with the help of the Iranian Statistics Research and Training Centre (ISRTC), has developed an electronic site in Farsi for the popularisation of statistics (www.mathhouse.org).

2   Challenges for the Pre- and In-Service Training of Teachers

Most of the statistics teachers in schools have a mathematical background but are unfamiliar with statistical concepts, methods, and reasoning. Many teachers do not have any feel for data. An attempt to change this situation with some recommendations was explained in a previous paper (Parsian & Rejali, 2008). Since there was no proper in-service programme for teachers who teach statistics, many of whom did not have a background in statistics, IMH developed lecture notes for mathematics teachers to help them understand the concepts of statistics and learn methods for teaching these concepts at the school level. These notes have been distributed to many mathematics teachers throughout the country.

IMH developed a programme in continuing education for mathematics teachers who want to teach statistics in schools or are already doing so. This programme was first developed for mathematics teachers of Isfahan in cooperation with MTSI and was implemented in the summer of 2004 as a workshop with active involvement of the participants. During the workshop, the volunteer teachers who took part in the programme worked in teams. The curriculum for the workshop was developed in cooperation with teachers and statisticians. After observing the effect of the first workshop, IMH in cooperation with the IRSS and the Iranian Association of Mathematics Teachers' Societies (IAMTS) ran workshops for volunteer mathematics teachers in eight provinces of Iran in the summer of 2005 and has announced its readiness to run the workshop at other sites.

3   Follow-Up Observations

A follow-up study in Isfahan province shows some successes in promoting statistics among Isfahan students as well as teachers. Every year, teams of high school students present statistical projects at the IMH annual festivals, and high school students seem interested in enrolling in undergraduate statistics programmes (pending entrance examination results to justify this improvement). High school teachers are actively participating in increasing numbers at the statistics education sessions of the Iranian Mathematical Education Conferences (IMECs). Many mathematics teachers, who, with little statistical knowledge, preferred to use any extra time to solve mathematics problems in their statistics classes, today organise discussions on statistical methods and reasoning. Teachers developed enough confidence so they now volunteer to coach teams of high school students for the annual statistics competition. Teachers are also more willing to deliver talks on statistical concepts and methods and on probability at IMECs and the weekly colloquiums of IMTS. These observations led the other mathematics teachers' societies to invite IMH to run workshops in their provinces.

4   Future Plans

To realise the statistical goals for teachers, a complete follow-up study on the impact of these workshops on teachers should be done by IAMTS; the impact of the workshop on students and general public should be studied by IMH; new resources for teachers and students should be published by IRSS; a forum for the teachers should by developed by IMH to continue their discussions on statistical teaching methods; teachers should be encouraged to participate in national statistics conferences; motivated statistics educators should be trained; and, finally, lessons from other projects should be studied to expand these efforts to enhance statistics education in Iran.

References

Batanero, C., Burrill, G., Reading, C., & Rossman, A. (Eds.). (2008). *Joint ICMI/IASE Study*: *Teaching Statistics in School Mathematics. Challenges for Teaching and Teacher Education. Proceedings of the ICMI Study 18 and 2008 IASE Round Table Conference*. Monterrey, Mexico: International Commission on Mathematical Instruction and International Association for Statistical Education. Online: www.stat.auckland.ac.nz/~iase/publications

Connor, D., & Davies, N. (2008). Technology in the teaching of statistics: Potentials and challenges in preparing the teachers. In C. Batanero, G. Burrill, C. Reading, & A. Rossman (2008).

Garfield, J. (2002). The challenge of developing statistical reasoning. *Journal of Statistics Education, 10*(3). Online: www.amstat.org/publications/jse/

Gattuso, L. (2006). Statistics and mathematics: Is it possible to create fruitful links? In A. Rossman, & B. Chance (Eds.), *Proceedings of the Seventh International Conference on Teaching Statistics*. Salvador (Bahia), Brazil: International Association for Statistical Education. Online: www.stat.auckland.ac.nz/~iase/publications

Jordan, J. (2007). The application of statistics education research in my classroom. *Journal of Statistics Education, 15*(2). Online: www.amstat.org/publications/jse/

Kahn, M. (2005). An excellent problem for teaching statistics. *Journal of Statistics Education, 13*(2). Online: www.amstat.org/publicatyions/jse/

Larsen, M. D. (2006). Advice for new and student lecturers on probability and statistics. *Journal of Statistics Education, 14*(1). Online: www.amstat.org/publications/jse/

Meletiou, M. (2003). On the formalist view of mathematics: Impact on statistics instruction and learning. In A. Mariotti (Ed.), *Proceedings of the Third European Conference in Mathematics Education*. Bellaria, Italy: European Society for Research in Mathematics Education. Online www.dm.unipi.it/~didattica/CERME3/proceedings

Melton, K. I. (2004). Statistical thinking activities: Some simple exercises with powerful lesson. *Journal of Statistics Education, 12*(2). Online: www.amstat.org/publications/jse/

Parsian, A., & Rejali, A. (1998). Statistics and probability in Iranian school curriculum. *ISI Newsletter, 22*(1), 21.

Parsian, A., & Rejali, A. (2008). A report on preparing mathematics teachers to teach statistics in high school. In C. Batanero, G. Burrill, C. Reading, & A. Rossman (2008).

Rejali, A. (1997). Teaching probability and statistics in schools. *Andishe-ye Amari, 2*, 10–18.

Sanchez, E., & Blancarte A. L. (2008). Statistical thinking as a fundamental topic in training the teachers. In C. Batanero, G. Burrill, C. Reading, & A. Rossman (2008).

Chapter 8
Reform Efforts in Training Mathematics Teachers to Teach Statistics: Challenges and Prospects

Enriqueta Reston and Lisa Grace Bersales

Abstract This chapter describes some reform efforts in the Philippines on the inservice training of mathematics and statistics teachers. It presents the Philippine experience as an example of how individuals, government, and private organisations work together to achieve reforms. In particular, some work developed through a government-aided project involving elementary mathematics teachers and the institutional efforts of the Philippine Statistical Association in the inservice training of statistics teachers throughout the country is described. The challenges encountered in these reform efforts are examined as a basis for recommendations towards improving the training and preparation of mathematics teachers to better succeed in teaching statistics.

1 Introduction

For most countries, the teaching of statistics at the school level is part of the mathematics curriculum and is therefore managed by mathematics teachers who frequently lack specific training and preparation in teaching statistics (Batanero, Godino & Roa, 2004; Arnold, 2008; Giambalvo & Gattuso, 2008). This is particularly true in the Philippine Educational System comprising 10 years of elementary and secondary education where mathematics is the basic subject in which statistical and probability concepts are taught. Based on the Revised Basic Education Curriculum of the Department of Education (DepEd), data organisation

E. Reston (✉)
Science and Mathematics Education Department, University of San Carlos,
Talamban Campus, Cebu City 6000, Philippines
e-mail: edreston@usc.edu.ph

L.G. Bersales
University of the Philippines, Diliman, Quezon City, Philippines
e-mail: lisa_grace.bersales@up.edu.ph

C. Batanero, G. Burrill, and C. Reading (eds.), *Teaching Statistics in School*   41
Mathematics-Challenges for Teaching and Teacher Education: A Joint ICMI/IASE Study,
DOI 10.1007/978-94-007-1131-0_8, © Springer Science+Business Media B.V. 2011

and graphical displays are introduced at the elementary level starting in Grade 3, while probability and averages are introduced in Grade 6. At the secondary level, the topics include probability and descriptive summary statistics. Only government science high schools and a few private schools offer statistics as a separate subject, usually an elective, in the secondary level.

As to preservice preparation of mathematics teachers, the teacher education curriculum requires only one three-unit statistics course in the Bachelor's programme in Education, major in Mathematics (Commission on Higher Education [CHED], 2004). Thus, individuals, private organisations, and the government initiate various reform efforts to enrich mathematics teachers' statistical content knowledge and pedagogical skills. This chapter examines and analyses two reform efforts directed towards the inservice training of mathematics teachers at the local and national levels.

2   Local and National Reform Efforts

2.1   A Certificate Programme for Elementary Mathematics Teachers

One reform effort launched by a local government unit in coordination with the DepEd was an inservice professional development programme for public school elementary teachers in Cebu City, Philippines. In 2007, some 200 elementary mathematics teachers were sent by the Cebu City Government to enroll in a customised 24-unit Certificate Programme for Elementary Mathematics Teachers delivered by five selected teacher education institutions in the city. The programme aimed to enhance elementary mathematics teachers' content mastery, pedagogical skills, assessment schemes, mathematics and communication proficiency, use of technology in teaching, and values (Department of Education, 2007).

Among the eight courses in this programme, *Teaching Statistics for Elementary Math Teachers* and *Investigatory Approaches in Elementary Math Instruction* were of particular interest to statistics educators. In the first course, teachers' statistical content knowledge and pedagogical skills related to teaching basic statistical concepts were enhanced through formal class discussion, interactions, and activities. The second course engaged the elementary mathematics teachers in activities that demonstrated investigatory approaches in teaching mathematics concepts, including basic statistical concepts within the elementary mathematics curriculum. The course further required the teachers to conduct an investigatory project that focused on selected mathematical or statistical concepts taught in the elementary level. Aside from coursework, teacher educators handling the courses conducted at least two class observations to monitor and evaluate how the teacher participants translated what they learned in their coursework into their actual teaching practices.

2.2   The Training of Statistics Teachers by the Philippine Statistical Association

At the national level, reform efforts were made by the Philippine Statistical Association (PSA) towards more intensive training of statistics teachers. PSA is the country's only national professional association in statistics, and its reform efforts are primarily done in collaboration with two government organisations, the Philippine Statistical System (PSS) and the Commission on Higher Education (CHED). The main agency involved in statistical capacity building is the Statistical Research and Training Center (SRTC), the PSS training and research arm. Aside from training, SRTC sponsored the writing of reference material that elementary teachers in various subjects can use to illustrate the use of statistics (Bersales & Patungan, 1999), a high school textbook for third and fourth year high school mathematics students (Bersales, 2003), and a tertiary level introductory statistics book.

The PSA also examined curricular resources for statistics instruction by conducting a nationwide evaluation of introductory statistics textbooks available locally. The results were presented at its 2005 annual conference with the theme "Are We Teaching Statistics Correctly to Our Youth?". Based on the results, PSA recommended improvements in (1) the availability and the selection process for better textbooks and (2) the competency of statistics teachers. In line with the second recommendation, PSA collaborated with SRTC and conducted a pilot training of statistics teachers in 2007 (Philippine Statistical Association, 2007). After a high satisfaction rating by the pilot training participants, in 2008 the PSA launched a CHED-funded nationwide course for statistics teachers named *PSA-CHED Training Course for College Teachers of Basic Statistics*. The course included topics in probability and probability distributions, sampling distributions, point and interval estimation, hypothesis testing and basic tests, and correlation and simple regression, and provided hands-on computer sessions with exercises using actual data. The course was conducted in 11 different venues nationwide with 298 teacher-participants from 53 colleges and universities. Course evaluation by participants yielded very positive results with a median score of 4.5 on a scale of 1–5 (Philippine Statistical Association, 2008).

An analysis of these reform efforts revealed that both programmes focused on strengthening teachers' content knowledge and pedagogical approaches in teaching statistics. Further, the development and review of curricular resources used in teaching statistics were also addressed. However, these two reform efforts were carried out by independent teacher training programmes at school and university levels and differed in their emphasis and orientation. The certificate programme for elementary math teachers was primarily aimed at developing teachers' pedagogical content knowledge in teaching statistics within the school mathematics curriculum, while the training programme for college statistics teachers was more content-oriented. Further, the teacher educators who taught the statistics course in the *Certificate Programme* are college statistics teachers who may not have benefited from *PSA-CHED Training Course*. Thus, there is a need for coherent instructional design framework for teacher training that addresses teachers' professional development needs in more integrated ways.

3   Conclusions and Future Directions

The Philippine experience provides an example of collaboration among government institutions, professional statistics associations, academic institutions offering statistics programmes, and teacher education institutions in the professional development of mathematics-trained teachers of statistics. However, the need for coherent instructional design framework for teacher training in statistics was also revealed.

It is important that future actions contribute towards aligning local reform efforts with the global reform movements in statistics education that shift focus from procedural knowledge in statistics to conceptual understanding, statistical thinking, and reasoning. Garfield and Ben-Zvi (2008), for example, described two types of professional development projects in Israel and in the United States for preparing knowledgeable and effective teachers of statistics based on six principles of instructional design described by Cobb and McClain (2004, cited in Garfield & Ben-Zvi, 2008). The application in the Philippines of similar research-based instructional frameworks may contribute towards better preparation of teachers.

Further, there is a need for more studies in the Philippines that inform education policy makers about the state of statistics education at the school level. A suitable approach might be the use of action research, similar to that described by Arnold (2008) where a community of learning was established in New Zealand to help secondary mathematics teachers improve their statistical content knowledge and pedagogical skills in response to change within the statistics strand of the new 2008 New Zealand school mathematics curriculum.

References

Arnold, P. (2008). Developing new statistical content knowledge with secondary school mathematics teachers. In C. Batanero, G. Burrill, C. Reading, & A. Rossman (2008).

Batanero, C., Burrill, G., Reading, C., & Rossman, A. (Eds.). (2008). *Joint ICMI/IASE Study: Teaching Statistics in School Mathematics. Challenges for Teaching and Teacher Education. Proceedings of the ICMI Study 18 and 2008 IASE Round Table Conference.* Monterrey, Mexico: International Commission on Mathematical Instruction and International Association for Statistical Education. Online: www.stat.auckland.ac.nz/~iase/publications

Batanero, C., Godino, J., & Roa, R. (2004). Training teachers to teach probability. *Journal of Statistics Education, 12*(1). Online: www.amstat.org/publications/jse/

Bersales, L. G. S. (2003). *What do these data tell me? Statistics for high school.* Quezon City, Philippines: Statistical Research Training Center.

Bersales, L. G. S., & Patungan, W. R. (1999). *The use of statistics in the enhancement of elementary education in the Philippines*. Quezon City, Philippines: Statistical Research and Training Center.

Commission on Higher Education. (2004). Memorandum Order No. 30. In *Revised policies and standards for undergraduate teacher education curriculum*. Pasig City, Philippines: Author.

Department of Education. (2007). *A brochure on the certificate program for elementary mathematics teachers*. Cebu City, Philippines: DepEd Cebu City Division.

Garfield, J., & Ben-Zvi, D. (2008). Preparing school teachers to develop students' statistical reasoning. In C. Batanero, G. Burrill, C. Reading, & A. Rossman (2008).

Giambalvo, O., & Gattuso, L. (2008). Teachers training in a realistic context. In C. Batanero, G. Burrill, C. Reading, & A. Rossman (2008).

Philippine Statistical Association. (2007). *PSA report on the PSA-SRTC pilot training course on teaching basic statistics at the tertiary level*. Manila: Author.

Philippine Statistical Association. (2008). *PSA report on PSA-CHED training course for college teachers of basic statistics*. Manila: Author.

Chapter 9
Statistical Training of Central American Teachers

M. Alejandra Sorto

Abstract The statistical preparation and training of primary and secondary teachers in two Central American countries, Panama and Costa Rica, are described and compared. Teachers in both countries that graduate with a college degree have several courses in statistics, but the purpose of the courses is to prepare the prospective teachers to do a senior research project and not necessarily to teach the subject. Primary teachers that graduate from a Normal School in Panama have no preparation in statistics.

1 Introduction

Central American countries have just recently started the process of developing and implementing new educational standards, which include, for the first time, the teaching and learning of statistics in primary and secondary levels of education. As a consequence, governments in some of these countries are starting to make important decisions with respect to the preparation of teachers. These decisions are usually made based on nonempirical evidence due to the lack of human capacity to carry out educational studies and to the fact that some developing countries do not participate in international comparison studies like the Third International Mathematics and Science Study (TIMSS) and Preparatory Teacher Education Study (PTEDS). However, a recent study (385 primary and secondary teachers) funded by a Panamanian private sector in collaboration with the Panama and Costa Rica governments found that Costa Rican teachers had statistically significantly higher scores than teachers from Panama in all eight items related to statistical knowledge (Sorto, 2008). About 92% of the Costa Rican teachers could correctly answer items measuring Grade 3 and Grade 7 content compared to about 73% in

M.A. Sorto (✉)
Texas State University, 601 University Drive, MCS 575, San Marcos, TX 78666-4616, USA
e-mail: sorto@txstate.edu

C. Batanero, G. Burrill, and C. Reading (eds.), *Teaching Statistics in School Mathematics-Challenges for Teaching and Teacher Education: A Joint ICMI/IASE Study*, DOI 10.1007/978-94-007-1131-0_9, © Springer Science+Business Media B.V. 2011

Panama. Further, about 40% of Costa Rican teachers correctly answered items measuring statistical knowledge for teaching compared to about 30% in Panama (Sorto, Marshall, Luschei, & Carnoy, 2009). To better understand these results this chapter describes and compares the preparation and training of teachers in statistics in both countries.

2 Teachers' Education

In comparing teacher preparation between Panama and Costa Rica (see Table 9.1), several general points can be made. First, all Costa Rican primary teachers (Grades 1–6) have university degrees, while in Panama, primary teachers can opt for a Normal School degree (which is equivalent to Grades 10–12, with an additional year of postsecondary training) or a university degree.

The statistical content course varies according to the different institutions that offer a degree in primary education. Slightly over half of prospective primary teachers in Panama opt for the Panama Normal School and take the Maestro degree, which includes no statistics training.

Table 9.1 Comparison of degree type and statistical training

	Teaching level	Teacher possible degrees (and years of university training)	Courses with statistical component
Panama	Primary school	*Maestro* (1 year)	None
		Licenciatura (4 years)	(a) Descriptive statistics, (b) Inferential statistics, and (c) Quantitative research methods
	Secondary school	*Licenciatura* (4 years)	(a) Statistics, (b) Probability, and (c) Quantitative research methods
Costa Rica	Primary school	*Bachillerato* (4 years)	(a) Research methods and (b) Teaching practice
		Licenciatura (5 years)	(a) Research course, (b) Quantitative methods I, and (c) Quantitative methods II
	Secondary school	*Profesorado* (3 years)	(a) Statistics and probability
		Bachillerato (4 years)	(a) Statistics and probability, (b) Inferential statistics, and (c) Quantitative methods
		Licenciatura (5½ years)	(a) Statistics and probability, (b) Inferential statistics, (c) Quantitative methods I, and (d) Quantitative methods II

A primary teacher with a 4-year degree (*Licenciatura*) in Panama receives one descriptive statistics course, one inferential course, one course in quantitative research methods, and two mathematics education courses. Costa Rican primary teachers in a 4-year degree programme (*Bachillerato*) study statistics as part of a classroom research course, and those who take a 5-year degree programme (*Licenciatura*) take two extra quantitative research methods courses. The rationale for the unusual, at least when compared to the typical United States primary teacher education, amount of statistics content in both countries is because the degree of *Licenciatura*, by definition, requires candidates to submit a research project for which, if a quantitative method is appropriate, students are expected to use their statistical knowledge.

At the secondary level (Grades 7–11), mathematics specialist teachers in both countries receive considerable preparation in mathematics and statistics content. In Panama, secondary teachers attend a 4-year programme (*Licenciatura*) with two separate courses, one in statistics (non-calculus–based) and another in probability (calculus-based). In addition, they take one quantitative method course.

In Costa Rica, there are up to three levels of degrees to train secondary school teachers depending on the institution of higher education. A 3-year programme (*Profesorado*) requires one course in statistics and probability; a 4-year programme (*Bachillerato*) requires in addition an inferential statistics course and one quantitative methods course; a 5½-year programme (*Licenciatura*) requires all courses of the previous level with an additional quantitative methods course. Courses for secondary teachers are more complete than those for primary school teachers. For example, the statistics course is calculus-based with a focus on concepts related to mathematical statistics and probability theory. The inference course focuses on sample techniques, sampling distributions, hypothesis testing, and linear regression. Educational research method courses also put an emphasis on quantitative analysis. In their senior year at the university, prospective secondary teachers are also expected to complete a research project (*Tesis de grado*) as a requirement for graduation.

An effort to improve the secondary teacher preparation in statistics in Costa Rica is a research project conducted by scholars in the Department of Mathematics at the National University of Costa Rica. They have examined the official curriculum programmes, textbooks, and students' beliefs about the learning of statistics at the upper secondary grades (Chaves, 2007). Chaves found that even though teachers and students believe statistics is important, there is little emphasis on statistics at the school level primarily due to lack of time and the absence of statistical content in national assessments. In order to help improve this situation the *Bachillerato* and *Licenciatura* programmes at The National University of Costa Rica offer an elective senior course on statistics and probability education. This course includes topics such as current issues and future perspectives, epistemological foundations, statistical reasoning and learning difficulties, statistical curricula in the secondary school, and teaching techniques. Many of the sources listed in the bibliography are from statistics educators from Spain and Latin America (e.g., Batanero & Godino, 2003).

3   Teachers' Professional Development

Both countries provide opportunities for professional development in all content areas, but the opportunities for Costa Rican teachers appear to be much greater. In Panama, training opportunities are concentrated during teachers' vacation time, and although required, only a small fraction of teachers attend.

In Costa Rica, however, a National Pedagogy Centre (Centro Nacional de Didáctica, CENADI) is in charge of the professional development of both primary and secondary teachers. Costa Rican teachers have opportunities to participate in extensive professional development both during vacations and the school year. For these trainings, CENADI contracts public universities and consultants in mathematics education to offer a series of courses for teachers. While courses held during teachers' vacations are voluntary, school year courses are mandatory for primary teachers who are selected by their directors to improve in certain areas. Secondary teachers from Costa Rica participate in courses based on their areas of specialty. Teachers participating in these courses receive a financial incentive. Finally, the existence of CENADI and the requirement of regional offices to design and submit a yearly professional development plan make the Costa Rican approach appear somewhat better coordinated than the approach in Panama. The professional development in statistics for secondary teachers in Costa Rica is designed and conducted by mathematics department faculty and focuses on training teachers to teach statistics in an integrated way. The specific topics for the professional development are derived from a survey of what the teachers perceive are the most difficult concepts for the students (M. Martinez, personal communication, January 27, 2010).

4   Conclusion

Costa Rican teachers have more opportunities to study statistics during their teacher preparation programmes and in their structured professional development activities than their counterparts in Panama. This could explain the different results on the survey conducted by Sorto et al. (2009) between teachers from the two countries with respect to their statistical knowledge. The low scores on statistical knowledge (about 40% correct for Costa Rican teachers and about 30% correct for Panama) could be due to the fact that their preparation focuses on learning statistics for educational research purposes and not necessarily for teaching. Primary teacher preparation programmes in Panama's Normal School need to add statistics content to their curriculum, while secondary teacher preparation programmes, in both countries, would benefit from offering a statistics education course like the one offered at the National University of Costa Rica.

References

Batanero, C., & Godino, J. D. (2003). *Análisis de datos y su didáctica (Data analysis and its didactics)*. Granada, Spain: Universidad de Granada.

Chaves, E. (2007). Inconsistencia entre los programas de estudio y la realidad de aula in la enseñanza de la estadística de secundaria (Inconsistency between the official program and the reality of classroom in the middle school statistical education). *Actualidades Investigativas en Educación, 7*(3), 1–35.

Sorto, M. A. (2008). The statistical performance in a comparative study between Central American teachers. In C. Batanero, G. Burrill, C. Reading, & A. Rossman (Eds.), *Joint ICMI/IASE Study: Teaching Statistics in School Mathematics. Challenges for Teaching and Teacher Education. Proceedings of the ICMI Study 18 and 2008 IASE Round Table Conference*. Monterrey, Mexico: International Commission on Mathematical Instruction and International Association for Statistical Education. Online: www.stat.auckland.ac.nz/~iase/publications

Sorto, M. A., Marshall, J. H., Luschei, T. F., & Carnoy, M. (2009). Teacher knowledge and teaching in Panama and Costa Rica: A comparative study. *Revista Latinamericana de Investigación en Matematica Educativa, 12*(2), 251–290.

Part II
Fundamentals for Teaching Statistics

Chris Reading

Statistics is becoming increasingly important to all levels of citizenry, with more and more data available to inform decision-making. How this data is utilised by those forming the decisions and those acting on the decisions is necessarily impacted by the statistical learning experiences made available. The variety of experiences in the teaching of statistics in countries across the world outlined in Part I shows that the importance of such teaching is being recognised. Align this importance with the changing focus in statistics from computation to inference and a reconceptualisation of the teaching of statistics becomes necessary. This reconceptualisation must include not just changes to teaching methods but changes to the fundamentals for teaching statistics.

Conversations in the preparation for the Joint Study are reflected in the discussion document that shared questions to frame the study. These questions addressed current problems in the teaching of statistics within school mathematics specificities such as teacher attitudes, current practices, empowering teachers, training teachers and building collaborations. The proposed research questions, organised into Joint Study Topics, provided a landscape for researchers to address in their conference presentations. As well as the planned topics that were addressed, what eventuated in these presentations was common underlying themes in relation to fundamentals that were seen to be impacting generally on statistics education and thus specifically on the teachers, teaching and teacher education. Some of these fundamentals, including technology, project work and assessment were the focus of specific Joint Study Topics but other fundamentals were not obvious in the Joint Study Conference programme. These fundamentals have been brought together in the chapters in Part II.

The ideas presented in Part II in relation to these fundamentals are relevant to all those involved in statistics education but have the potential to have most impact on the work of those developing curriculum, planning teaching or training teachers.

C. Reading (✉)
SIMERR National Centre, University of New England, Education Building,
Armidale, NSW 2351, Australia
e-mail: creading@une.edu.au

Authors in this part propose certain fundamentals that impact on the way that statistics is approached: appreciating the different perspectives on statistics that can be taken as a foundation for teaching statistics (Chap. 10); strengthening the role that probability plays in the statistics curricula (Chap. 11); and recognising the differences between mathematical thinking and statistical thinking (Chap. 15). Authors also give consideration to how these fundamentals can be supported during teaching: taking a modelling approach for learning statistics (Chap. 12); using technology to support new approaches to teaching statistics (Chap. 13); using a project-based approach to better support statistical thinking (Chap. 14); and revamping assessment approaches to better measure statistical thinking (Chap. 16).

As a guide to a general approach to teaching statistics, Burrill and Biehler (Chap. 10) present four different perspectives that can be taken on the teaching of statistics: a framework for statistical thinking; statistics as a process different from mathematics; statistical literacy and stochastics. Teachers are advised to choose a specific perspective to suit the needs of the context and the learners. Based on four criteria for deciding whether an idea is fundamental, Burrill and Biehler present seven statistical ideas that are fundamental to teaching in the mathematics classroom. They describe the way in which mathematics and statistics approach each of the fundamental ideas and suggest how each fundamental idea should be taught.

Focusing in particular on probabilistic thinking, Borovcnik (Chap. 11) explains that probability is a complex concept and is needed to deal with statistics. A detailed interpretation of probability is provided from different perspectives that have been taken over time: Laplacian, Frequentist and Subjectivist. Explaining the present and predicting the future have been critical aspects of life for centuries and Borovcnik elaborates on how these have been achieved by relying on divination, causality and creationism. More recently probability has developed as another tool to aid in this process fuelling the debate of randomness versus divination, causality and creationism. Thinking probabilistically involves conflict between strategies and intuition and is unfortunately often neglected in teaching. Various arguments are provided for challenging those who hold the view that randomness does not exist and thus probability does not have a role in school curricula. These culminate in suggestions to support the development of a significant role for probability in the mathematics curricula.

Continuing the theme of different perspectives on probability, Chaput, Girrard and Henry (Chap. 12) propose that using a modelling perspective to teach statistics can achieve a synthesis between approaches. More experimental activities and more statistical software available in schools have led to choice when introducing the concept of probability but Chaput, Girrard and Henry warn that students must still understand the difference between models and reality. They propose a three-step model for teaching modelling. However, there are difficulties in using models in statistics when compared to using models in other areas such as geometry. For example, the learning of statistics begins after naïve conceptions have set in while geometry learning begins earlier; and when learning statistics, models are more removed from reality, that is, more abstract, than in geometry. Teachers must be aware that simulations are an artificial reproduction of the theoretical model of a situation and that there will always remain the problem with justifying to students the equivalence of a computer simulation with real random experiments or pseudo-concrete descriptions.

Impacting critically on the way that both teachers and students approach statistics is the difference between mathematical and statistical thinking. Gattusso and Ottaviani (Chap. 15) compare mathematical and statistical thinking but maintain that despite the differences statistics is currently taught as part of mathematics. Students need chances to develop statistical thinking and understand how it is different from mathematical thinking. In fact, there are certain skills that are utilised in learning statistics that contribute to more effective learning of mathematics, including: working within contexts; posing good questions; "rerouting" logic during the analysis process; constructing representations; and communicating results. Important implications for teaching are provided including the fact that while students are doing statistics they are also doing mathematics and that assessment needs to be revamped to include methods of revealing understanding.

With the increasing focus on statistics education MacGillivray and Pereira-Mendoza (Chap. 14) point out that projects provide an investigative context that nurtures the learning of statistical thinking. They advocate the use of a practical framework to demonstrate and learn statistical thinking, namely a data investigative cycle that incorporates stages: defining problem/planning; collecting data; analysing data; and interpreting data is proposed based on various other cycles. Samples of projects are provided and MacGillivray and Pereira-Mendoza maintain that projects that are suitable for use with students should be used when training pre-service teachers and in-service teachers so that the teachers will be more likely to use such projects with their students.

The use of technology can provide strong support in implementing new and different approaches to teaching statistics, especially the modelling approach, and Pratt, Davies and Connor (Chap. 13) examine the affordances of technology in teaching statistics. There are new, interesting and as yet under-utilised ways that technology can be used to facilitate learning in statistics. These include using large data sets that were previously unmanageable, and using dynamic digital graphical representations to aid analysis. Various issues that discourage the use of technology when teaching statistics are explained. For example, even if a technology-focused curriculum encourages the use of technology, the assessment regime will ultimately have a strong impact on whether technology is used in teaching. One aim of teacher training programmes should be to increase the technological pedagogical content knowledge (TPCK) of pre-service teachers so that they focus on technology as a teaching tool and not just on technology as part of course content. Various ideas for encouraging such an approach are explored.

With the changing focus on what is important in the learning of statistics there needs to be changes in the way that this learning is assessed. Garfield and Franklin (Chap. 16) recommend that teachers broaden the way they view assessment by considering assessment of learning, assessment for learning and assessment as learning when designing assessment tasks. They propose three foundational pillars of assessment that should underlay all assessment design: cognition, observation and interpretation. Principles are provided to assist educators to design statistics assessment that complies with the requirements of each of these three pillars.

Recommendations include collaborative activities for assessment and opportunities for pre-service teachers and in-service teachers to learn how to assess student learning. Some issues, such as high-stakes tests and the use of technology, that hinder the broadening of one's view of assessment are considered.

Those responsible for teaching statistics need to consider the fundamentals presented in Part II as a foundation for reconceptualising how statistics might be presented to learners. While it is recognised that changing those very ideas that underpin one's teaching is challenging, Part II also presents some supportive strategies to facilitate such change. With these fundamentals as a backdrop those involved in statistics education should use the detailed information on teacher knowledge, teaching and teacher training presented in Parts III and IV to advance a revolution in the teaching of statistics.

Chapter 10
Fundamental Statistical Ideas in the School Curriculum and in Training Teachers

Gail Burrill and Rolf Biehler

Abstract This chapter considers several perspectives on approaches to teaching statistics and summarises some of the literature related to these perspectives, in particular looking at the relationship between probability and statistics. Adapting criteria from the literature, each perspective is examined to identify statistical ideas that seem to be fundamental for understanding and being able to use statistics in the workplace, in personal lives, and as citizens. The chapter next considers the possible tensions between mathematics and statistics in the way each discipline approaches these fundamental ideas and finishes with implications for training teachers.

1 Introduction

Documents such as the *Principles and Standards for School Mathematics* (National Council of Teachers of Mathematics [NCTM], 2000), *New Zealand Mathematics and Statistics Curriculum* (New Zealand Ministry of Education, 2006), and the *National Standards for Mathematics in Grades* 5–10 in Germany (Kultusministerkonferenz, 2004) provide convincing rationales for why statistics is important and are explicit about the content that should be in school instructional programmes. Many educators agree on the increasing importance for students to gain competence in using and interpreting data as part of critical citizenship and the need for statistical reasoning and sense making in personal decisions, in the workplace, and in supporting progress in other fields and disciplines (Franklin et al., 2005).

G. Burrill (✉)
Michigan State University, 240 Erickson, East Lansing, MI 48824, USA
e-mail: burrill@msu.edu

R. Biehler
University of Paderborn, Warburger Str. 100, 33098 Paderborn, Germany
e-mail: biehler@math.upb.de

C. Batanero, G. Burrill, and C. Reading (eds.), *Teaching Statistics in School*   57
Mathematics-Challenges for Teaching and Teacher Education: A Joint ICMI/IASE Study,
DOI 10.1007/978-94-007-1131-0_10, © Springer Science+Business Media B.V. 2011

In general, a vast difference in approaches to the statistics curriculum exists among different countries, as can be seen by the discussion in the chapters in Part I of this book (for example, Newton, Dietiker, & Horvath; Opolot & Opyene; Reston & Bersales). The preparation of teachers to teach statistics also varies widely. Usually school subjects are aligned with university subjects in teacher training, and teachers study the respective university subject in some depth. Statistics, however, is often taught by mathematics teachers who have not had a specific statistics education themselves.

This chapter looks closely at what statistics educators deem important, from several perspectives, in describing a set of fundamental ideas in statistics that should be taught in school mathematics and that every student should know by the time he or she leaves secondary school. The chapter makes the case that part of this learning is clarifying the distinction between mathematics and statistics, and highlights issues related to these fundamental ideas that need to be addressed in training both beginning and practising teachers to teach statistics.

To establish a base for the work, four perspectives, which represent diverse ways of thinking about teaching statistics, are described in the next section.

2   Perspectives on Statistics Education

2.1   A Framework for Statistical Thinking

One perspective is provided by Wild and Pfannkuch's (1999) often quoted framework that focused on the thought processes involved in solving problems in statistics. The framework has four dimensions: investigative cycle, interrogative cycle, types of thinking, and dispositions. Within the types of thinking, those specific to statistics are recognition of the need for data, transnumeration (changing representations of data to increase understanding), reasoning with statistical models, consideration of variation, and integrating statistics and context. The framework was not intended to illustrate how concepts develop across grade levels.

The Wild and Pfannkuch framework considers variability as the defining ingredient in statistical reasoning. They quote Snee (1990, p. 118), who defined statistical thinking as "thought processes, which recognise that variation is all around us and present in everything we do, all work is a series of interconnected processes, and identifying, characterising, quantifying, controlling, and reducing variation provide opportunities for improvement".

2.2   Statistics as a Process Different from Mathematics

Another perspective that views statistics as a *process* for dealing with variability in data is described in the *Guidelines for Assessment and Instruction in Statistics*

(GAISE) *K-12 Report* (Franklin et al., 2005). GAISE is based on two beliefs the authors claim distinguish statistics from mathematics: the centrality of random variability or variability in data in statistics as opposed to the deterministic nature of mathematics, and the role of context; in statistics context provides meaning whereas in mathematics context provides the opportunity for applications.

The GAISE framework has four components, each formulated in terms of variability:

- Formulate a question – anticipating variability without which the question is not statistical;
- Collect data – acknowledging variability by designing for differences;
- Analyse data – accounting for variability using distributions; and
- Interpret results – allowing for variability and looking beyond the data.

The conceptual structure of the framework is composed of two dimensions: the first is described in terms of problem-solving processes, where the key processes include posing questions, random sampling, designing experiments, comparing variability among individuals and groups, associating two variables, generalising from sample to population, and distinguishing between association studies and experiments.

The second dimension of the framework is comprised of three developmental levels, reflecting an increasing sophistication in the ability to understand and operate within each level. The levels are not tied to particular grades; each level builds on concepts from the lower levels as the depth of sophistication in using statistical methods is increased. The nature and focus of variability is portrayed in increasingly complex ways in each of the levels. The first level describes variability inherent in a context, measurement variability, and induced variability with a focus on variability within a group. The second level adds sampling variability, an abstract concept of variability, with a focus on variability between groups and on covariation. The third level deals with chance variability, and the focus is variability in model fitting.

2.3 Statistical Literacy

A third perspective is that of statistical literacy, where learners are seen as users instead of producers of data or statistical results. Schield (1999) defined statistical literacy as the study of arguments that use statistics as evidence. Watson (1997) suggested a three-tiered hierarchy that set goals for statistical literacy: the ability to (1) understand basic statistical terminology, (2) understand it in context, and (3) question claims made without proper statistical justification. Gal (2002) claimed that the statistical literacy needs of adults are to be able to (a) interpret and critically evaluate statistical information, data-related arguments, or stochastic phenomena that they may encounter in diverse contexts, and, when relevant, (b) discuss their reactions to statistical information, such as their understanding of the meaning of information, their opinions about the implications of this information, or their concerns regarding the acceptability of given conclusions. This work included the identification of five necessary components for statistical literacy:

1. Knowing why data are needed and how data can be produced;
2. Having familiarity with basic terms and ideas related to descriptive statistics;
3. Having familiarity with basic terms and ideas related to graphical and tabular displays;
4. Understanding basic notions of probability;
5. Knowing how statistical conclusions or inferences are reached.

2.4   Stochastics

A fourth perspective is conveyed in the context of "stochastics teaching", which is described in Germany as a sub-domain of mathematics comprising probability and statistics. Heitele (1975) offered a set of fundamental ideas for teaching stochastics that included probability; sample space; addition and product rules for probability; independence and compound/conditional probability; equidistribution; combinatorics; random variable and probability distributions; simulation; sampling; and the Law of Large Numbers. These were selected as fundamental because among other things, they (a) are powerful as each helped situate probability as a mathematical theory; (b) can be taught at different levels in the curriculum, from primary school to university, and students can progress in formalisation and completeness; and (c) appear in most random situations. A major difference between the frameworks for statistics and stochastics is that Heitele does not describe thinking processes but does describe concepts.

In general, stochastics as a perspective has implications for what is taught in the sense that probability has not only the role of a servant of statistics but also is the mathematical branch that models nondeterministic relationships, random phenomena, and decisions under uncertainty (see Borovcnik, 2006 for a comparison of stochastics and statistical thinking). The stochastics framework is influential in Germany and some other European countries.

Each of the above perspectives refers in some way to probability or chance, as well as statistics. The next section elaborates in more detail on the relationship between statistics and probability.

3   Connecting Statistics and Probability

Some statistics educators tend to say "not more probability than is needed for statistics" (e.g., Moore, 1997a). This point of view may be too narrow compared to those in general education if we take elements of probability literacy into account (see Borovcnik, this book). What counts is that inference in statistics is based on probability, and many curricula make a distinction between statistics without probability (descriptive statistics, exploratory data analysis) and statistics with probability (inferential statistics). The latter is often taught at upper levels after probability has been introduced. Both the stochastics and statistical literacy perspectives identify probability, chance, as central in the work of statistics. GAISE (Franklin et al., 2005) suggests that probability should only be emphasised in the ways it is used in statistical thinking.

A fundamental idea in statistical inference is that empirical distributions have to be interpreted and seen from the perspective of hypothetically assumed theoretical distributions. But a probability model is more than just a static description of a probability distribution over a sample space. Probability models signify "data-generating processes" or data-producing "chance setups" as Hacking (1965) called them. According to the Law of Large Numbers the theoretical distribution is fictitiously identical to the empirical one when the sample size equals infinity. Due to finite sample sizes the theoretical distribution can never be established for certain, and several models consistent with empirical data may emerge. This position has two implications.

1. Probability should not be taught "data-free" but with a view towards its role in statistics. Probability models should be introduced as models to predict real data from random experiments and how empirical data may randomly differ from the theoretical distribution even if this distribution is assumed to be true. Schupp (1982, p. 210) once formulated an allusion to a famous sentence of the philosopher Kant: "Statistics without probability is blind, and probability without statistics is empty" [authors' translation]. Fischbein (1990) and Freudenthal (1961) offered similar arguments.
2. Data analysis should not be taught completely "model-free" but with a view towards theoretical distributions and underlying processes.

Inspired by the work in the 1970s with Exploratory Data Analysis (EDA), data analysis without probability took root at the school level and opened the road for more genuine data analysis in the classroom without the straightjacket of inferential statistics (Tukey, 1972). The emergence of EDA supported a view that "data analysis" without probability is important because looking at data only with a probability lens may distort the message in the data. Tukey (1972, p. 51) stated … "'data analysis' instead of 'statistics' is a name that allows us to use probability where it is needed and avoid it where we should". On the other hand, Tukey considered inferential statistics as the next step after EDA.

The philosophical differences between EDA and other types of statistics received little attention (Biehler, 1994). However, the question of how early data and chance should and can be connected in school has been tabled at curricular discussions. Some researchers (e.g., Pfannkuch, 2006; Rubin, Hammerman, & Konold, 2006) argued for the need to develop a sense of informal inference prior to a more formal approach.

4   Fundamental Statistical Ideas

Given the above considerations as starting premises, how can ideas fundamental to developing an understanding of statistics be identified? Any claim about what is fundamental should be grounded in a rationale for the claim. For example, Heymann (2003) asserted that main ideas about fundamental mathematical concepts should allow the relationship between the mathematical and nonmathematical culture to become perceptible, express universal features in ways that are comprehensible to students, and describe ideas that are both meaningful for individual mathematical topics and something other than basic mathematical concepts.

Adapting Heymann's criteria for fundamental ideas in mathematics and Heitele's criteria for fundamental ideas in stochastics, the authors of this chapter suggest that fundamental concepts in statistics should share some commonality within the different perceptions or ways of thinking about teaching statistics, be able to connect the discipline to other experiences in the world and to aspects of culture, illustrate the structure of the discipline perhaps clarifying specific characteristics and features important in the discipline, and allow for deepened understanding across time as students mature in their knowledge of statistics.

Based on the different perspectives on teaching statistics described above and using these four criteria, the following concepts are suggested (with references to appropriate chapters in this volume) as fundamental ideas in statistics.

1. *Data* – including types of data, ways of collecting data, measurement, respecting that data are numbers with a context (Ridgway, Nicholson, & McCusker, this book).
2. *Variation* – identifying and measuring variability to predict, explain, or control (Sánchez, Borim, & Coutinho, this book). The term "variability" is used for the general phenomenon of change and "variation" for describing the total effect of the change.
3. *Distribution* – including notions of tendencies (Jaccobe & Carvalho, this book) and spread (Sánchez, Borim, & Coutinho, this book) that are foundational for reasoning about statistical variables from empirical distributions, random variables from theoretical distributions, and summaries in sampling distributions (Reading & Canada, this book).
4. *Representation* – graphical or other representations that reveal stories in the data including the notion of transnumeration (González, Espinel, & Ainley, this book).
5. *Association and modelling relations between two variables* – nature of the relationships among statistical variables for categorical and numerical data (Engel & Sedlmeier, this book) including regression for modelling statistical associations.
6. *Probability models for data-generating processes* – modelling hypothetical structural relationships generated from theory, simulations, or large data set

approximations, quantifying the variability in data including long-term stability (Borovcnik, this book; Girard, Chaput, & Henry, this book).
7. *Sampling and inference* – the relation between samples and the population and the essence of deciding what to believe from how data are collected to drawing conclusions with some degree of certainty (Harradine, Batanero, & Rossman, this book).

Given that these ideas are critical for teachers to know and convey in their instruction, the next step is to consider how teachers, grounded in mathematical ways of thinking, will come to understand them.

5 Teaching Fundamental Statistical Ideas in the Mathematics Classroom

Many statistics educators (Franklin et al., 2005; Rossman, Chance, & Medina, 2006; Scheaffer, 2006; Gattuso, 2008) contend that mathematics and statistics differ in their essential defining characteristics: role of context, methods of reasoning, precision, role of data, and data collection. Because statistics is often taught in the mathematics classroom, the discussion below suggests some ways within the current culture of school mathematics in which the two disciplines, statistics and mathematics, are alike and then points out tensions and differences (see also Gattuso & Ottaviani, this book).

5.1 Bridging Between Mathematics and Statistics

Opportunities exist within the mathematics curriculum to build bridges to the fundamental ideas in statistics, in particular with respect to variation, association, and modelling, and developing informal notions of inference. One such opportunity exists with the topic bivariate data to expand the concept of mathematical function to model random dependencies. Batanero, Godino, and Estepa (1998) suggested that one of the three settings in which judging association is important is scatter plots (the other two being contingency tables and comparison of samples) and recommend that association be described in terms of intensity varying from independence to functional relationships. If an association has been identified between two or more variables, regression methods can be used to fit different types of mathematical functions to predict one of the variables (dependent variable) as a function of the other. The adequacy of a mathematical model for a situation can bear similar uncertainties to those in statistics. Categorical bivariate data analysis (gender and participation in athletics, for example) might be related to the probabilistic notion of independence and possible associations explored.

5.2   Tensions Between School Mathematical and Statistical Thinking

Many topics that may on the surface look the same in mathematics and statistics require very different types of thinking. The following relates tensions as relevant to each fundamental statistical idea.

1. *Data*: Data are typically used in mathematics classrooms in the context of "visualisation of numbers" and the study of functions, but the work rarely reaches the level of context-related reading between and beyond the data in the sense of Friel, Curcio, and Bright (2001). Measurement is done with standard magnitudes, often without regard for error, and little consideration is given to measuring categorical attributes. Probability is developed from rules, and data enter the picture as an application of the rules rather than as a way to develop the notions of probability.

2. *Variation*: Variation has a different nature in the two disciplines. Mathematics is often taught in school as being exact and precise. Statistics is about "noise", that is, how to measure and control variability. Real data in statistics are contextual, containing uncertainty and error while data in many school mathematics classrooms are typically assumed to perfectly fit a mathematical model. The teaching of functions in particular often undermines statistical concepts, for instance, when data lie exactly on a function graph.

3. *Distribution*: Distributions are developed in the context of teaching statistics only and do not evoke specific tensions with concepts taught in mathematics.

4. *Representation*: Statistics and mathematics differ in approaches to representations of data in several ways including the following: most statisticians begin with a graph; many mathematics students and teachers "crunch numbers" without paying attention to a visual representation of the data; and while in statistics different graphs or representations are used to identify different aspects of the same data (transnumeration), graphs in mathematics are often used in showing the same relationship in different representations (tables, graphs, and symbols).

5. *Association and modelling relations between two variables*: Cartesian coordinate plots are typically used in mathematics classrooms only to draw graphs of functions and not as scatter plots for bivariate data. As alluded to earlier, the potential for using mathematical modelling as a bridge between mathematics and statistics, in particular to bivariate quantitative data analysis, is seldom exploited; on the contrary, the mathematics educators who do research and development in mathematical modelling (Blum, Galbraith, Henn, & Niss, 2007) often pay little attention to statistical aspects, in particular to the central aspects of data in the process of modelling. Data collection plays no systematic role in going from a real situation to the mathematical model nor does comparing mathematical results to empirical data. In such models there is no need for a statistical lens, for example, to check residuals or think about how the context might relate to the choice of a model.

6. *Probability models for data-generating processes*: Drawing random samples from population data and sample-to-sample variation can be modelled with

probability models. Moore (1997b) argued that when students study probability with a formal approach, they will learn formalisms without understanding the phenomena described by this mathematics. Therefore teaching probability has to be enriched by broad phenomenological experiences, in which simulation can play a prominent role. However, this modelling applies only to those samples "randomly drawn" from a population or from a random allocation and assignment. This fundamental aspect is often neglected in a typical mathematical treatment of probability, which then undermines statistical understanding. This modelling will depend on assumptions, such as independence or equiprobability, which do not always hold, and are often taken as given and not to be considered or checked.

7. *Sampling and inference*: As Freudenthal (1974) pointed out with regard to sampling: what is important for statistics is sample-to-sample variation and how this variation decreases as the sample size increases. An intuitive understanding of this property can prevent students from believing in the law of small numbers, an unrealistic stability of samples with "small" sample sizes (Tversky & Kahnemann, 1971). The mathematical approach to proportional reasoning, however, often undermines the statistical approach for reasoning from samples. Percentages in mathematics are often applied in simple contexts, where the reference is set and the units are clear and constant. Careful statistical statements made about margin of error and confidence intervals are replaced by simplistic "inferences" from "sample" to "population", assuming a perfect proportional relationship. Ignoring uncertainty and variability, sample results are reported in point estimates rather than interval estimates in many media reports. Preparing students for statistical thinking requires that discussions in mathematics classrooms make this difference explicit.

With regard to inference there are the following tensions. In mathematics, deciding what to believe is straightforward: conclusions follow deductively from definitions and agreed-on principles. In statistics, reasoning is partly inductive, and conclusions always uncertain. The degree of faith in a statistical conclusion depends on the integrity of the entire investigative process, while in mathematics a proof makes you certain. In statistics, how the data were collected and the role of randomness determines how you can interpret the results, while in (pure) mathematics, the reasoning is independent of the data. However, justifying the validity of mathematical models requires reasoning more akin to statistical reasoning than to reasoning in pure mathematics.

6   Preparing Teachers to Teach the Fundamental Statistical Ideas

In nearly all countries, statistics is not a separate school subject but is taught by teachers of mathematics, and training for teaching statistics occurs, if at all, as a "catch-up" in the form of professional development for practising teachers

(Coutinho, 2008; North & Schieber, 2008). Prospective teachers often receive their mathematics education training in mathematics departments and, consequently, the education of teachers of statistics at the school level is oriented towards the scientific discipline of mathematics. Without some direct intervention in their training, teachers' "philosophy of mathematics" may create a tension with an adequate "philosophy of statistics". Cuoco, Goldenberg, and Mark (1996) described mathematical habits of mind that included multiple points of view, mix of deduction and experiment, emphasis on language, and willingness to conjecture, tinker, search for patterns, guess, and visualise. What are the statistical habits of mind teachers and students should develop as they grow in their understanding of the fundamental concepts in statistics? Resources to help think about this question might include the principles that informed the work of the Quantitative Literacy Series (Scheaffer, 1990; Cobb, 1992; Rossman & Chance, 2004). Possible statistical habits of mind are listed below:

- Use real data (call attention to variation and noise, pay attention to the source of the data in deciding what to believe);
- Build intuitions (use simulations to generate sampling distributions, predict before calculating, ask questions about chance based on data);
- Begin with a graph (investigate associations, analyse different representations of distributions, emphasise visualisation as a tool for learning about relationships – both data-driven and mathematical functions);
- Explore alternate representations of data (contrast what can be learned about shape, centre, and spread of distributions from different representations to understand relationships and connections among variables);
- Investigate and explore before introducing formulas (use simulations to model probability distributions, allow students to play with chance events and to experience variability);
- Use student projects and experiments to engage students in doing statistics (collect data to investigate questions, consider ways to reduce variability).

The preparation of teachers should not only include the fundamental statistical ideas described above but should also help teachers move interchangeably between the two disciplines, conscious of the differences and making links between the two that enhance student learning in both.

Several questions emerge for the statistics education community: The criteria for selecting the fundamental statistical ideas were given earlier in Sect. 4. Do the set of fundamental statistical ideas meet those criteria? Do these ideas give a sense of the structure and characteristics of statistics? Do these ideas share some commonality within the different perceptions or ways of thinking about statistics? Do these ideas connect statistics to other experiences in the world and to aspects of culture? Do these ideas develop and deepen throughout the school curriculum? Are these criteria the right criteria? What do we look for and what do we measure as evidence of progress?

The bottom line, however, is how can we enable those who train teachers to make visible the fundamental ideas in statistics, given the variations across countries in the constraints and conditions in which teachers are prepared? Part IV in this book is about the challenges and experiences in training teachers to teach statistics, and can begin to provide a framework to address this crucial challenge.

References

Batanero, C., Burrill, G., Reading, C., & Rossman, A. (Eds.). (2008). *Joint ICMI/IASE Study*: *Teaching Statistics in School Mathematics. Challenges for Teaching and Teacher Education. Proceedings of the ICMI Study 18 and 2008 IASE Round Table Conference.* Monterrey, Mexico: International Commission on Mathematical Instruction and International Association for Statistical Education. Online: www.stat.auckland.ac.nz/~iase/publications

Batanero, C., Godino, J., & Estepa, A. (1998). Building the meaning of statistical association through data analysis activities. In A. Olivier & K. Newstead (Eds.), *Proceedings of the 22nd Conference of the International Group for the Psychology of Mathematics Education* (pp. 221–236). Stellenbosch, South Africa: University of Stellenbosch.

Biehler, R. (1994). Probabilistic thinking, statistical reasoning, and the search for causes—Do we need a probabilistic revolution after we have taught data analysis? In J. Garfield (Ed.), *Research papers from the Fourth International Conference on Teaching Statistics (ICOTS 4)* (pp. 20–37). Minneapolis, MN: University of Minnesota.

Blum, W., Galbraith, P. L., Henn, H.-W., & Niss, M. (Eds.). (2007). *Modelling and applications in mathematics education. The 14th ICMI Study.* Berlin: Springer.

Borovcnik, M. (2006). Probabilistic and statistical thinking. In M. Bosch (Ed.), *Proceedings of the Fourth Congress of the European Society for Research in Mathematics Education* (pp. 484–506). Sant Feliu de Guitxols, Spain: European Research in Mathematics Education. Online: ermeweb.free.fr/CERME4/

Cobb, G. (1992). Teaching statistics. In A. S. Lynn (Ed.), *Heading the call for change suggestions for curriculum action* (pp. 3–43). Washington, DC: Mathematical Association of America.

Coutinho, C. (2008). Teaching statistics in elementary and high school and teacher training. In C. Batanero, G. Burrill, C. Reading, & A. Rossman (2008).

Cuoco, A., Goldenberg, P., & Mark, J. (1996). Habits of mind: An organizing principle for mathematics curricula. *Journal of Mathematical Behavior, 15,* 375–402.

Fischbein, E. (1990). Training teachers for teaching statistics. In A. Hawkins (Ed.), *Training Teachers to Teach Statistics. Proceedings of the International Statistical Institute Roundtable Conference* (pp. 48–57). Voorburg, The Netherlands: International Statistical Institute.

Franklin, C., Kader, G., Mewborn, D. S., Moreno, J., Peck, R., Perry, M., & Scheaffer, R. (2005). *Guidelines for assessment and instruction in statistics education (GAISE) report: A pre-K-12 curriculum framework.* Alexandria, VA: American Statistical Association. Online: amstat.org/education/gaise/

Freudenthal, H. (1961). Models in applied probability. In B. H. Kazemier & D. Vuysje (Eds.), *The concept and the role of the model in mathematics and natural and social sciences* (pp. 78–88). Dordrecht: Reidel.

Freudenthal, H. (1974). The crux of course design in probability. *Educational Studies in Mathematics, 5,* 261–277.

Friel, S. N., Curcio, F. R., & Bright, G. W. (2001). Making sense of graphs: Critical factors influencing comprehension and instructional implications. *Journal for Research in Mathematics Education, 32*(2), 124–158.

Gal, I. (2002). Adults' statistical literacy: Meanings, components, responsibilities. *International Statistical Review, 70,* 1–51.

Gattuso, L. (2008). Mathematics in a statistical context? In C. Batanero, G. Burrill, C. Reading, & A. Rossman (2008).

Hacking, I. (1965). *Logic of statistical inference*. Cambridge, UK: Cambridge University Press.
Heitele, D. (1975). An epistemological view on fundamental stochastic ideas. *Educational Studies in Mathematics, 6*(2), 187–205.
Heymann, H. (2003). *Why teach mathematics: A focus on general education*. Dordrecht: Kluwer Academic Publishers.
Kultusministerkonferenz (2004). *Bildungsstandards im fach mathematik für den mittleren schulabschluss (Educational standards for secondary school mathematics (grade 10))*. München, Germany: Wolters Kluwer.
Moore, D. (1997a). New pedagogy and new content: The case of statistics. *International Statistical Review, 65*, 123–165.
Moore, D. (1997b). Probability and statistics in the core curriculum. In J. Dossey (Ed.), *Confronting the core curriculum* (pp. 93–98). Washington, DC: Mathematical Association of America.
National Council of Teachers of Mathematics. (2000). *Principles and standards for school mathematics*. Reston, VA: National Council of Teachers of Mathematics.
New Zealand Ministry of Education. (2006). *The New Zealand Curriculum: Mathematics and statistics*. Online: nzcurriculum.tki.org.nz/
North, D., & Schieber, J. (2008). Introducing statistics at school level in South Africa: The crucial role played by the National Statistics Office in training in-service teachers. In C. Batanero, G. Burrill, C. Reading, & A. Rossman (2008).
Pfannkuch, M. (2006). Informal inferential reasoning. In A. Rossman & B. Chance (Eds.), *Proceedings of the Seventh International Conference on Teaching Statistics*. Salvador, Bahia, Brazil: International Statistical Institute and International Association for Statistical Education. Online: www.stat.auckland.ac.nz/~iase/publications
Rossman, A., & Chance, B. (2004). A data-oriented, active learning, post-calculus introduction to statistical concepts, methods, and theory. In G. Burrill & M. Camden (Eds.), *Curricular Development in Statistics Education: International Association for Statistical Education 2004 Roundtable*. Voorburg, The Netherlands: International Statistical Institute. Online: stat. auckland.ac.nz/~iase/publications
Rossman, A., Chance, B., & Medina, E. (2006). Some key comparisons: statistics and mathematics, and why teachers should care. In G. Burrill (Ed.), *Thinking and reasoning with data and chance, 68th NCTM Yearbook (2006)* (pp. 323–334). Reston, VA: National Council of Teachers of Mathematics.
Rubin, A., Hammerman, J., & Konold, C. (2006). Exploring informal inference with interactive visualization software. In A. Rossman & B. Chance (Eds.), *Proceedings of the Seventh International Conference on Teaching Statistics*. Salvador, Bahia, Brazil: International Statistical Institute and International Association for Statistical Education. Online: www.stat. auckland.ac.nz/~iase/publications
Scheaffer, R. (1990). The ASA-NCTM quantitative literacy project: An overview. In D. Vere-Jones (Ed.), *Proceedings of the Third International Congress on Teaching Statistics*. Dunedin, New Zealand: International Statistical Institute. Online: www.stat.auckland.ac.nz/~iase/ publications
Scheaffer, R. (2006). Statistics and mathematics: On making a happy marriage. In G. Burrill (Ed.), *Thinking and reasoning with data and chance, 68th NCTM Yearbook (2006)* (pp. 309–321). Reston, VA: National Council of Teachers of Mathematics.
Schield, M. (1999). Statistical literacy: Thinking critically about statistics. *Of Significance, 1*(1), 15–20.
Schupp, H. (1982). Zum Verhältnis statistischer und wahrscheinlichkeitstheoretischer Komponenten im Stochastikunterricht der Sekundarstufe I (On the relationship between statistical and probabilistic components in lower secondary stochastics teaching). *Journal für Mathematik-Didaktik, 3*(3/4), 207–226.
Snee, R. D. (1990). Statistical thinking and its contribution to total quality. *American Statistician, 44*, 116–121.

Tukey, J. W. (1972). Data analysis, computation and mathematics. *Quarterly of Applied Mathematics, 30*, 51–65.
Tversky, A., & Kahneman, D. (1971). Belief in the law of small numbers. *Psychological Bulletin, 76*, 105–110.
Watson, J. M. (1997). Assessing statistical literacy using the media. In I. Gal & J. B. Garfield (Eds.), *The assessment challenge in statistics education* (pp. 107–121). Amsterdam: IOS Press and International Statistical Institute.
Wild, C., & Pfannkuch, M. (1999). Statistical thinking in empirical enquiry. *International Statistical Review, 67*, 223–265.

Chapter 11
Strengthening the Role of Probability Within Statistics Curricula

Manfred Borovcnik

Abstract This chapter illustrates probability as a type of thinking, which has its own existence even without a theoretical study. While such thinking is usually omitted in teaching, it is deep-rooted. The success of probabilistic models as compared to other, possibly primitive approaches is difficult to judge. This might hinder learners in accepting and applying – not to speak of understanding – the concepts. Probabilistic models seem to resemble scenarios more directly than fit perfectly to real situations. This runs contrary to current trends to reduce the link between probability and data to relative frequencies. A wider framework for the interpretation seems to be required. Consequences of such views on randomness and probability for teaching statistics are described.

1 Introduction

In this chapter, philosophical and psychological issues related to the understanding of chance and probability are analysed and implications for teaching statistics are discussed. The deliberations encompass official *and* private conceptions. The latter embed the concepts of randomness and probability in a wider context, which may account for their peculiar features.

As humans we think about what we could have done better yesterday, and we also care about what we can do for tomorrow. Like the child in Doris Day's "Que sera", we refuse to accept that "the future is not ours to see". We look for anyone who can dismantle the future: astrologists, gurus, and statisticians. Consistently, any conceptual offer will be judged by the extent to which it contributes to predicting the future. Hereby, "prediction of the future" comprises predicting the

M. Borovcnik (✉)
Department of Statistics, Alpen-Adria-Universität Klagenfurt, Universitätsstraße 65–67,
Klagenfurt 9020, Austria
e-mail: manfred.borovcnik@uni-klu.ac.at

C. Batanero, G. Burrill, and C. Reading (eds.), *Teaching Statistics in School
Mathematics-Challenges for Teaching and Teacher Education: A Joint ICMI/IASE Study*,
DOI 10.1007/978-94-007-1131-0_11, © Springer Science+Business Media B.V. 2011

unknown, that is, making inference about possible causes, or events in the past; it also includes inference about an unknown population.

Teaching focuses on the interpretation of probability as relative frequencies, as seen in curricula worldwide or in a famous discussion by Berry (1997), Albert (1997), Moore (1997), and Witmer, Short, Lindley, Freedman, and Scheaffer (1997) in the *American Statistician*. This approach is easier to understand and more objective (Moore, 1997). Endeavours centre on the relationship between data and probability reducing probability to a subsidiary concept, which is replaced by data analysis of suitable (simulated) data-sets.

Randomness is a reservoir of phenomena, including unpredictability, lack of patterns, lack of control over outcomes, and fairness. Probability is but one concept for dealing with randomness; as such it faces competing perceptions. To ignore private notions and to try to reach sound concepts as soon as possible hardly convinces learners that their efforts will pay off in the end. Stochastic notions are also intertwined with philosophical (Hacking, 1990) and psychological (Fischbein, 1975) components, since intuitions play an eminent role in the understanding of concepts (Kapadia & Borovcnik, 1991).

The philosophical Sect. 2 summarises the foundation of probability and characterises randomness by three debates: randomness–divination, randomness–causality, and random evolution–creationism. Section 3, on thinking probabilistically, elaborates five general features of such thinking. Finally, the educational situation is reviewed in Sect. 4, leading to arguments for a strong role of probability in statistics curricula.

2 Philosophy of Probability and Paradigms of Science

The concept of probability has developed only relatively recently and has received different interpretations that even today are subject to controversy (Batanero, Henry, & Parzysz, 2005). In this section, philosophical interpretations of randomness are summarised from the foundations debate and complemented by the concept's overlap with three further "dimensions": divination, causality, and creationism.

2.1 *Frequentist and Other Interpretations of Probability*

The frequentist interpretation of probability has grown in importance since the "golden" law of Bernoulli (1713/1987), which described a "theoretical" convergence of relative frequencies to the underlying probability. However, there are more officially accepted interpretations; the following are two of the different types of information for which probability may stand.

- *Objectivist information* where probability is described by (1) proportions of (equiprobable) cases, favourable to an event (Laplacian view); or (2) frequencies of an event in independent identical repetitions of a random experiment (Frequentist view).

- *Subjectivist information* (Subjective view) where probability is described by (1) qualitative knowledge of the event, perhaps of experts (possibly transferred from similar contexts, or gained by assumptions beyond any scrutiny); or (2) private degree of confidence in the statement (event).

Since the debates on the foundations between Bayesians (subjectivists) and objectivists (Barnett, 1973; Hacking, 1990), subjectivist views have found their way into applications with moderate positions (like Berger, 1993). If frequentist information is available, it has a priority; if it is missing (or too costly), qualitative knowledge is used instead.

2.2 Randomness–Divination

Early endeavours to take control of the future were connected to the course of the sun. Astronomy has been accompanied by astrology, which is connected with fate. Since the origins of astrology, predicting the future has been connected to devices, which bear an element of randomness. In *Games, Gods and Gambling*, David (1962) discriminated between two aspects of divination: to explore god's will and to surrender crucial decisions to god.

In ancient times, randomness was personified by female gods like the Greek Týche who changed the course of action with her moods. Fortuna, her Roman counterpart, is complemented with Iustitia: with sceptre and sword, later with a scarf to blind her eyes, she is the allegory of justice and fairness. Blindly drawing balls from an urn, Iustitia was revived to symbolise random samples as fair, when statistics bureaus justified random sampling to replace complete census (Kiaer, 1899).

2.3 Randomness–Causality

With the Renaissance the causal paradigm emerged from astronomical issues to explain the present and to predict the future, and probability (with an objective view) started to emerge as a branch of mathematics. At times chance was connected to deist determinism (see the following references in Batanero et al., 2005):

1. "It is written up there" (Diderot, 1796/1983, p. 103);
2. "[…] if […] future would not arrive with certainty, we cannot see how the supreme Creator could preserve […] his […] omnipotence" (Bernoulli, 1713/1987, p. 14); and
3. "Present events are connected with preceding ones by a link based upon the evident principle that a thing cannot occur without a cause which produces it". (Laplace, 1814/1995, p. vi).

Causal approaches have the advantage that once one recognises the mechanism of how a specific cause establishes the effect one can predict the future. Laplace gave the first definition (based on equally probable cases) of probability; yet for him probability is merely a substitute for ignorance in an otherwise purely deterministic world. Probability gains a genuine position by the success in thermodynamics in the nineteenth century when macroscopic causal laws were explained by random models at the microscopic level. This encouraged theoretical physicists of the twentieth century to eliminate causality by referring it back to randomness (Styer, 2000). However, the dispute in the foundations is ongoing; recent approaches allow a *deterministic* view on quantum mechanics (Dürr, Goldstein, Tumulka, & Zanghi, 2004).

2.4 Random Evolution–Creationism

Another philosophical question is the relationship between randomness and creationism. If an all-knowing god exists, the question arises as to whether this may be reconciled with randomness and free will. As randomness means unpredictability it conflicts with omniscience. If god knows every action of any individual in advance, how may the individual exert its free will? If no free will exists, no responsibility can remain with the individual. The debate between Darwin's evolutionary theory and creationism is similar to the causality–randomness debate in physics. Is nature created by a supernatural being or has it evolved solely due to random effects? Here and there the question can be put within a scientific debate but has to remain unanswered.

3 Thinking Probabilistically: Competing Intuitions and Strategies

While mathematicians define thinking probabilistically in terms of adequate use of probabilistic models, individuals are often faced with the *context* of situations to be modelled, which possibly leads them in directions different from standard models. To characterise thinking probabilistically is a genuine didactical task. Five features of probabilistic thinking are now elaborated.

3.1 Probability as an Index of Surprise

Randomness may be linked to surprise: more surprising events are usually judged as less probable. When a coincidence happens like meeting a neighbour during an overseas conference one might be surprised. With highly surprising events, however, individuals are inclined to think about alternative explanations such as "God's interference". Similarly, "rare events" serve as a basis for rejection in statistical tests.

To consider probability as an index of surprise may explain some well-known paradoxes. According to the conjunction fallacy (Tversky & Kahneman, 1983), people wrongly judge the conjunction of two "events" as more probable than each of the single events. This fallacy will not be elaborated here but serves only to illustrate matters. An index of surprise might lead wrongly to a higher probability judgement for the *conjunction* if it is less surprising to see the two events acting *together* than a single event, which might happen if one event seems to strongly suggest the other one.

3.2 Feedback from Probabilistic Situations is Indirect

Faced with the spinners in Fig. 11.1 – labels representing prizes – one might reason that it is better to choose the left. However, when playing (a) one loses quite often with both spinners, and (b) playing only a few times one might well lose all the games with the left spinner.

Feedback from this activity is indirect and valid only in a *series* of trials; but people feel confronted with a one-off situation. So how should one develop intuitive thought on the merits of measuring success by relative frequencies on the long run? Even with the "best" choice one is prone to lose when playing only once. The player is in the same situation as in the divination: Why did *he* lose? This prompts speculation about underlying reasons and might lead to a preference for magical thinking rather than knowledge.

If selecting the better option were always rewarding, then learning by trial and error could correct misleading conceptions. However, with random situations such learning is missing. The concepts are perceived as artificial, as are the possibilities to measure success. Alternative concepts might also fail; they might, however, have once proven to be successful. Who could convince such a person that his conception is wrong? It is remarkable that people tend to return to their previous unaltered private conceptions when they feel free from the demand to "answer tests". Fischbein (1975, 1987) developed the notion of primary (raw) intuitions, which are present before or develop without formal education, and secondary intuitions, which emerge from learning:

> Conflicts appear also between intuitive interpretations and formal ones (acquired by instruction). In children, such contradictory interpretations may annihilate the formal conception. [...] It is recommended that the student should be made aware of his tacit mental conflicts in order to strengthen the control of the taught conceptual structures over the primary intuitive ones (1987, p. 205).

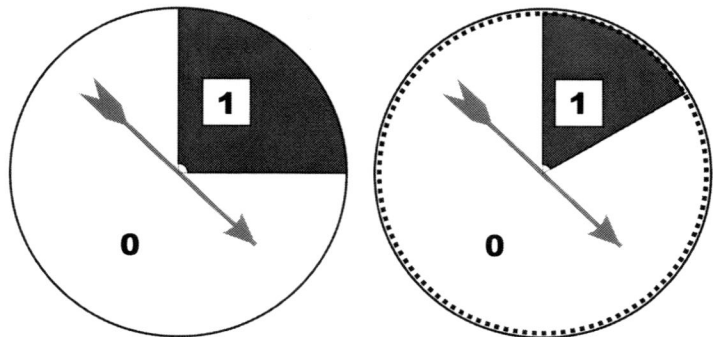

Fig. 11.1 Falk spinners

3.3   The Causal Alternative to Randomness

A causal view of a situation is always an alternative to a probabilistic perception of it. Especially if conditional probabilities are involved, causal perceptions are reinforced. If one of two events may be interpreted as cause C and the other as effect E, two sorts of confusion are conceivable: either the "cause" C is thought to be independent of the "effect" E, or the conditional probabilities are "reversed": $P(C|E) = P(E|C)$. While the first is prompted in time-bound contexts, reversing is popular in diagnostic contexts. For example, in the experiment of drawing twice – without replacement – from the "Falk urn" with two white and two black balls, possible questions (Falk & Konold, 1992) are:

(a) The first marble drawn is white. What is the probability that the second marble drawn is also white?
(b) The first marble drawn is hidden from you and laid to the side. Then the second marble drawn is white. What is the probability that the first marble is white?

Causal thinking is deep-seated and time-bound; in (b) people are convinced that what happened later lacks any relevance for previous events. They equate missing causal influence to independence and answer 1/2; causal schemes seem to attract more trust than "anaemic" probabilistic rules; the results were replicated by Borovcnik and Bentz (1990/2003).

In discussing conditional probabilities $P(X|D)$, Tversky and Kahneman (1980) distinguished two *types* of situations: D as a causal event if it is perceived as cause for X; D as a diagnostic event if X is a possible cause for D. Faced with the causal information $P(+|D) = 0.99$ of a medical test being positive given that a person has an illness D, people tend to equate the diagnostic probability $P(D|+)$ to 0.99; of course, this is much lower depending on the prevalence of the disease (Gigerenzer, 2002). Bayes' formula enhances the difference between causal and diagnostic probabilities but it is hard to learn. Moreover, the usual development of statistics leads to a violent clash as "prior" probabilities in this formula are genuinely subjective and thus are beyond any frequency interpretation.

3.4   The Conflict Between Actions and Reflections

Operations and (future) actions are on a different cognitive level from reflections. To ask a person to give an *estimate* of an unknown probability elicits different answers than to ask which of the options to *choose* if, for example, a bet is offered. This can be illustrated by the Kahneman and Tversky (1972) example: Which of the two following sequences of coin tossing is the more probable: (a) HTTHTH, or (b) HHHHTH? In in-depth interviews, mathematics students – at the level of *reflection* – recognised the possibilities as equally likely (Borovcnik & Bentz, 1990/2003). But, when they were offered a bet – at the level of *action* – they strongly favoured sequel (a), accepting unfair stakes of up to 2:1 (which amounts to a probability of 2/3).

The conflict between actions and reflections, discussed in Borovcnik and Peard (1996), is connected to the outcome approach of Konold (1989), who explains a tendency for people to focus on such information from the context of tasks, which allows a direct choice of an action, while neglecting other information: "This [outcome approach] can be characterised as involving two general features: (a) the tendency to predict the outcome of *a single trial*, and (b) the reliance on causal as opposed to stochastic explanations of outcome occurrence and variability" (Konold, 1989, p. 65).

3.5   Non-probabilistic Criteria for Decisions

Divination, free will, and self-responsibility act together and underlie the Monty Hall problem where a prize is hidden behind one of three doors (Gigerenzer, 2002). The question remains why such a comparatively easy task is disputed so much, even by mathematicians (vos Savant, n.d.). Why are people so emotionally laden? Of course, experts understand the solution with Bayes' formula; yet some still *refute it*! Again, here is circumstantial evidence that stochastic concepts are hierarchically ranked below other concepts. This statement is in accord with research findings, which show that mathematical competencies of adults "reduce" to those they develop in the first 6 years of education (Peters, 2008).

Reluctance to accept probabilistic reasoning may be explained by the following argument: If we stay with our first choice, randomness has given us our part – divination – fate. ... If we change our choice and reject what fate has attributed to us, we have to take *responsibility* if we lose. We are inclined to surrender our own responsibility to fate – or to randomness. Since ancient times, crucial decisions have been outsourced to divination. Now the outsourcing in financial markets has changed to computer programs with poorly understood algorithms.

4 Implications for Teaching

Implications for teaching include review of some issues from the educational debate on the role of probability, listing of reasons for a strong role of *probability*, and suggestions for addressing probability in teaching.

4.1 The Role of Probability in School Curricula

An investigation among researchers within the didactics of stochastics (Nemetz, 1997) revealed the following reasons for probability vanishing from curricula internationally: probability is (a) orientated too much towards mathematics; (b) too tightly connected to games of chance and possibly amoral; and (c) only required to *justify* the methods of inferential statistics. Reason (a) may be addressed by using simulation and interactive animations when teaching probability. Reason (b) neglects the influence of gambling and games of chance; to clarify and understand such games is better than trying to ignore them. Essentially, probability concepts originate from games and insurance enterprises (which are "games"). Reason (c) ignores probability as a tool to investigate reality (see Sect. 4.2). In fact, inference is included at high school level in some countries (see Harradine, Batanero, & Rossman, this book).

4.2 Joint Study Discussion Points About Probability

The ICMI/IASE Joint Study Conference focused on teaching *statistics*; yet the presentations in Panel 2 "The Interplay of Probability and Statistics in Teaching and in Training the Teachers" as well as the presentations of Carranza and Kuzniak (2008), and Chaput, Girard, and Henry (2008) "invited" the audience to explore an agenda on *probability*:

- Crucial issues of a modelling approach and the role of pseudo-concrete models;
- Problems within a frequentist approach towards probability;
- Disadvantages of approaches based on a reduced interpretation of probability;
- Subjective probability and its potential role within frequentist approaches;
- Relative merits of real applications and artificial situations;
- Features of sampling including real sampling and simulation.

A tentative response to the first point relates also to the other points. Probability is a complex concept. To facilitate understanding, didacticians focus on models which (a) embody the theory-like urns, or (b) come as close to the real situation as possible and that are taken into reality via simulation and attain the status of real existence. In both approaches the difference between model and reality collapses; resulting data is taken as factual, neglecting the circumstance that it is restricted to the *model* used (see Chaput et al., 2008 or 2011, this book).

Here, a peculiarity of probability models is worth mentioning: their *scenario character*. Even if models miss a perfect fit, some characteristics may be derived from them for making decisions *transparent*. The case of taking out an insurance policy for a car, discussed in Borovcnik (2006) is but one example. While the frequency interpretation is useful for the company, because it always makes money in the long run, it is irrelevant for the policy taker, because it is a one-off situation. Interestingly, there is a solution to this problem without eliciting the person's exact subjective probabilities.

In technical systems, another example concerns the reliability of units and complex systems (Borovcnik, 2006). Two issues make it doubtful whether there is a link between probability of failure and data. First, data is missing to estimate the probability; if there is data, they result from accelerated life tests and it is speculative as to how to transfer to normal conditions. Second, the assumption of independence between units in systems is unreasonable and there is hardly a way to check it. Moreover, how does one interpret a risk of 10^{-12}? Beyond doubt, such a value has more of a figurative character in the sense that it is better than 10^{-6}. Such risks lack a frequency interpretation. Despite the clash between the real situation and the "model", conclusions about the real situation may be derived.

4.3 Supporting the Role of Probability in the Curriculum

The following theses are meant to explain probability's role when opposed by people who suggest that *randomness does not exist*. Randomness is a concept that allows one to think about the world. We have a strong affinity with other kinds of thinking, which might lead us in directions different from those predesigned by probability theory. This creates special challenges for teaching.

- Only in rare cases do results from data handling speak for themselves and allow a clear message without referring to probability.
- The peculiarity of probabilistic thinking in contrast to logical, causal, or mystic thinking is important. To clarify the abundance of personal thought will help to build stable intuitions about probability and its potential.
- Clarification of the mutual dependencies between frequentist, Bayesian, and mathematical conceptions and intuitive thought makes the concept of probability flexible and robust.
- Basic notions of expected value, risk, and variability rely on sound conceptions of probability.
- The emergence of probability concepts is embedded in games of chance. This context helps to anchor mental images in the learners' cognitive system, which could serve both as prototypes for modelling and for clarifying intuitions.

- A restricted notion of probability as frequency in the long run ignores valuable *qualitative* sources of information. Moreover, it hinders applications in contexts of reliability, or risk, and impedes a sound interpretation of methods of inferential statistics.
- Probability is only *one* kind of approach to a problem, which might alternatively be solved by, for example, differential equations. This corroborates the idea that probability models have more the character of *scenarios*.
- Clarification and demystification of conditional probability from concurrent causal ideas is a prerequisite for understanding concepts of inferential statistics like type I and II errors, or *p* values.
- Probability enhances the role of *random* samples for generalising findings from samples to populations. Moreover, inferential statistical methods are intermingled with conditional probabilities and a sound understanding thereof.

4.4 Perspectives on Teaching

Following are some arguments for a strong role of probability within stochastic curricula (a) probability is indispensable for *understanding* the methods of inferential statistics; (b) probability offers a type of tool for modelling and/or "creating" reality; and (c) probability offers a type of thinking, with which one can reflect on reality. From these arguments one may see that to abandon probability within the curricula means reducing the repertoire of intellectual possibilities.

The position of probability within mathematics curricula is disputed as probability seems to be ambiguous and prone to private (and therefore officially unacceptable) conceptions. Most mathematicians who teach probability strive for a direct path to "clean" concepts.

- The didactical approach has always challenged such a view as inappropriate. By learning only *mathematics*, some misconceptions remain unaltered, as we may see from studies such as that of Díaz and Batanero (2009).
- Chaput et al. (2008, p. 6) expressed: "The construction of mental images relative to randomness is delicate". To meet this problem, they suggested *activities* of *simulating* experiments at *earlier* stages than now.
- The emotional component of private thought is touched by Garuti, Orlandoni, and Ricci (2008, p. 5), who explicitly referred to the existing "magic view of random events". They suggested "to tame it" by a *mixture* of classical probability *and* real applications.
- Personal intuitions about probability may be classified into objective and subjective conceptions. This establishes a further source of problems as Carranza and Kuzniak (2008, p. 3) noted a tendency to avoid subjective views in teaching: "Thus the concept […] is truncated: the frequentist definition is the only one approach taught, while the students are confronted with frequentist and Bayesian problem situations". They argued for a duality between the two aspects of probability, to reduce to one aspect provokes special problems of comprehension on the side of the learners.

- Yet another source of complexity for learning is the overlap between probabilistic and deterministic reasoning. Ottaviani (2008, p. 1) referred to Fischbein in accentuating that probability and statistics belong to a line of thought which is essentially different from deterministic reasoning and continued to state: "It is not enough to show random phenomena. To enrich the child's probabilistic experience, it is necessary to draw the distinction between what is random and what is chaos".

Modelling and simulation are the usual suggestions to cope with the multifaceted character of probability. The arguments of Carranza and Kuzniak (2008) confirm that we need wider approaches if teaching is to be successful. Moreover, Girard (2008, p. 2) remarked that "teaching probability by modelling and simulation is not easy [...]. The link between statistics and probability is still to be clarified [...]". Ottaviani (2008, p. 2) considers the need for further research, "as the risk could be that [statistics and probability] continue not to converge in school and, worse still, in our learners' minds".

Simulation or interactive animations may be used to reduce the need for mathematical sophistication. Simulation should be accompanied by the idea of a scenario, a potential counterpart of the real situation, which helps to explore it. There is a need for a reference concept wider than the frequentist approach. In the early phases of teaching, the attitude of empirical research contributes to clarifying the more archaic private conceptions and how formal concepts of probability may substitute some of our strategies (Lysø, 2008).

The mathematics to continue from there could well be impregnated with *fundamental ideas*. The ideas of Heitele (1975) are undisputed as the discussion by Batanero et al. (2005) showed. A description of the named ideas clearly reveals their didactic value. Yet, Heitele's list reads like the contents of a textbook reflecting the aim to replicate the inner-mathematical structure. An outside perspective addressing "What purpose do the concepts serve, what other concepts are available for the task, and what answers do the concepts fail to give?" is Borovcnik (1996), who centres his "programme" on *information* as key idea.

Probability is signified by a peculiar kind of *thinking*, which is not preserved in its mathematical conception. Concurrent and overlapping modes of thought interfere with it. This chapter shows that modelling and simulation are only two steps towards the direction to disclose probability as a *cultural* phenomenon that may clear our thought and clarify our expectations when we apply related methods to problems under "uncertainty".

References

Albert, J. (1997). Teaching Bayes' rule: A data-oriented approach. *American Statistician, 51*(3), 247–253.
Barnett, V. (1973). *Comparative statistical inference*. New York: Wiley.
Batanero, C., Burrill, G., Reading, C., & Rossman, A. (Eds.) (2008). *Joint ICMI/IASE Study: Teaching Statistics in School Mathematics. Challenges for Teaching and Teacher Education.*

Proceedings of the ICMI Study 18 and 2008 IASE Round Table Conference. Monterrey, Mexico: International Commission on Mathematical Instruction and International Association for Statistical Education. Online: www.stat.auckland.ac.nz/~iase/publications

Batanero, C., Henry, M., & Parzysz, B. (2005). The nature of chance and probability. In A. G. Jones (Ed.), *Exploring probability in school: Challenges for teaching and learning* (pp. 15–37). New York: Springer.

Berger, J. O. (1993). *Statistical decision theory and Bayesian analysis.* New York: Springer.

Bernoulli, Jacques (1987). *Ars conjectandi–4ème partie* (N. Meunier, Trans.). Rouen, France: Institut de Recherche sur l'Enseignement Mathematique (Original work published in 1713).

Berry, D. A. (1997). Teaching elementary Bayesian statistics with real applications in science. *American Statistician, 51*(3), 241–246.

Borovcnik, M. (1996). Fundamentale Ideen als Organisationsprinzip in der Mathematik-Didaktik (Fundamental ideas as organizing logic in the didactics of mathematics). In K. P. Müller (Ed.), *Beiträge zum Mathematikunterricht* (pp. 106–109). Hildesheim: Franzbecker.

Borovcnik, M. (2006). Probabilistic and statistical thinking. In M. Bosch (Ed.), *Proceedings of the Fourth Congress of the European Society for Research in Mathematics Education* (pp. 484–506). Barcelona: European Society for Research in Mathematics Education. Online: ermeweb.free.fr/ CERME4/

Borovcnik, M., & Bentz, H. J. (1990/2003). *Intuitive Vorstellungen von Wahrscheinlichkeitskonzepten: Fragebögen und Tiefeninterviews.* (Intuitive conceptions of probabilistic concepts: Questionnaire and in-depth interviews.) (Technical Reports). Klagenfurt am Wörthersee: Klagenfurt University.

Borovcnik, M., & Peard, R. (1996). Probability. In A. Bishop, K. Clements, C. Keitel, J. Kilpatrick, & C. Laborde (Eds.), *International handbook of mathematics education* (pp. 239–288). Dordrecht, The Netherlands: Kluwer.

Carranza, P., & Kuzniak, A. (2008). Duality of probability and statistics teaching in French education. In C. Batanero, G. Burrill, C. Reading, & A. Rossman (2008).

Chaput, B., Girard, J. C., & Henry, M. (2008). Modeling and simulations in statistics education. In C. Batanero, G. Burrill, C. Reading, & A. Rossman (2008).

David, F. N. (1962). *Games, gods and gambling.* London: Charles Griffin.

Díaz, C., & Batanero, C. (2009). University Students' Knowledge and Biases in Conditional Probability Reasoning. *International Electronic Journal of Mathematics Education, 4*(3), 131–162. Online: www.iejme.com/

Diderot, D. (1983). *Jacques le Fataliste* (*Jacques the Fatalist*). Paris: Le Livre de Poche (Original work published 1796).

Dürr, D., Goldstein, S., Tumulka, R., & Zanghi, N. (2004, August 24). Bohmian mechanics and quantum field theory. *Physical Review Letters, 93*(9). Online: arxiv.org/abs/quant-ph/ 0303156v2.

Falk, R., & Konold, C. (1992). The psychology of learning probability. In F. Sheldon & G. Sheldon (Eds.), *Statistics for the twenty-first century* (MAA notes 26, pp. 151–164). Washington, DC: The Mathematical Association of America.

Fischbein, E. (1975). *The intuitive sources of probabilistic thinking in children.* Dordrecht, The Netherlands: Reidel.

Fischbein, E. (1987). *Intuitions in science and mathematics. An educational approach.* Dordrecht, The Netherlands: Reidel.

Garuti, R., Orlandoni, A., & Ricci, R. (2008). Which probability do we have to meet? A case study about statistical and classical approach to probability in students' behaviour. In C. Batanero, G. Burrill, C. Reading, & A. Rossman (2008).

Gigerenzer, G. (2002). *Calculated risks: How to know when numbers deceive you.* New York: Simon & Schuster.

Girard, J. C. (2008). The Interplay of probability and statistics in teaching and in training the teachers in France. In C. Batanero, G. Burrill, C. Reading, & A. Rossman (2008).

Hacking, I. (1990). *The taming of chance.* Cambridge, UK: Cambridge University Press.

Heitele, D. (1975). An epistemological view on fundamental stochastic ideas. *Educational Studies in Mathematics, 6*, 187–205.

Kahneman, D., & Tversky, A. (1972). Subjective probability: A judgement of representativeness. *Cognitive Psychology, 3*, 430–454.

Kapadia, R., & Borovcnik, M. (1991). *Chance encounters: Probability in education*. Dordrecht, The Netherlands: Kluwer.

Kiaer, A. N. (1899). Die repräsentative Untersuchungsmethode (The representative method of investigation). *Allgemeines Statistisches Archiv, 5*, 1–22.

Konold, C. (1989). Informal conceptions of probability. *Cognition and Instruction, 6*(1), 59–98.

Laplace, P. S. (1995). *Théorie analytique des probabilités* (*Analytical theory of probabilities*). Paris: Jacques Gabay. (Original work published 1814).

Lysø, K. (2008, July). Strengths and limitations of informal conceptions in introductory probability courses for future lower secondary teachers. Paper presented at the *Eleventh International Congress on Mathematics Education* (*ICME 11*), *Topic Study Group 13 "Research and development in the teaching and learning of probability"*. Monterrey, Mexico. Online: tsg.icme11.org/tsg/show/14

Moore, D. S. (1997). Bayes for beginners? Some reasons to hesitate. *American Statistician, 51*(3), 254–261.

Nemetz, T. (1997). An overview of the teaching of probability in secondary schools. In B. Phillips (Ed.), *Papers on statistical education presented at ICME 9* (pp. 75–86). Hawthorn: Swinburne University.

Ottaviani, M. G. (2008). The interplay of probability and statistics in teaching and in training the teachers. In C. Batanero, G. Burrill, C. Reading, & A. Rossman (2008).

Peters, E. (2008). Numeracy and the perception and communication of risk. *Annals of the New York Academy of Science, 1128*, 234–267.

Styer, D. F. (2000). *The strange world of quantum mechanics*. Cambridge, UK: Cambridge University Press.

Tversky, A., & Kahneman, D. (1980). Causal schemas in judgment under uncertainty. In M. Fishbein (Ed.), *Progress in social psychology*. Hillsdale, NJ: Lawrence Erlbaum.

Tversky, A., & Kahneman, D. (1983). Extensional versus intuitive reasoning: The conjunction fallacy in probability judgment. *Psychological Review, 90*(4), 293–315.

vos Savant, M. (n.d.). *Game show problem*. Online: www.marilynvossavant.com/articles/gameshow.html

Witmer, J., Short, T. H., Lindley, D. V., Freedman, D. A., & Scheaffer, R. L. (1997). Teacher's corner. Discussion of papers by Berry, Albert, and Moore, with replies from the authors. *American Statistician, 51*(3), 262–274.

Chapter 12
Frequentist Approach: Modelling and Simulation in Statistics and Probability Teaching

Brigitte Chaput, Jean-Claude Girard, and Michel Henry

Abstract In this chapter, a question is posed about the link between two traditional approaches to the notion of probability, classical (or Laplacian) and frequentist, in secondary teaching. Different conceptions of probability, objective and subjective, are considered, some didactical difficulties of the frequentist option are underlined, and the modelling view point is presented. A critical description of a modelling process of a random situation for teachers' training in secondary teaching is proposed and it is developed for the example of a queue. Finally, the status of simulations on computers in classrooms is clarified and their didactical relevance is highlighted.

1 Introduction

During the last 30 years, teaching of probability in secondary schools has evolved considerably in many countries. Traditionally, this teaching was based on the classical definition of probability that had emerged in the early eighteenth century and was notably defined by De Moivre (1967/1756, p. 1) in his Doctrine of Chances (original text):

> If we constitute a Fraction whereof the Numerator be the Number of Chances whereby an Event may happen, and the Denominator the Number of all the Chances whereby it may either happen or fail, that Fraction will be a proper designation of the Probability of happening.

B. Chaput
École Nationale de Formation Agronomique (ENFA), 2 route de Narbonne, BP 22687,
31326 Castanet-Tolosan, France
e-mail: brigitte.chaput@educagri.fr

J.-C. Girard
Université Claude Bernard, Lyon I, IREM, 43 Boulevard du 11 Novembre 1918,
69622 Villeurbanne, France
e-mail: jcg.girard@laposte.net

M. Henry (✉)
Université de Franche-Comté, UFR ST, IREM La Bouloie, 25030 Besançon, France
e-mail: michel.henry@univ-fcomte.fr

C. Batanero, G. Burrill, and C. Reading (eds.), *Teaching Statistics in School*
Mathematics-Challenges for Teaching and Teacher Education: A Joint ICMI/IASE Study,
DOI 10.1007/978-94-007-1131-0_12, © Springer Science+Business Media B.V. 2011

At the end of the twentieth century, the development of data analysis, the evolution of mathematics teaching towards more experimental activities, and the allocation to primary and secondary schools of computers with statistical software has led, especially in France, to another fundamental epistemological choice for the introduction of the notion of probability. This choice is based on observations of the well-known phenomenon of the stabilisation of the relative frequencies of an event associated with a random experiment when it is possible to repeat this experiment a sufficiently large number of times. This "stabilised" value is then considered as an objective measure of the probability of this event. Some authors of textbooks propose this approach to give a definition, called "frequentist", to probability (Renyi, 1992/1966, p. 25): "We will call probability of an event the number around which the relative frequency of the considered event fluctuates…"

Thus, the notion of probability can be introduced through these two approaches, which are not independent of philosophical options, objectivist and subjectivist. The modelling perspective achieves a synthesis between these two approaches. Recent didactic studies have led to teaching based on this modelling process using simulations of models in statistics. Such a choice necessarily involves some implications for teacher training.

2 The Modelling Perspective to Link the Classical and Frequentist Approaches in Probability Teaching

The link between the classical and frequentist approaches must be clarified for secondary teaching (Garuti, Orlandoni, & Ricci, 2008). In the twenty-first century the French curriculum, like that of other countries, opted for a modelling perspective: probability is a theoretical value of the degree of confidence that one can give to a random outcome. This probability can be either worked out a priori or issued a posteriori from observing its relative frequencies or estimated subjectively. In this framework, to represent a random drawing of subjects from a population, a probability distribution of some observable characteristics has the status of an interpretative model of the data which can come from a real observation of this population. This perspective concurs with the general process of contemporary statistical thinking (Wild & Pfannkuch, 1999), which allows the application of powerful theoretical results in correct mathematical conditions, notably in statistical inferences or in data analysis.

Such a didactic choice links the classical and frequentist approaches of the notion of probability. At the same time, it contributes to the learning of a modelling process. The confrontation between theoretical results obtained from an accurately constructed model and an experimental reality is important in a scientific process and must therefore be present in a school curriculum. It is pointed out that in statistics, more so than in other fields, it is possible to provide the students with models which are simple enough to be easily understood. Teachers have to ensure that students make a clear distinction between model and reality.

Presenting probability with a modelling approach for mathematics teacher training requires clarification of the modelling process which leads from an observation of a random situation and its *perception*, to its *description* and *comprehension*, and finally to its *signification*, according to the three stages described for primary teaching by Biembengut (2007).

Finally, from a probabilistic model built to interpret a set of statistical data, powerful computers available in a classroom can be used to explore the working effects of the model or to make, for instance, numerous simulations of random samples. But what pedagogical advantage can we draw from using such simulations?

3 Various Conceptions of Probability

Historically, the notion of probability of a random event is the subject of different conceptions, objectivist and subjectivist, in a duality for which Hacking (1975, p. 12) pointed out:

> It is notable that the probability that emerged so suddenly is Janus-faced. On the one side it is statistical, concerning itself with stochastic laws of chance processes. On the other side it is epistemological, dedicated to assessing reasonable degrees of belief in propositions quite devoid of statistical background.

The objectivist conception of probability has two main approaches:

- Classical or logical, based on physical considerations of symmetries in the random generator or on similarities in the wording that describe the random situation, which enables the same probability to be given to such symmetric cases, while admitting that these symmetries are subjective in nature.
- Frequentist, which considers that the probability is included in the random experiment itself and appears as a stabilised relative frequency when the same random experiment is repeated independently in the same conditions, but this independence cannot be clearly defined in reality and must be subjectively accepted.

On the other hand, the subjectivist or Bayesian conception of probability considers that the probability of a random event, or a distribution of probabilities, is a personal degree of belief a priori put on a random situation, evolving with the experimental data towards a conditional probability. During the Joint ICMI/IASE Study Conference in Monterrey, Carranza and Kuzniak (2008) presented the duality between frequentist and Bayesian approaches in French education.

Thus, the subjectivists of the twentieth century, when developing the Bayesian methods, declared as De Finetti (1974, preface p. X) claimed in uppercase: "PROBABILITY DOES NOT EXIST". As Keynes (1921) had remarked previously, this statement leads to the fact that the probability of an event depends on the field of knowledge of the observer and could vary from one person to another. It could also vary for the same observer acquiring new knowledge, which allows the modification of this a priori probability by referring to Bayes' theorem.

Even if it is not ignored by teachers, the subjective conception of probability is rarely taken into account in basic probability teaching. However, Albert (2006) proposed activities based on this conception and gave references for teaching about the debate between objectivists and subjectivists. Moreover, mathematicians adopt an axiomatic standpoint defining the probability as a mathematical abstract object used to model random situations (Batanero, Henry, & Parzysz, 2005). Thus, probability can be understood either as an intrinsic value in a random experiment, independent of the observer, or as a subjective value depending on the observer's knowledge, or as a theoretical value in an interpretative model.

So, we come to a fundamental didactical question: Can we introduce these historical and philosophical controversies in the secondary teaching of probability?

4 Which Conception of Probability Should Be Used for Secondary Teaching?

In the 1990s, the teaching of probability had to take into account the rapidly growing use of computers in data handling (Biehler, 1991), and statistical thinking became an object of study that could not be ignored in order to enhance citizenship education, as stated by the official Reflection on Mathematics Education Committee, presided over by J. P. Kahane, in its report to the French Ministry of National Education (2002, p. 53, authors' translation):

> Some statistical concepts … are useful to understand public debates in every country; then, statistical language with its proper rules, syntax and semantics has to be learned; the teaching of statistics is, by nature, associated with the teaching of probability, it is actually a teaching about randomness.

Understandably, these recommendations link statistics and probability. The tendency to teach probability through a frequentist approach has grown in many countries, including France since 1991. This objective conception leads to definitions like the one given by Alfred (Renyi, 1992/1966, p. 26, authors' translation): "The mathematical theory of probability does not deal with subjective judgments; it relates to objective probabilities that can be measured as physical magnitudes".

As a result, two different interpretations of the notion of probability are now presented in secondary teaching:

- A theoretical a priori value corresponding to the idea of *chance*, computable as long as one makes an equally likely assumption *somewhere*; and
- An experimental measurement obtained by the observation of a *stabilised relative frequency* when the same random experiment is repeated a large number of times under the same conditions.

These two interpretations should not be separated if students are to develop a good understanding of probability, and to be able to apply it in practical situations. The French choice in secondary teaching is to adopt an objective conception, with both classical and frequentist approaches inseparable, because focusing on

- The classical approach only leads to reduce the understanding of probability to the counting of elementary events and to the overuse of combinatorics;
- The frequentist approach only, because of its empirical nature, generates didactic issues like the confusion between the observation of reality (stabilised frequency) and theoretical knowledge (probability).

In fact, the definition of probability given in some books or curricular guidelines as a *stabilised relative frequency* raises serious epistemological problems, because one can only estimate a value of a probability through empirical frequencies and cannot take this value for a definition of a mathematical object. In this approach, reality and mathematics domains are confused.

A meticulous wording of the Law of Large Numbers, even in the simplest form of Bernoulli's theorem, presupposes a *mathematical definition* of probability and should not be introduced in a context in which model and reality are mixed up.

So, which approach would be the best to use when teaching at the secondary level? This question is all the more tricky. If a modelling approach is adopted to teach probability, these different conceptions can be overcome by presenting the probability of a random event as a theoretical model of a ratio of cases or a stabilised relative frequency or a degree of belief.

5   A Modelling Approach for Teaching Statistics and Probability

Through scientific reasoning, the modelling perspective moves the debate between subjectivists and objectivists back to the choice of the most relevant probabilistic conception. Probability is then defined in an axiomatic way as a theoretical object (Steinbring, 1991), ideally quantifying the possibility of a given event, intuitively estimated, a priori calculated, or experimentally measured in its practical uses.

The 14th International Commission on Mathematical Instruction Study (Blum, Galbraith, Henn, & Niss, 2007) reviewed the situation about the use of modelling and applications in mathematics education.

A presentation of a model accounts for various levels of abstraction and formalism. First for a didactic purpose, some basic models may be presented in relation to reality by using everyday words. Then, the objects from reality are idealised by selecting relevant characteristic properties. In this process, one obtains what we call *pseudo-concrete models*. This is the case for the two colours Bernoulli's urn model (Henry, 2001a) which can be applied to a population (for example, in situations of random sampling for polls) in which individuals have one characteristic in the proportion p. A draw at random of one individual from the population is represented by a draw of a marble at random from the urn. By definition, this ideal urn contains marbles "undistinguishable by touch" (as described in some problems), which implies the implicit hypothesis: an equal probability of all the marbles in a draw at random.

Among different types of representations, mathematical language and mathematical symbolism allow strong descriptions on which general properties and algorithms can operate. These representations will be called *mathematical models*. Below are described three steps in the didactic analysis of teaching a modelling process: pseudo-concrete model, mathematisation and validation (Henry, 2001a).

5.1   *Pseudo-concrete Model and Description of Reality*

The first step of a modelling process consists of describing the concrete situation in usual language and building up an experimental protocol containing a set of instructions to be followed in order to carry out an experiment and to reproduce it under the same conditions, if necessary. This description leads to the creation of hypotheses which are intended to interpret the situation. For instance, if students use an urn model, then they have to decide that the marbles represent events which have the same chance of occurring.

Consider the example of a queue. Suppose that the arrival of a customer is isolated and the frequency of such an event in a given period of time depends neither on the time chosen for the observation nor on the previous arrivals. A minimum theoretical knowledge enables the transformation of these empirical observations into *working hypotheses*. From a didactic point of view, this stage is called *a contextualisation of previous knowledge*. In the queue example, it can be assumed that the probability of a customer's arrival in a very short time interval is quasi-null (rare event), that the arrival of a customer in a given period of time depends only on its duration (homogeneous phenomenon through time), and that two arrivals in disjoint intervals are independent (phenomenon without memory).

5.2   *Mathematisation and Formalisation*

Next comes the second step of the modelling process: translating the *working hypotheses* into *model hypotheses*. Students are required to translate the pseudo-concrete model into a simplified and mathematical symbolic system, and to choose the characteristics of the real objects which are idealised in order to design a relevant *probabilistic model*. For instance, for the urn model, the discrete uniform distribution on an abstract set is used to model the situation.

In the queue example, the work hypotheses are interpreted by formal hypotheses: one supposes that the probability of a customer's arrival in the time interval $[t, t+\Delta t]$ does not depend on t (homogeneous phenomenon); let $P(\Delta t)$ be this probability, one supposes that $P(\Delta t)/\Delta t$ tends towards a constant c (the rhythm of the arrivals) when Δt tends towards 0 (rare event); for this P, the customers' arrivals in two disjoint time intervals are independent events (phenomenon without memory). These hypotheses lead to a Poisson model for the number of customers showing up in a time interval [0, T] and to an exponential distribution for the duration of the waiting between two customers (Henry, 2001b; Borovcnik, Bentz, & Kapadia, 1991).

5.3   Validation and Interpretation in the Context

The third, and final, step consists first in translating the mathematical results according to the previous pseudo-concrete model, then giving them a meaning to create answers to the original problem in the real world, and then again comparing these answers with the *model hypotheses*. Finally, the answers have to be put into perspective to estimate whether the model was adequate for the real problem. In the previous example of the queue, one has to check whether the statistical data fit the Poisson model in a relevant way.

6   Didactic Difficulties Linked to a Modelling Process

The French curriculum of the twenty-first century in statistics and probability suggests the teaching of probability via modelling. This perspective is similar to the construction process of the Euclidean model in geometry at compulsory education level (Henry, 1999; Girard, 2004), because in both cases the starting point is the observation and the description of real objects. In geometry, more so than in statistics, it is easy to find modelling activities in real situations that children can understand. In this case, young children discover real geometric objects, represent them by geometric drawings (or on computer screens with dynamic geometric software), then the properties of the objects are progressively drawn out and idealised, and finally the mathematical objects can be defined as *figures*. This conceptual jump generates difficulties for the students who gradually find out about the scientific process. This modelling process is usual in mathematics education, for example, the progressive building of different sets of numbers, or equations written to solve concrete problems in which students have to choose unknown quantities.

  Statistics and probability belong to a different context compared to geometry, and have been treated differently over time. In France, the first difference is that geometry teaching starts at elementary school and lasts 10 years, while statistics and probability teaching (unlike in other countries such as United States of America, United Kingdom, Spain, Brazil, and Australia) starts only at the end of middle school and in the first 2 years of senior secondary level, when students have reached the ages of 15–16. Naive conceptions have been settled already in their minds as the perception of randomness which is not univocal and is linked to many different beliefs (Kahneman, Slovic, & Tversky, 1982; Lecoutre & Fischbein, 1998). A late introduction of probability in secondary teaching conflicts with these náive conceptions. So secondary school students often have misconceptions about probability (Batanero & Sanchez, 2005).

The second difference, in France, is that most of the geometry problems are posed and solved in the Euclidean model context. A model is seldom used to treat practical problems; at best it concerns pseudo-concrete problems and most of the time the associated modelling is already completely detailed for students. On the contrary, in almost all statistics and probability applications, the modelling stage has to be present. Exercises are presented in a concrete packaging and their contexts are close enough to the learning situations involved to enable the students to transfer them to usual probability distributions. If the learning period is too short in time, there is a strong risk that such an approach may not make sense to the students. In spite of these difficulties, the French curriculum has taken up the modelling perspective. For example, the official guideline for Grade 11 (16–17-year-olds) edited by the national French experts group on mathematics school curricula (Groupe d'Experts sur les Programmes Scolaires) gives this definition (GEPS, 2001, p. 68): "One will make the list of the elementary mathematical properties of the object 'distributions of relative frequencies' and one will define a probability distribution as a mathematical object which has the same properties". The following comment is also provided:

> Modelling a random experiment means associating a probability distribution with it. A frequency is empirical: it is calculated from experimental data, whereas the probability of an event is a 'theoretical value'… Thanks to practical examples, students should understand that modelling means choosing a probability distribution.

7 Simulations of Models in the Teaching of Statistics and Probability

Modelling is also present in various simulation activities. Simulation is an artificial reproduction of a theoretical model of the situation (Girard, 2004; Parzysz, 2009), analysing its behaviour in response to input variations, and eventually planning the consequences of similar changes within a real context. In the past, when building a sailing boat a scale model was made some months in advance to follow during the different building stages and to anticipate any possible failures. Similarly, for manufacturing an aircraft wind tunnel tests are made on a scale model of a plane. The training of astronauts and plane pilots makes use of flight simulators that include theoretical models of foreseeable incidents.

Currently in statistics teaching, computer tools allow, for example, the creation of numerous sample simulations based on probabilistic models of populations and the determination of various parameters through a frequentist approach, or the testing of some theoretical models by comparing their behaviour with the real observed data.

But the status of a computer simulation in class must be analysed, according to Mills (2002), and its didactical pertinence must be discussed, according to others (Burrill, 2002; Zieffler & Garfield, 2007).

In many countries the statistics curriculum for students of different ages introduces simulations, but the term *simulation* involves more than what is actually required from the students: representing the outcomes of a concrete experiment. Accepting this restriction does not raise the problem of a subjacent theoretical model necessary to achieve the task or even reveal the absence of this model in the students' minds (Parzysz, 2009). Thus, in simple situations such as throwing dice, equal probability is implicitly accepted and associated with a uniform discrete distribution which is supposed to control the random numbers used in the simulation. However, no theory of what uniform distribution means has previously been developed with students. In fact, the designing of a *simulation* requires a minimal knowledge of probability, particularly about probability distributions, that students do not actually have.

For example, a set of values of a physical quantity is generally modelled by a normal distribution, as long as the variations spread around a central value and are coming from measurement subject to errors or other random causes. The normal curve is then a model for a histogram of the simulated data.

The question is how to justify to students the equivalence of real random experiments or pseudo-concrete descriptions with a computer simulation, judiciously programmed from a theoretical model. The equivalence is ensured by the fact that both experiments are relative to the same probabilistic model, a concept not yet available to the students. Another question is how to interpret the sampling fluctuations observed in the repetition of the simulation. Without answers to these two questions, teachers are in a difficult didactic situation. This didactic inconsistency is identified in the French curriculum guideline (GEPS, 2001, p. 72, authors' translation), which gives teaching tips for Grade 11: "The respective positions of modelling and simulation will be briefly clarified: modelling consists of associating a model with experimental data while simulating consists of producing data from a defined model".

In modelling, the Law of Large Numbers plays a decisive role. Use of the computer allows students to work quickly on large statistical series, and thus helps them to understand this law. Yet, using a computer for its mere power and speed for the sake of presenting a wide range of new random experiments is not satisfactory. The didactic interest of simulation lies elsewhere: in the analysis of the random situation, the design of model hypotheses, and their translation into computer instructions that are necessary before simulations (Parzysz, 2009).

8 Implications for Teacher Training

As for other mathematical notions, statistics knowledge and probability knowledge are rooted in everyday life. Modelling is an essential process and the introduction of basic probabilistic notions raises specific issues as described above. Students meet a new difficulty when they have to link probabilistic notions to reality. The probability theory taught in a finite context is very simple but its abstract model component is not direct, particularly in the somewhat artificial situations presented in schools. Modelling is a critical stage in the use of the probability theory, especially in the different statistics applications.

The construction of mental images relative to randomness is delicate. In order to create these images, it is necessary to present activities in a random context to students right from the start of primary school. Later, links between statistics and probability can be established. As for geometry, the sequence could be observation, description, model building, reproduction, and representation of random experiments. The techniques of descriptive statistics are used to analyse and communicate the results of these experiments. The notions of sampling fluctuation and model could then be progressively constructed.

Teaching statistics and probability at secondary level still faces various obstacles (Girard, 2001), which teachers will only be able to overcome through a more in-depth focus in postgraduate and in-service training. This training should deal with different approaches to randomness, the epistemological basis of the concept of probability and current research about didactic questions raised by the management of modelling and computer simulations in the analysis of statistical populations in the classroom (Batanero, Burrill, Reading, & Rossman, 2008).

References

Albert, J. (2006). Interpreting probabilities and teaching the subjective viewpoint. In G. Burrill (Ed.), *Thinking and reasoning with data and chance: 68th NCTM Yearbook (2006)* (pp. 417–433). Reston, VA: National Council of Teachers of Mathematics.

Batanero, C., Burrill, G., Reading, C., & Rossman, A. (Eds.) (2008). *Joint ICMI/IASE Study: Teaching Statistics in School Mathematics. Challenges for Teaching and Teacher Education. Proceedings of the ICMI Study 18 and 2008 IASE Round Table Conference.* Monterrey, Mexico: International Commission on Mathematical Instruction and International Association for Statistical Education. Online: www.stat.auckland.ac.nz/~iase/publications

Batanero, C., Henry, M., & Parzysz, B. (2005). The nature of chance and probability. In G. A. Jones (Ed.), *Exploring probability in school: Challenges for teaching and learning* (pp. 20–42). New York: Springer.

Batanero, C., & Sanchez, E. (2005). What is the nature of high school students' conceptions and misconceptions about probability? In G. A. Jones (Ed.), *Exploring probability in school: Challenges for teaching and learning* (pp. 241–266). New York: Springer.

Biehler, R. (1991). Computers in probability education. In R. Kapadia & M. Borovcnik (Eds.), *Chance encounters: Probability in education* (pp. 169–212). Dordrecht, The Netherlands: Kluwer.

Biembengut, M. S. (2007). Modelling and applications in primary education. In W. Blum, P. L. Galbraith, H. W. Henn, & M. Niss (Eds.), *Modelling and applications in mathematics education: The 14th ICMI Study* (pp. 451–456). New York: Springer.

Blum, W., Galbraith, P. L., Henn, H. W., & Niss, M. (Eds.). (2007). *Modelling and applications in mathematics education: The 14th ICMI Study.* New York: Springer.

Borovcnik, M., Bentz, H. J., & Kapadia, R. (1991). A probabilistic perspective. In R. Kapadia & M. Borovcnik (Eds.), *Chance encounters: Probability in education* (pp. 27–73). Dordrecht, The Netherlands: Kluwer.

Burrill, G. (2002). Simulation as a tool to develop statistical understanding. In B. Phillips (Ed.), *Proceedings of the Sixth International Conference on Teaching Statistics.* Cape Town, South Africa: International Statistical Institute and International Association for Statistics Education. Online: www.stat.auckland.ac.nz/~iase/publications

Carranza, P., & Kuzniak, A. (2008). Duality of probability and statistics teaching in French education. In C. Batanero, G. Burrill, C. Reading, & A. Rossman (2008).

De Finetti, B. (1974). *Theory of probability. A critical introductory treatment.* New York: Wiley.

De Moivre, A. (1967). *The Doctrine of chances: or a method of calculating the probabilities of events in play*. New York: Chelsea Publishing Company (Original work published in 1756, London: A. Millar).

Garuti, R., Orlandoni, A., & Ricci, R. (2008). Which probability do we have to meet ? A case study about statistical and classical approach to probability in students' behaviour. In C. Batanero, G. Burrill, C. Reading, & A. Rossman (2008).

Girard, J. C. (2001). Quelques hypothèses sur les difficultés rencontrées dans l'enseignement des probabilités (Some hypotheses on difficulties coming across probability teaching). In *Autour de la modélisation en probabilités* (pp. 189–200). Besançon, France: Presses Universitaires de Franche-Comté.

Girard, J. C. (2004). La liaison statistique-probabilités dans l'enseignement (The link statistics-probability in teaching). *Repères-IREM, 57*, 83–91.

Groupe d'Experts sur les Programmes Scolaires, Direction de l'Enseignement Scolaire (2001). *Accompagnement des programmes de mathématiques, classe de Première des séries générales* (Support of mathematics curricula, First class of the general series). Paris: Centre National de Documentation Pédagogique (CNDP). Online: www.sceren.fr/produits/detailsimp. asp?ID=84491

Hacking, I. (1975). *The emergence of probability*. Cambridge, UK: Cambridge University Press.

Henry, M. (1999). L'introduction des probabilités au lycée: Un processus de modélisation comparable à celui de la géométrie (Introduction of *probability* in high school: A modelling process comparable with geometry teaching). *Repères-IREM, 36*, 15–34.

Henry, M. (2001a). Notion d'expérience aléatoire, vocabulaire et modèle probabiliste (Notion of random experiment, vocabulary and probabilistic model). In *Autour de la modélisation en probabilités* (pp. 225–232). Besançon, France: Presses Universitaires de Franche-Comté.

Henry, M. (2001b). Construction d'un modèle de Poisson (Construction of a Poisson's model). In *Autour de la modélisation en probabilités* (pp. 161–172). Besançon, France: Presses Universitaires de Franche-Comté.

Kahane, J. P. (2002) (Ed.), *L'enseignement des sciences mathématiques*: *Rapport au Ministre de l'Éducation Nationale* de la Commission de Réflexion sur l'Enseignement des Mathématiques (Mathematics science teaching: Report to the National Education Minister from the Commission of Reflection on the Teaching of Mathematics). Paris: Odile Jacob.

Kahneman, D., Slovic, P., & Tversky, A. (1982). *Judgement under uncertainty, heuristics and biases*. Cambridge: Cambridge University Press.

Keynes, J. M. (1921). *A treatise on probability*. London: Macmillan.

Lecoutre, M.-P., & Fischbein, E. (1998). Évolution avec l'âge de "misconceptions" dans les intuitions probabilistes en France et en Israël (Evolution with the age of probabilistic intuitively based misconceptions in France and in Israel). *Recherches en Didactique des Mathématiques, 18*(3), 311–332.

Mills, J. D. (2002). Using computer simulation methods to teach statistics: A review of the literature. *Journal of Statistics Education, 10*(1). Online: www.amstat.org/publications/jse/

Parzysz, B. (2009). Des expériences au modèle, via la simulation (From experiences to a model, via simulation). *Repères-IREM, 74*, 91–103.

Renyi, A. (1992). *Calcul des probabilités (Probability calculus)*. Paris: Jacques Gabay (Original work published in 1966).

Steinbring, H. (1991). The theoretical nature of probability in the classroom. In R. Kapadia & M. Borovcnik (Eds.), *Chance encounters: Probability in education* (pp. 135–168). Dordrecht, The Netherlands: Kluwer.

Wild, C., & Pfannkuch, M. (1999). Statistical thinking in empirical enquiry. *International Statistical Review, 67*(3), 221–248.

Zieffler, A., & Garfield, J. B. (2007). Studying the role of simulation in developing students' statistical reasoning. In *Proceedings of the 56th Session of the International Statistical Institute*, Lisboa, 22–29 August 2007. Voorburg, The Netherlands: International Statistical Institute. Online: www.stat.auckland.ac.nz/~iase/publications

Chapter 13
The Role of Technology in Teaching and Learning Statistics

Dave Pratt, Neville Davies, and Doreen Connor

Abstract In this chapter the merits, or otherwise, of using technology in teaching and learning statistics are considered. The many affordances that technological advances offer to teachers of statistics and the issues that hinder their widespread use in classrooms are summarised. When statisticians do statistics they get involved with far deeper concepts and carry out activities that require a wider range of cognitive skills compared with just applying techniques. It seems that pedagogic developments have not kept pace with those in software design, in that the opportunity to use computers to engage students in the full statistical enquiry cycle is not being exploited. The authors believe that beginning teachers must be exposed to such opportunities if they are to appreciate the key role that technology could have in facilitating the development of students' understanding of statistics.

1 Introduction

We are all familiar with the increasing importance of technology within all realms of our existence but what role does technology have in the teaching and learning of statistics? In responding to this question, this chapter focuses on the potential of

D. Pratt (✉)
Institute of Education, University of London, 20 Bedford Way, London WC1H 0AL, UK
e-mail: d.pratt@ioe.ac.uk

N. Davies
Royal Statistical Society Centre for Statistical Education,
University of Plymouth, Plymouth, UK
e-mail: neville.davies@rsscse.org.uk

D. Connor
Clifton Campus, Nottingham, NG11 8NS, UK
e-mail: doreen.connor2@ntu.ac.uk

C. Batanero, G. Burrill, and C. Reading (eds.), *Teaching Statistics in School Mathematics-Challenges for Teaching and Teacher Education: A Joint ICMI/IASE Study*, DOI 10.1007/978-94-007-1131-0_13, © Springer Science+Business Media B.V. 2011

technology in the teaching and learning of statistics, recognising however that those potentials are mediated inevitably by many factors, including inter alia software, tasks and teaching approaches. Harradine (2008) has noted how in the late 1990s statistical investigations began to incorporate real and large data sets as computers managed the number crunching. However, he has expressed concern that even now pedagogy in statistics education appears to have not moved on. This chapter explores some of the opportunities technology offers teachers of statistics while remaining alert to the issues surrounding their use. The authors' emphasis is similar to that of Goldstein (2003), namely, that it is not good enough to only consider which technology to use, but that, in order for effective learning to take place, it is how the technology is integrated into the curriculum and learning process and how the teacher uses it that are vital. In this chapter, we focus both on the use of computers themselves and some specific software tools.

2   What Does Technology Offer Teachers of Statistics?

This section discusses the special opportunities for teaching statistics that technology offers teachers who aim to provide rich learning experiences for their students. Ben-Zvi (2000) proposed the following categories of software: statistical packages (tools), microworlds, tutorials, resources (including Internet resources) and teachers' metatools. The discussion below considers five important affordances for teaching that might accrue should these categories of software be more widely adopted.

2.1   Using Representations as Dynamic Tools for Analysis

Traditionally, graphs are used to report data, often through displays and presentations. Exploratory Data Analysis (EDA) encourages deep interaction with data typically supported by the sort of immediate graphical representation of data that can be generated by computers. Computers thus enable representations to be used as analytical tools during an investigation rather than only as presentational tools at the end of the investigation. When graphs are used to try to make sense of data during analysis, the representations need to appeal to an intuitive sense of position, spread and outlying values. One example is the hat plot in Tinkerplots (www.keypress.com/x5715.xml), designed to appeal to students' intuitive notion of a modal clump (Konold et al., 2002). Similarly, Cobb, Gravemeijer, Bowers, and McClain (1997) developed their mini-tools as part of an intuitive infrastructure within a learning trajectory for statistical ideas based around the affordance of computer software to enable dynamic manipulation of images and numerical data. McClain (2008) reported teacher use of these mini-tools to develop ways of thinking about distribution.

    There are forms of representation that are as yet underdeveloped. For example, the manipulation of multivariate data might be better supported through digital technology. Ridgway, Nicholson, and McCusker (2008) reported on the development of new web browser tools (www.dur.ac.uk/smart.centre/) for displaying up to six variables for the easy exploration of complex data sets.

2.2   Expressing Personal Models

Statistical analysis involves the creation and use of models that describe the data arising from the phenomenon in question. Teachers of statistics might therefore involve their students in the activity of expressing personal models that attempt to capture the inherent structure in a situation. In this way, students might embrace the full cycle of statistical enquiry (discussed in more detail later) through the stages of reality description and pseudo-concrete model, mathematisation-formalisation and validation (Henry, Girard, & Chaput, 2008).

Computers offer flexible tools that empower the levels of expressiveness needed to develop models that fit data. In EDA, students express their own informal models for the data by searching for trends and patterns in the data, a process often referred to as expressive modelling (Doerr & Pratt, 2008). New developments in Tinkerplots promise to provide a graphical probabilistic language to model the generation of data sets (Konold, Harradine, & Kasak, 2007). Teachers could use the software as an authoring tool in which they build models for students to explore or as an expressive tool in which students build their own models of phenomena.

It is only by engaging with models that students might become tuned towards the uncertainty in the model (Chatfield, 1995). There are opportunities here to compare the logical necessity of mathematics and the vagaries of statistics. Biehler (2008), for example, has argued that the automatic graph plotting, now available through computers and graphing calculators fitted with data logging devices, offers a natural platform for discussing idealised mathematical functions alongside processes containing noise as in statistical situations. Early work in this area was done by Ainley, Pratt, and Nardi (2001), who proposed a pedagogic technique called *active graphing*, in which students carried out experiments to generate noisy data. Comparison between the data arising from the manual data collection and the smooth mathematical functions arising from digital representations raises the issue for discussion about the difference between mathematical and statistical data and the role of mathematical functions as models underpinning data. However, in teaching statistics we should always bear in mind that, in building models for data, the insightful comment of Box (1979) that all models are wrong, but some are useful is highly relevant.

2.3   Exploring Models

There are a number of key statistical concepts with which teachers expect students to engage. Expressive modelling might provide opportunities for students to appreciate the utility of those concepts in specific contexts but expressive modelling is unpredictable and cannot guarantee engagement with any specific concept. An alternative is for models to be built into the computer software and used to generate simulations that can be explored by the student, who might see the real-world phenomenon through the mathematical model (rather than see the model through the data). Integrating simulation into teaching can have positive benefits by allowing students to experiment with data and statistical distributions (Garfield & Ben-Zvi, 2007, 2008; Engel, Sedlmeier, & Worn, 2008). Such approaches are referred to as exploratory modelling (Doerr & Pratt, 2008). The computational power is directed towards providing feedback according to the in-built model in response to the action of the student.

Simulations have been used to help students bridge ambiguous statistical concepts. Some examples are Engel et al. (2008) on explained and unexplained variation; Kadijevich, Kool-Voljic, and Lavicza (2008) on sampling distributions; Abrahamson and Wilensky (2007) on different epistemological perspectives for probability; Prodromou and Pratt (2006) on determined and stochastic causality; and the Winton programme for the Public Understanding of Risk (www.understandinguncertainty.org) on absolute and relative risk.

2.4   Storing and Processing Real Data

Digital technology facilitates the use of large data sets through its capacity for data storage, easy retrieval and universal availability thanks to the increasing use of idealised data formats. Data sets enable the analysis of data drawn from situations that are meaningful to students.

One of the most significant developments for schools in this respect has been the development of CensusAtSchool (Connor, Davies, & Holmes, 2000; Davies, Connor, & Spencer, 2003). On the Royal Statistical Society Centre for Statistical Education (RSSCSE) web site (www.censusatschool.org.uk), there are over 30 databases containing 1.3 million responses of real data, collected from learners in five countries, available for sampling.

Hall (2008) described her use of CensusAtSchool with elementary teachers. She noted that although the data sets were powerful, it was equally important to have available a dynamic statistical analysis software, such as Tinkerplots. Web-based tools that allow the intuitive graphing of data sets, such as the tools in CensusAtSchool (www.censusatschool.org.uk/get-data/datatool), are increasing in availability.

Real data sets present issues that are often not present in sanitised data. For example, difficult numbers, errors in data and missing values are all qualities of data that might be avoided in carefully prepared situations. At some point in a student's education, these issues need to be confronted since they raise important questions about the limitations, scope and reliability of inferences that can be made, as well as techniques for handling the problems (for further discussion of using real data in statistics education, see Hall, this book).

2.5 Sharing and Communicating

Ben-Zvi (2007) has argued for the creation of statistics courses using a wiki to promote collaborative learning. He has researched several types of wiki activities: collaborative writing; glossaries; discussion and review; statistical projects; self-reflective journals; and assessment. The creation of such resources could be beneficial to teacher trainees, both for their own learning and for the future benefit of their students.

Nolan and Temple Lang (2007) showed how a large meaningful data set (9,000 email messages containing 30 variables for each message) was used to teach statistical practice in a dynamic document environment. Actions allowed by the live worksheet include interacting with the data, modifying inputs, changing computations and exploring a range of analyses. By allowing the electronic document to be interactive, the authors demonstrated how it could be used to reflect what a statistician might do in carrying out a statistical investigation. In essence they aimed to provide a problem-solving environment in which practising statisticians and statistical researchers work so that educators/teachers could more effectively teach students.

3 Issues Regarding the Use of Technology in Teaching and Learning Statistics

It is true that technology has changed the way people consume statistics, the speed and efficiency with which researchers produce statistics and how statisticians do statistics. We might also expect the power of technology as described above to change the way people teach statistics and students learn statistics (Chance, Ben-Zvi, Garfield, & Medina, 2007). Rubin (2007) reflected on experience in using technology in statistics education between 1992 and 2007 and commented that "... as amazing and inspiring as these technologies may seem, none of them have any educational effect without carefully constructed curriculum and talented teaching" (p. 2). In fact, teaching and learning is such a complex process that the very power that creates potential for dramatic change comes hand-in-hand with a set of issues that threaten to curtail the possible development. Five of these issues are elaborated.

3.1 Teachers May Not Prioritise the Use of Technology

In the end, what matters is what students learn but there is little statistics education research about teachers using such technological tools (Chance et al., 2007). In practice, teachers have to make decisions about what is appropriate to incorporate into already-written curricula and have little time to research such matters. They also have competing demands on classroom time and are forced to prioritise. Placing more emphasis on enquiry and investigation, and using new forms of technology appear, on the face of it, to be time-consuming and so demand a commitment to a pedagogical approach that many teachers may not share.

3.2 Curriculum Specifications May Not Support the Use of Technology

Teachers' priorities might be influenced by curriculum specification. However, there is much variation in the use of technology between the statistics curricula that schools follow within and between countries. In Australia, for example, all curriculum documents include emphasis on the use of technology, but the extent to which this enables statistical data investigations depends on the assessment regime. A common technology used in Australia is the graphics calculator, but the flexibility in the Queensland system is seeing more schools making use of a combination of computers and graphics calculators than in other states.

3.3 Assessment Methods May Not Encourage the Use of Technology

Even if the curriculum encourages the use of computers, teachers realise that examinations may be more important to their students and the students' parents than anything else. The actual curriculum as experienced by students will always be driven by the assessment regime, irrespective of pedagogic guidelines. In many countries it remains the case that examination boards do not allow the use of technology. For example, in Queensland, the assessment for senior school is school-based centrally moderated assessment; it is mandatory in mathematics (including statistics) to include alternative assessments such as investigations. In contrast, the emphasis in New South Wales is almost entirely on a final statewide exam and this has probably contributed to the lack of progress in developing students' statistical conceptual understanding. Teachers who consider the use of technology in their classrooms may be concerned that their students will not develop practices that transfer to such traditional examination contexts.

3.4 Teachers May Not Re-skill

Experienced teachers may not have benefited from the use of computers in their own learning. Their own experience of being taught is unlikely to have included the changes in pedagogy that Chance et al. (2007) have claimed flow from technological advances, including balancing the role of computers and non-technological environments for teaching how to "unlock stories in data" (Pfannkuch, 2008). Garfield and Ben-Zvi (2007) have supported exposing teachers to innovative software, such as *TinkerPlots* and *Fathom*, to enable them to explore data. The intention is that exposure during teacher education might motivate teachers to pass their experience on and engage their students in statistical investigations through using similar software.

3.5 Technology May Not Be Used to Teach Statistical Concepts but Instead May Reinforce Emphasis on Computation

Technology as a productivity tool has been concerned with developing faster and faster ways of processing data to calculate routine and/or sophisticated statistics and, more recently, with processing very large databases. However, the way statisticians do statistics also involves planning the investigation, collecting data, processing, analysing results and drawing appropriate conclusions (Marriott, Davies, & Gibson, 2009; Stuart, 1995). Therefore it is important to stress that, when teaching the subject, processing and analysing results is just *one* part of the statistical enquiry cycle. The way that computers can support an emphasis on modelling as well as on analysis, as discussed above, could ensure that students do not lose touch with the statistical thinking necessary to do statistics when they use technology.

4 Challenges for Teacher Education

It has been argued that technology offers a range of affordances that could revolutionise statistics education but there are reasons why that transformation has not yet happened and indeed might never happen, except in certain pockets of good practice. For a number of years, statistics educators have proposed that the way statistics is taught should be changed (Stuart, 1995; Chance et al., 2007; Marriot et al., 2009), and that the place to begin is in teacher education.

4.1 Immersing Beginning Teachers in the Use of Technology for Teaching Statistics

Da Ponte (2008) commented on the need for a better vision of how teachers learn. Studies have reported on the difficulties teachers face due to their own lack of exposure and knowledge of the best use of technology (Reston & Bersales, 2008) and yet beginning teachers are often heavily influenced in their teaching by the way that they themselves were taught (Ball, 1988). When new teachers teach areas in which their own conceptual knowledge is weak, they have been shown to revert more readily to those methods they themselves experienced as a learner (Sedlmeier & Wassner, 2008). The use of a wide range of technology can present problems for beginning teachers, who are accustomed to a certain style of instruction (Healy & Hoyles, 2001). Immersion in the use of technology for statistics education should therefore be a major feature of any teacher education course. However, teacher educators should first explore the many ideas being put forward by researchers and developers (Batanero, Godino, & Roa, 2004).

4.2   How Might Teacher Educationalists Deploy Technology?

Immersion, though, must involve the full statistical enquiry cycle as discussed above. Using modern software to express and explore models of real data sets, such as those available through CensusAtSchool, would expose weaknesses in the beginning teachers' own statistical knowledge in a non-threatening environment. Since Lee and Hollebrands (2008) reported that teachers' decisions about using computer tools were often based on knowledge gained during their teacher education courses, it is reasonable to suppose that a modelling approach in teacher education courses might enable the development of a more sophisticated understanding from the exposed weaknesses. This immersion in the use of statistics software in teacher education courses is of vital importance due to the gearing effect on eventual student learning of statistical ideas and because teacher educators have potential influence over many students who later become teachers.

4.3   What Might Beginning Teachers Learn About Statistics Through Using Technology?

The aim of such immersion in teacher education would be to enhance teachers' technological pedagogical content knowledge (TPCK) (Lee & Hollebrands, 2008), distinguishing between technology as course content and technology as a teaching tool (Habre & Grundmeier, 2007). Learning how the special characteristics of technology might be integrated into teaching to support learning of statistical concepts can be considered a key aspect of TPCK. Pfannkuch (2008) stated that teacher educators need both to build teachers' statistical concepts and to make teachers aware of how students' conceptual understanding may develop.

The focus on EDA, supported by technology, can aid beginning teachers themselves engage with statistical concepts and the whole statistical investigation cycle, helping them to reach a higher level of conceptual understanding of statistics before considering how they might use technology in their own teaching, thus also enabling students to understand statistical concepts.

5 The Way Forward

The development of technologies, especially in the form of new software tools, will of course continue and no doubt so will the careful analysis of how their design and implementation impacts on teaching and learning. Nevertheless, the way forward is to place emphasis on teacher education so that the affordances of technology, as identified in Sect. 2 of this chapter, can be understood by teachers, the issues alerted in Sect. 3 can be shared and avoided, and finally so that the challenges in the final section above can be met. Research is needed to examine and evaluate such developments in teacher education and to help understand and measure the presumed consequent gearing effect that results from the way that teachers influence many students in their courses.

References

Abrahamson, D., & Wilensky, U. (2007). Learning axes and bridging tools. *International Journal of Computers for Mathematics Learning, 12*(1), 23–55.

Ainley, J., Pratt, D., & Nardi, E. (2001). Normalising: children's activity to construct meanings for trend. *Educational Studies in Mathematics, 45*, 131–146.

Ball, D. L. (1988). *Unlearning to teach mathematics.* (Issue Paper 88–1). East Lansing: Michigan State University, National Center for Research on Teacher Education.

Batanero, C., Burrill, G., Reading, C., & Rossman, A. (Eds.) (2008). *Joint ICMI/IASE Study: Teaching Statistics in School Mathematics. Challenges for Teaching and Teacher Education. Proceedings of the ICMI Study 18 and 2008 IASE Round Table Conference.* Monterrey, Mexico: International Commission on Mathematical Instruction and International Association for Statistical Education. Online: www.stat.auckland.ac.nz/~iase/publications

Batanero, C., Godino, J., & Roa, R. (2004). Training teachers to teach probability. *Journal of Statistics Education, 12*(1). Online: www.amstat.org/publications/jse/

Ben-Zvi, D. (2000). Toward understanding the role of technological tools in statistical learning. *Mathematical Thinking and Learning, 2*(1 & 2), 127–155.

Ben-Zvi, D. (2007). Using wiki to promote collaborative learning in statistics education. *Technology Innovations in Statistics Education, 1*(4). Online: repositories.cdlib.org/uclastat/cts/tise/

Biehler, R. (2008). From statistical literacy to fundamental ideas in mathematics: How can we bridge the tension in order to support teachers of statistics. In C. Batanero, G. Burrill, C. Reading, & A. Rossman (2008).

Box, G. E. P. (1979). Robustness in the strategy of scientific model building. In R. L. Launer & G. N. Wilkinson (Eds.), *Robustness in statistics.* New York: Academic Press.

Chance, B., Ben-Zvi, D., Garfield, J., & Medina, E. (2007). The role of technology in improving student learning in statistics. *Technology Innovations in Statistics Education, 1*(1). Online: repositories.cdlib.org/uclastat/cts/tise/

Chatfield, C. (1995). Model uncertainty, data mining and statistical inference. *Journal of the Royal Statistical Society A, 158*, 419–466.

Cobb, P., Gravemeijer, K. P. E., Bowers, J., & McClain, K. (1997). *Statistical minitools.* Nashville, TN/Utrecht: Vanderbilt University and Freudenthal Institute/Utrecht University.

Connor, D., Davies, N., & Holmes, P. (2000). CensusAtSchool 2000. *Teaching Statistics, 22*, 66–70.

da Ponte, J. P. (2008). Preparing teachers to meet the challenges of statistics education. In C. Batanero, G. Burrill, C. Reading, & A. Rossman (2008).

Davies, N., Connor, D., & Spencer, N. M. (2003). An international project for the development of data handling skills of teachers and pupils. *Journal of Applied Mathematics and Decision Sciences, 7*, 75–83.

Doerr, H., & Pratt, D. (2008). The learning of mathematics and mathematical modeling. In M. K. Heid & G. W. Blume (Eds.), *Research on technology in the teaching and learning of mathematics: Syntheses and perspectives. Mathematics learning, teaching and policy* (Vol. 1, pp. 259–285). Charlotte, NC: Information Age.

Engel, J., Sedlmeier, P., & Worn, C. (2008). Modelling scatter plot data and the signal-noise metaphor: towards statistical literacy for pre-service teachers. In C. Batanero, G. Burrill, C. Reading, & A. Rossman (2008).

Garfield, J., & Ben-Zvi, D. (2007). How students learn statistics revisited: A current review of research on teaching and learning statistics. *International Statistical Review, 75*(3), 372–396.

Garfield, J., & Ben-Zvi, D. (2008). Preparing schoolteachers to develop students' statistical reasoning. In C. Batanero, G. Burrill, C. Reading, & A. Rossman (2008).

Goldstein, R. (2003). Integrating computers into the teaching of mathematics. In L. Haggarty (Ed.), *Teaching mathematics in secondary schools – A reader* (pp. 143–159). Milton Keynes: Open University.

Habre, S., & Grundmeier, T. (2007). Prospective mathematics teachers' views on the role of technology in mathematics education. *Issues in the Undergraduate Mathematics Preparation of School Teachers: The Journal, 3*. Online: www.k-12prep.math.ttu.edu

Hall, J. (2008). Using CensusAtSchool and Tinkerplots to support Ontario elementary teachers' statistics teaching and learning. In C. Batanero, G. Burrill, C. Reading, & A. Rossman (2008).

Harradine, A. (2008). Birthing big ideas in the minds of babes. In C. Batanero, G. Burrill, C. Reading, & A. Rossman (2008).

Healy, L., & Hoyles, C. (2001). Software tools for geometrical problem solving: Potentials and pitfalls. *International Journal of Computers for Mathematical Learning, 6*(3), 235–256.

Henry, M., Girard, J., & Chaput, B. (2008). Modelling and simulations in statistics education. In C. Batanero, G. Burrill, C. Reading, & A. Rossman (2008).

Kadijevich, D., Kool-Voljic, V., & Lavicza, Z. (2008). Towards a suitably designed instruction on statistical reasoning: Understanding sampling distribution with technology. In C. Batanero, G. Burrill, C. Reading, & A. Rossman (2008).

Konold, C., Harradine, A., & Kasak, S. (2007). Understanding distributions by modeling them. *International Journal of Computers for Mathematical Learning, 12*(3), 217–230.

Konold, C., Robinson, A., Khalil, K., Pollatsek, A., Well, A., Wing, R., & Mayr, S. (2002). Students' use of modal clumps to summarize data. In B. Phillips (Ed.), *Proceedings of the Sixth International Conference on Teaching Statistics*. Cape Town, South Africa: International Association for Statistics Education. Online: www.stat.auckland.ac.nz/~iase/publications

Lee, S., & Hollebrands, K. (2008). Preparing to teach data analysis and probability with technology. In C. Batanero, G. Burrill, C. Reading, & A. Rossman (2008).

Marriott, J. M., Davies, N., & Gibson, E. (2009). Teaching, learning and assessing statistical problem solving. *Journal of Statistics Education, 17*(1). Online: www.amstat.org/publications/jse/

McClain, K. (2008). The evolution of teachers' understandings of distribution. In C. Batanero, G. Burrill, C. Reading, & A. Rossman (2008).

Nolan, D., & Temple Lang, D. (2007). Dynamic, interactive documents for teaching statistical practice. *International Statistical Review, 75*(3), 295–321.

Pfannkuch, M. (2008). Training teachers to develop statistical thinking. In C. Batanero, G. Burrill, C. Reading, & A. Rossman (2008).

Prodromou, T., & Pratt, D. (2006). The role of causality in the coordination of two perspectives on distribution within a virtual simulation. *Statistics Education Research Journal, 5*(2), 69–88. Online: www.stat.auckland.ac.nz/serj/

Reston, E., & Bersales, L. G. (2008). Reform efforts in training statistics teachers in the Philippines: Challenges and prospects. In C. Batanero, G. Burrill, C. Reading, & A. Rossman (2008).

Ridgway, J. Nicholson, J., & McCusker, S. (2008). Reconceptualising 'statistics' and 'education'. In C. Batanero, G. Burrill, C. Reading, & A. Rossman (2008).

Rubin, A. (2007). Much has changed; little has changed: Revisiting the role of technology in statistics education 1992–2007. *Technology Innovations in Statistics Education, 1*(1). Online: repositories.cdlib.org/uclastat/cts/tise/

Sedlmeier, P., & Wassner, C. (2008). German mathematics teachers' views on statistics education. In C. Batanero, G. Burrill, C. Reading, & A. Rossman (2008).

Stuart, M. (1995). Changing the teaching of statistics. *The Statistician, 44*, 45–54.

Chapter 14
Teaching Statistical Thinking Through Investigative Projects

Helen MacGillivray and Lionel Pereira-Mendoza

Abstract Projects involving investigations are ideal vehicles for student engagement, for learning problem-solving in context and for synthesising components of learning. They are also a natural environment for learning statistical thinking through experiencing the process of carrying out real statistical data enquiries from first thoughts, through planning, collecting and exploring data, to reporting on its features. In addition, they foster collaborative learning, provide learning opportunities for students of all abilities and educational levels, and can facilitate rich information for teachers as they assist and observe students' work. This chapter considers the benefits of the data investigative process in a project-based approach.

1   Introduction

Statistics and statistical thinking have become increasingly important in a society that relies more and more on information and demands for evidence. Hence the need to develop statistical skills and thinking across all levels of education has grown and is of core importance in a century which will place even greater demands on society for statistical capabilities throughout industry, government and education.

The past two decades have seen considerable discussion, research and developments across all levels of education to meet the challenges of facilitating the learning of statistical thinking and reasoning. These have included data-driven approaches, more emphasis on data production and the measuring and modelling of

H. MacGillivray (✉)
Mathematical Sciences, Queensland University of Technology, Gardens Point Campus,
GPO Box 2434, Brisbane Q 4001, Australia
e-mail: h.macgillivray@qut.edu.au

L. Pereira-Mendoza
Educational Consultant, 3366 Drew Henry Drive, Osgoode, Ottawa K0A 2W0,
Ontario, Canada
e-mail: lionel@iammendoza.com

C. Batanero, G. Burrill, and C. Reading (eds.), *Teaching Statistics in School*
Mathematics-Challenges for Teaching and Teacher Education: A Joint ICMI/IASE Study,
DOI 10.1007/978-94-007-1131-0_14, © Springer Science+Business Media B.V. 2011

variability (Moore, 1997), real data and contexts, and generally a more holistic approach that reflects the practice of statistics. The emphasis in creating environments for learning has been on active learning, hands-on experience and problem-solving.

Although gathering and interpreting data, and statistical thinking pervade everyday living, disciplines, workplaces and research, they are remarkably challenging to learn and teach. Data are inherently messy, and interpretations of models and analyses, from the very simple to the most complex, require judgement and understanding dependent on assumptions and context, but avoiding susceptibility to context "intuitions". Whether considering development of school curricula and resources, or pre-service and in-service programmes for teachers, the learning and teaching of statistical thinking require gradual building up of concepts, understanding and skills, in a coherent, consistent and cumulative way that engages students in real contexts and authentic learning experiences. This is an ongoing challenge requiring cooperation and contributions from statisticians, educators and teachers.

The term "statistical thinking" will be used here with the broad meaning of making sense of information in which variation and/or uncertainty is present. It is thus inclusive of chance and data, which should be regarded as intertwining and interacting elements of statistical thinking. Section 2 outlines some of the commonalities in frameworks that have been advanced for statistical thinking and how it is learnt experientially. Some of the shared elements include focus on what statisticians do in the process of data investigations, and on articulating this. The statistical investigative process is often described as the data investigative or enquiry cycle, and this is increasingly emphasised as a vehicle for both statistical problem-solving and learning statistical thinking. The key elements of the process and some articulations of it are described in Sect. 2 and the stages of two articulations provide the structure for Sects. 4–7 in discussing the value of its explicit use in learning and teaching statistical thinking.

One context in which a description of the data investigative cycle was advanced was the United Kingdom (UK) National School Curriculum in the mid-1970s. Holmes (1997) described the introduction and development in the final 2 years of school study in the UK, of a compulsory project to encourage a more holistic and practical approach to statistics, reflecting what statisticians do. The project-based approach is ideal for the statistical investigative process, and Sect. 3 outlines the approach of learning statistical thinking within the vehicle of the data investigative cycle. Reference is made to this approach throughout Sects. 4–7. Section 8 provides a small selection of examples of practical and accessible projects that have proved of value in engaging students and teachers in developing statistical thinking within particular stages or in the full process of the investigative cycle.

2   Statistical Thinking and the Data Investigative Cycle

As stated, the term "statistical thinking" is used here in an inclusive and encompassing sense as envisaged by the analysis of Wild and Pfannkuch (1999). The framework of their model for statistical thinking is that of empirical enquiry, and aims to generalise and synthesise from a combination of the literature and qualitative research into the approach and thinking of professional statisticians and statistics students during the process of investigating and solving real, vaguely described problems. The authors comment that they are "not concerned with finding some neat encapsulation of statistical thinking" (Wild & Pfannkuch, 1999, p. 224), nor do they address the full spectrum of statistical thinking. Their focus is on what professional statisticians do in solving real problems involving the need for modelling and analysing context information, uncertainty and data. Dimension one of their model is an articulation of the data investigative cycle.

This is also the focus of Cameron (2009) in a paper written from the viewpoint of training collaborative research statisticians. Cameron commented on Chambers' (1993) description of what statisticians do as "greater" statistics involving three components: preparing data (including planning, collecting, organising and validating); analysing data; and presenting information from data. In adding an initial stage of formulating a problem so that it can be tackled statistically, and another possible stage of researching the interplay of observation, experiment and theory to develop new methods, Cameron's (2009) model of what professional statisticians do in the practice of statistics is not only consistent with dimension one (the investigative cycle) of Wild and Pfannkuch's (1999) model of statistical thinking, but also reflects dimension two of their model in the types of thinking fundamental to statistics. These include recognition of the need for data; changing the representation to assist understanding and problem-solving; investigating variation; reasoning with statistical models; and incorporating statistics and context.

Thus an expression describing the data investigative cycle provides a practical framework for demonstrating and learning statistical thinking. Exact descriptions of the cycle vary slightly but all share common concepts and structure. Cameron's (2009) description was based on descriptions by professional statisticians. Wild and Pfannkuch's (1999) description is the Problem, Plan, Data, Analysis, Conclusion (PPDAC) cycle adapted from MacKay and Oldfield (1994) that reflects the statistical process (see, for example, Shewhart & Deming, 1986). The description of the data-handling cycle that featured in the UK National School Curriculum since at least the mid-1970s (Holmes, 1997) has become the Plan, Collect, Process, Discuss (PCPD) cycle that is at the heart of the extensive pedagogies and resources produced by the Royal Statistical Society's Centre for Statistical Education (www.rsscse.org.uk/). Marriott, Davies, and Gibson (2009) included a mapping of the problem-solving approach of the PCPD cycle onto Bloom's taxonomy of the cognitive skills of educational objectives as revised by Anderson and Krathwol (2001). A mapping of the learning objectives of this form of the cycle onto a two-way classification that combines the cognitive and the knowledge dimensions of Anderson and Krathwol (2001) is given in www.rsscse. org.uk/qca/doc/PSAtwowaymap.pdf

Statisticians and statistical educators are increasingly emphasising the importance in statistical education of including all stages of the investigative cycle, particularly those that produce data to be investigated – those that involve identifying the problem or issue, planning the investigation and collecting the data – and the stage of interpreting the results of analysis or exploration of the data in context. That is, the stage described as "analysis" in the PPDAC description, or as "process" in the PCPD description, should be taught as part of the overall process of statistical thinking. Such emphasis requires not just real contexts and real data, but placement of components of learning within an overall framework representing whole and complex problems needing the full gamut of the knowledge dimension in statistics of Anderson and Krathwol (2001): factual, conceptual, procedural and metacognitive.

3 Projects and the Investigative Cycle

Thus statisticians and statistical educators advocate enquiry and investigation approaches in the development of statistical thinking, and use of the investigative cycle as a framework. Learning experiences, small or large, can be couched in terms of investigating real problems with real data, and can be explicitly embedded in the framework recommended for the development of statistical thinking. That is, learning experiences can target parts or all of the investigative cycle – the key pedagogies are emphasis on investigation and identification of the stages of the investigative cycle in problem-solving.

Holmes (1997) identified and discussed the advantages of projects in statistics as natural vehicles for the data investigative cycle and holistic experiential learning. Although the context is senior school, the comments could equally well apply to all levels of education across and beyond schooling. Projects may vary in size and in the time allocated to them, but are characterised by incorporating a whole process from identifying a problem or issue of interest through to presenting a report. Holmes (1997, p. 156) described a project as a piece of work "that would start with defining a problem, collecting the appropriate data, analysing the data and drawing appropriate inferences. All this was to be presented in a written project report of about 15 pages". Statistical projects, whether large or small, provide experiential learning of statistical investigations. Such learning brings together concepts, knowledge and skills in contexts that can engage and motivate students as well as teach them about the nuances of statistical thinking, the vagaries of data and the challenges of communicating interpretations in context.

This also applies within teacher education, and as the use of projects is also accepted as an integral part of school learning experiences, the use of data investigation projects is ideal to develop both statistical understanding and pedagogy for teachers. Associating the statistical investigative cycle with projects will assist in educating teachers to teach statistics, as the cycle may be used to capture statistical thinking within a pedagogical framework of active learning through projects. For example, the PPDAC form of the data investigative cycle is being used in an ongoing study designed to understand primary school teachers' experiences as they develop confidence in teaching statistical enquiry (Makar, 2008).

The projects discussed in Holmes (1997) are free-choice data investigations in which students identify their topic to be investigated, plan and implement a data collection strategy to investigate the topic, explore and analyse their data, and produce a written report. This is often a group project because the task needs a group at all stages, particularly in free-choice data investigations in identifying the topic, planning, and collecting the data. The strong sense of ownership also facilitates teamwork as the project moves through the full process of data handling, exploration, analysis (if appropriate), interpretation and reporting in context. At all educational levels, such projects are advocated for experiential learning of the process of statistical enquiry, because they capture the challenges of turning ideas and questions into plans for investigation, the practicalities and messiness of data collection and handling, the essentials of choosing and using statistical tools, and the synthesis of statistical interpretations in real and authentic contexts. As students move from primary school through the secondary school levels and beyond, the same general approach is followed, but the level of statistical sophistication increases and the level of teacher direction decreases.

A significant impetus for learning is student ownership – of the ideas, the data and therefore, the analysis (MacGillivray, 1998, 2002; Chance, 2005; Lee, 2005). If students do not choose the topic to be investigated or if the topic and data are supplied, then teaching strategies need to address student engagement in the problem and all the stages of the data investigative cycle. No matter what the size or restrictions of a learning experience, what makes it a statistical data investigation is its placement within the process of statistical enquiry. That is, if a learning experience focuses on part of the data investigation cycle, it is important for students to understand where it fits and, if possible, at least consider the other aspects of the cycle relevant to the topic.

Sections 4–7 below consider learning experiences within the stages of the statistical investigative cycle as discussed in Sect. 2, referring to the descriptions PPDAC and PCPD of the data investigative cycle, as these are probably the best known. Section 4 considers the Problem and Plan stages of the PPDAC description and the Plan stage of the PCPD description. Section 5 considers the Data stage of PPDAC and the Collect stage of PCPD. Section 6 considers the Analysis stage of PPDAC and the Process stage of PCPD. Finally, Sect. 7 considers the Conclusion stage of PPDAC and the Discuss stage of PCPD.

4 Identifying the Problem and Planning

Whether data are to be collected, selected or provided, identification of the problem or topic to be investigated and the plan for investigation are essential and significant aspects of statistical thinking and problem-solving (Wild & Pfannkuch, 1999; Nolan & Lang, 2007) and need much more emphasis than has traditionally been given. Once a general topic or aspects of a topic are identified for investigation, other questions can be asked to assist the planning, irrespective of whether data are to be collected, selected or provided. These include:

- What do we want to find out about? What can we find out about?
- What can we measure or observe? Can we measure what we want?
- Is there anything else we should observe or record … just in case?
- Should we do a preliminary experiment?
- How can we collect data that are representative?

Whatever the school level, statistical projects and statistical learning experiences should involve a number of variables. This provides authenticity in that almost all real problems are complex and involve – actually or potentially – many variables. This also provides experience in identifying questions, selecting data and method, and reporting interpretations in context. Even when illustrating methods involving just one or two variables, selecting from within a wider context or a more complex dataset facilitates a more holistic and realistic approach, and therefore more statistical thinking. This applies across educational levels, particularly as students mature. Ridgway, McCusker, and Nicholson (2005), MacGillivray (2005), and Schield (2005) all advocated the value of working within contexts with more than two variables in developing statistical thinking as students progress. Examples of problems and datasets that can involve many variables may be found in CensusAtSchool data (for example, Turner, 2006).

The Problem and Plan stage(s) of the cycle include identifying variables, identifying the subjects of the study, and considering the practicalities. When students are familiar with appropriate software, considering what the resultant spreadsheet of raw data will look like is an excellent aid in planning – if the spreadsheet cannot be visualised or described, then the planning is not complete.

Whether data are to be collected (primary) or provided (secondary), the Planning stage must consider the question of representativeness of data. Data can be used to make inferences about a larger group or a more general situation if the data can be considered to be representative of that larger group or more general situation *with respect to the question(s) of interest*. As students mature and come to consider the concept of inferring, the challenges of planning data collections to ensure such representativeness, or identifying the representativeness of secondary data, are greatly assisted through experiencing the Problem/Plan stage of the cycle. Each type of statistical investigation – experiment, observational study, survey or a mixture of types – has its own challenges in planning to achieve data relevant to, and representative of, the problem.

The thinking involved in considering the topic and its context, what variables to use, what data to obtain and how to obtain representative data, is profoundly statistical and incorporates almost all of the types of thinking fundamental to statistical thinking of dimension two of Wild and Pfannkuch's (1999) model. It can be seen why statisticians and statistical educators are emphasising the importance of inclusion in teaching statistics of these aspects of data investigations.

Progressive development of statistical concepts and methods should gradually proceed in types of data, and therefore types of variables, as well as number of variables to be considered, and complexities of contexts and topics. The simplest data are categorical, and early learning is of simple categorical variables. A delightful paper by du Feu (2005) demonstrated how statistical thinking and projects can commence at an early age with categorical data, and motivate and help develop the earliest concepts of presentation of data and commenting on features.

5   Collecting the Data

Data can be generated through surveys, experiments, observations, or from pre-existing datasets such as those available through the Internet and other sources. Offices of National Statistics may also provide a useful source of data for projects at all levels, although such data often tend to be in at least some form of summary; for example, it can be quite difficult to access original data or data on more than two variables at once. The CensusAtSchool data aim to use students' interest in data about themselves to engage them in statistical questions and explorations. The Internet is a rich source of data of interest to older students, and also encourages them to choose their own topics to explore. Topics of interest include sport, weather, music charts, movies, as well as more serious topics of social issues relevant to teenagers. Examples of work involving the analysis of media data can be found in Watson (1997).

The full Collecting the Data stage of the data investigation cycle involves collection, handling and cleaning of the data. In projects with data either provided or from secondary sources, these key elements can still be discussed, with the emphasis on understanding the need for well-collected, representative data. Pilot studies help in planning collection of good data. For simple projects at primary and even secondary school, such preliminaries may be as straightforward as trialling questions with each other, but no matter how simple the project, identifying the issue to be investigated, and planning the acquisition of, or access to, good data are essential in learning statistical thinking. In both collecting and preparing data for exploration or analysis, the representation of the problem and of the variables must be considered, whether we are dealing with simple or complex categorical variables, or measurement challenges. Data cleaning and data entry, whether students are using spreadsheets or summarising data themselves, have many aspects of challenge and fun. Is there really a student who lives 5 km from school but takes 40 min to reach school? Is there really a student who estimated 5 s by 3 s but 10 s by 15 s? We recorded 16 different colours of cars – how should we group them?

6   Analysing and Processing

At the school level, this stage refers mainly to choosing and using data representations and summaries for data exploration. In many ways, the word "process" of the PCPD version of the data investigation cycle is more appropriate for school levels, and the word "analysis" for post-school levels. It is this component that is most closely dependent on the student cohort's level and the details of the curriculum. But at all levels, this stage involves not only investigating variation but also reasoning with statistical models and incorporating statistics and context.

Types of data representations, summaries and commenting on features of data are underpinned by considerations of types and number of variables; these are essential building blocks of statistical models. Thus there is naturally a gradual development from categorical to count to continuous, and from single variables to two variables and more. There is also a need for a gradual development of awareness of variation, including the important learning development from consideration of variation *within* a dataset to variation *between* groups of data to variation *across* datasets from the same or similar situations or contexts. If this is done in association with simulations – produced and demonstrated by the teacher is sufficient – strong foundations can be laid in students' understanding of variation and sampling variation. Also important is the key concept of representativeness of data with respect to the questions or topics under investigation.

An emphasis on exploring data through graphical representations includes development of summary statistics with associated discussion of both the strengths and weaknesses of single-valued quantities in representing features of data. Early introduction and ongoing use of words such as "estimate" assist in providing a lasting foundation for future statistical learning. Data should be linked with chance at every opportunity, not only to reinforce concepts of estimation but also to assist in embedding understanding of probability in real and everyday contexts.

7   Interpreting and Discussing

Projects embedded in the data investigative cycle provide a natural environment for developing both verbal and written communication skills within each stage of the cycle, and facilitate coherent and gradual development of such skills as students mature. However, as commented in Forster, Smith, and Wild (2005), there is a need to systematically teach and develop skills in communicating, particularly in the Conclusions/Discuss stage of PPDAC and PDPC. Benefits of the integration of verbal and written communication within statistical projects go beyond specific development of these skills. Lipson and Kokonis (2005) pointed out that report writing in statistics is a metacognitive activity that facilitates the learning of statistical literacy and thinking.

It is also at this stage that students can learn about the nuances and pitfalls of commenting on variation, and allowing for variation in commenting on features of data, as well as commenting in context. It is of great importance in developing statistical thinking to emphasise commenting, interpreting and discussing, and NOT the definiteness of "answering the question". There are certainly incorrect comments and interpretations that can be made, but the focus should be on appropriate rather than "right" comments. There should also be emphasis on distinguishing between what the data are telling us and what might be the reasons. Interpreting data in context does not mean drawing conclusions based on contextual intuition.

Wherever possible, the word "estimate" should be used. If syllabi and school level permit the study of the concept of error of estimate, then it is the concept and understanding that are vital, not the names or jargon. Introduction of interval estimates through proportions avoids the messiness and complications inherent in estimating means. The worst misconceptions in interval estimates for means are those that interpret the interval as one in which most *individual* values lie. In contrast, such misinterpretation is almost impossible in estimating a proportion; this is just one reason statisticians are increasingly suggesting introducing formal inference through inference for proportions.

8 Some Examples of Projects

The following three examples, selected from free-choice projects conducted by students, have proved popular in workshops for teachers conducted in Australia, South Africa, and New Zealand. They illustrate some of the variety and learning potential in real and accessible contexts for statistical projects; Sects. 8.1 and 8.2 have been particularly popular for hands-on experience in planning investigations and trialling data collection, while Sect. 8.3 illustrates connecting chance and data. The Royal Statistical Society Centre's ExperimentsAtSchool also provide a number of well-constructed project activities.

8.1 An Experiment Involving Measurement Choices

Jelly snakes are a confectionery that appeals to the consumer because of its stretchiness. However, the apparently simple idea of investigating the stretchiness of jelly snakes can produce a wide range of ideas and designs for experimentation. Factors can include one or all of colour, brand and temperature, with the latter lending itself to linking with science discussion. But the most challenging aspect that leads to the greatest variation in ideas and some very interesting planning discussions is the question of how to measure the stretchiness, and what measures of the unstretched snake to include in the investigation. Some examples of suggestions of how to measure the stretch have arisen from students and from teachers, including: stretch to break and record length at breaking; stretch to a selected length, let go and measure length to which snake returns; stretch at constant speed; stretch vertically; and remove head of snake and stretch remainder. This is an example of a topic for investigation that can be made as simple or as complicated as desired. Because of its appeal and potential for diversity in approach, Conker Statistics (www.conkerstatistics.co.uk) have chosen it as an activity for the development of resources.

Table 14.1 Numbers of commuters at a bus station

Off peak	Down	Up	Total	Peak	Down	Up	Total
Lift	35	62	97	Lift	148	16	164
Stairs	26	23	49	Stairs	592	8	600
Total	61	85	146	Total	740	24	764

8.2 A Survey that Involves Human Characteristics

Human characteristics are always a fascinating topic for students. Surveys are popular choices in free-choice projects but the design of questions is usually far harder than it first appears. An example of a survey without question design problems and which can also include experimental design aspects is investigating how people clasp their hands. Some reports say people tend to place the left thumb on top (see, for example, http://humangenetics.suite101.com/article.cfm/dominant_human_genetic_traits). Is it related to how people fold their arms? Because the data are categorical, a reasonable amount of data is required for meaningful discussion based on plots and tables, but the data are also quick to collect. One such investigation (MacGillivray, 2007) found that key aspects included the importance of a pilot study and the randomisation of the order in which subjects were asked to do the clasping and folding.

8.3 An Observational Investigation that Links with Chance

Many aspects of human behaviour provide categorical data, and relationships can be explored through two-way tables and side-by-side or segmented bar charts. These also lead naturally to estimating conditional probabilities without the need for any theoretical concepts or jargon. Table 14.1 shows data from an observational study of the use of stairs or lifts at a bus station during a peak period and an off-peak period.

The probability that a person going up uses the lift can be estimated by $62/85 = 0.73$ during off-peak times, and $16/24 = 0.67$ during peak periods. The probability that a person using the lift is going up can be estimated by $62/97 = 0.64$ during off-peak periods, and by $16/164 = 0.1$ during peak periods. These and other estimates lead to a wealth of discussion, and key questions about the context and the data collection methods – questions that are core to the data investigation cycle.

9 Conclusion

Statistics is a very challenging area to teach. In addition, many teachers have limited statistical content knowledge as well as little, if no, exposure to any specific pedagogy related to the teaching of statistics. While one can hope that this situation might change over time, the question facing statistics educators is how to educate teachers to be effective teachers of statistics. The word educate, rather than train, is used because to be effective, teachers need to learn about the appropriate pedagogy as well as update their knowledge and understanding. Governments should work towards a significant increase in the number of teachers specifically educated to teach statistics but this requires time and planning to achieve. The authors make the following specific recommendations:

1. Statistical projects should be included in the mathematics education components of both pre-service and in-service teacher education programmes. This is feasible since the development and discussion of projects is already an integral part of most mathematics education teacher training.
2. The statistical projects used should emphasise the data investigative cycle as a vehicle to teach statistical thinking and to develop teachers' own statistical understanding and knowledge. The PCPD or PPDAC description can provide a framework. Burgess (2008) demonstrated how teachers' statistical knowledge for teaching statistics could be usefully benchmarked using the PPDAC framework.
3. The projects used as part of both pre-service and in-service teacher education should, wherever possible, be projects that can be adapted to school use. Many statistical questions can be approached at different levels of sophistication. By using projects that can be used with their students it is more likely that the pre-service and in-service teachers will utilise the materials and ideas in their teaching.
4. Teachers need to undergo the same learning experiences as their students (Burgess, 2008; Pfannkuch, 2008), and teachers of teachers may also need to be trained in substantive statistical content and pedagogical knowledge.

More research should be undertaken into frameworks that will help to structure the teacher "learning" of statistics. Very little is known about effective mechanisms. Much of the research thus far has been on the nature of statistical understanding and thinking; a focus on effective pedagogy in the statistical education of teachers, per se, is a key area of what needs to be undertaken.

References

Anderson, L., & Krathwohl, D. (2001). *A taxonomy for learning, teaching and assessing: A revision of Bloom's educational objectives*. New York: Longman.

Batanero, C., Burrill, G., Reading, C., & Rossman, A. (Eds.) (2008). *Joint ICMI/IASE Study: Teaching Statistics in School Mathematics. Challenges for Teaching and Teacher Education. Proceedings of the ICMI Study 18 and 2008 IASE Round Table Conference*. Monterrey, Mexico: International Commission on Mathematical Instruction and International Association for Statistical Education. Online: www.stat.auckland.ac.nz/~iase/publications

Burgess, T. (2008). Teacher knowledge for teaching statistics through investigations. In C. Batanero, G. Burrill, C. Reading, & A. Rossman (2008).

Cameron, M. (2009). Training statisticians for a research organisation. *Proceedings of the International Statistical Institute 57th Session*. Durban, South Africa: International Statistical Institute. Online: http://isi.cbs.nl/

Chambers, J. (1993). Greater or lesser statistics: A choice for future research. *Statistics and Computing, 3*, 182–184.

Chance, B. (2005). Integrating pedagogies to teach statistics. In J. Garfield (Ed.), *Innovations in teaching statistics* (pp. 101–112). Washington, DC: Mathematical Association of America.

du Feu, C. (2005). Bluebells and bias, stitchwort and statistics. *Teaching Statistics, 27*(2), 34–36.

Forster, M., Smith, D., & Wild, C. (2005). Teaching students to write about statistics. In L. Weldon & B. Phillips (Eds.), *Proceedings of the ISI/IASE Satellite on Statistics Education and the Communication of Statistics*. Voorburg, The Netherlands: International Statistical Institute. Online: www.stat.auckland.ac.nz/~iase/publications

Holmes, P. (1997). Assessing project work by external examiners. In I. Gal & J. Garfield (Eds.), *The assessment challenge in statistics education* (pp. 153–164). Amsterdam: IOS Press.

Lee, C. (2005). Using the PACE strategy to teach statistics. In J. Garfield (Ed.), *Innovations in teaching statistics* (pp. 13–22). Washington, DC: Mathematical Association of America.

Lipson, K., & Kokonis, S. (2005). The implications of introducing report writing into an introductory statistics subject. In L. Weldon & B. Phillips (Eds.), *Proceedings of the ISI/IASE Satellite on Statistics Education and the Communication of Statistics*. Voorburg, The Netherlands: International Statistical Institute. Online: www.stat.auckland.ac.nz/~iase/publications

MacGillivray, H. (1998). Developing and synthesizing statistical skills for real situations through student projects. In L. Pereira-Mendoza (Ed.), *Proceedings of the 5th International Conference on Teaching Statistics* (pp. 1149–1155). Voorburg, The Netherlands: International Association for Statistical Education.

MacGillivray, H. (2002). One thousand projects. *MSOR Connections, 2*(1), 9–13.

MacGillivray, H. (2005). Coherent and purposeful development in statistics across the education spectrum. In G. Burrill & M. Camden (Eds.), Curricular development in statistics education. Voorburg, The Netherlands: International Association for Statistics Education. Online www. stat.auckland.ac.nz/~iase/publications

MacGillivray, H. L. (2007). Clasping hands and folding arms: A data investigation. *Teaching Statistics, 29*(2), 49–53.

MacKay, R. J., & Oldfield, W. (1994). *Stat 231 Course Notes Fall 1994*. University of Waterloo, Waterloo, Canada

Makar, K. (2008). A model of learning to teach statistical enquiry. In C. Batanero, G. Burrill, C. Reading, & A. Rossman (2008).

Marriott, J., Davies, N., & Gibson, L. (2009). Teaching, learning and assessing statistical problem solving. *Journal of Statistics Education, 17*(1). Online: www.amstat.org/publications/jse

Moore, D. (1997). New pedagogy and new content: The case of statistics (with discussion). *International Statistical Review, 65*(2), 123–137.

Nolan, D., & Lang, D. T. (2007). Dynamic, interactive documents for teaching statistical practice. *International Statistical Review, 75*(3), 295–321.

Pfannkuch, M. (2008). Training teachers to develop statistical thinking. In C. Batanero, G. Burrill, C. Reading, & A. Rossman (2008).

Ridgway, J., McCusker, S., & Nicolson, J. (2005). Uncovering student statistical competencies via new interfaces. In G. Burrill & M. Camden (Eds.), *Curricular Development in Statistics Education: International Association for Statistical Education 2004 Roundtable*. Voorburg, The Netherlands: International Association for Statistics Education. Online: www.stat. auckland.ac.nz/~iase/publications

Schield, M. (2005). Statistical literacy curriculum design. In G. Burrill & M. Camden (Eds.), *Curricular Development in Statistics Education. International Association for Statistical Education 2004 Roundtable*. Voorburg, The Netherlands: International Statistical Institute. Online: www.stat.auckland.ac.nz/~iase/publications

Shewhart, W., & Deming, W. (Eds.). (1986). *Statistical method from the viewpoint of quality control*. New York: Dover (Original work published 1939).

Turner, C. (2006). Height, foot length and threat to woodland: Positive learning from pupil relevant data. *Teaching Statistics, 28*(1), 22–25.

Watson, J. (1997). Assessing statistical thinking using the media. In I. Gal & J. Garfield (Eds.), *The assessment challenge in statistics education* (pp. 107–121). Amsterdam: IOS Press.

Wild, C. J., & Pfannkuch, M. (1999). Statistical thinking in empirical enquiry (with discussion). *International Statistical Review, 67*(3), 223–265.

Chapter 15
Complementing Mathematical Thinking and Statistical Thinking in School Mathematics

Linda Gattuso and Maria Gabriella Ottaviani

Abstract The introduction of statistics into school curriculum within the mathematics subject poses multifaceted problems to mathematics teachers. This chapter first discusses the relevance of developing mathematical and statistical literacy in schools, and secondly reflects on some current recommendations to teach statistics in the school mathematics and challenges faced in the training of teachers. Then the chapter underlines differences between mathematical and statistical thinking and suggests that, taking account of their specificities, it is possible to generate teaching strategies that allow the harmonious development of both mathematical and statistical thinking in school. Some implications for teacher training are finally included.

1 Introduction

In the last few decades of the twentieth century unprecedented innovation in society and, in particular, globalisation catalysed by modern telecommunications did justify a new perceived complexity of reality, enhancing "the central importance of mathematics and its applications in today's world with regard to science, technology, communications, economics and numerous other fields" (United Nations Educational, Scientific and Cultural Organisation [UNESCO], 1997). The basic aim of mathematics in this changing society had already been formulated, in the United States of America, within the National Council of Teachers of Mathematics

L. Gattuso (✉)
Université du Québec à Montréal, 3902 rue Drolet, Montreal, QC H2W 2L2, Canada
e-mail: gattuso.linda@uqam.ca

M.G. Ottaviani
Sapienza, Università di Roma, Rome, Italy
e-mail: mariagabriella.ottaviani@uniroma1.it

C. Batanero, G. Burrill, and C. Reading (eds.), *Teaching Statistics in School Mathematics-Challenges for Teaching and Teacher Education: A Joint ICMI/IASE Study*, DOI 10.1007/978-94-007-1131-0_15, © Springer Science+Business Media B.V. 2011

(NCTM) Standards (1989), where the concept of "mathematical power" was presented as the notion of empowering the students to be able to apply their mathematical knowledge, concepts, and ability in problem solving, communication and reasoning.

In 2001, "use mathematics to solve problems and communicate" (Stein, 2001, p. 17) was listed as one of the 16 Equipped for the Future (EFF) standards needed by adults to effectively carry out their different roles in society. The use of knowledge in concrete situations – where the active participation of individuals is required – stresses the concept of "competence". For mathematics the curriculum shift from content topics to competences has many consequences from the pedagogical point of view both for teachers and for students, and implies a new teaching/learning style relying particularly on problem-posing and problem-solving teaching methods. In particular, giving evidence of workplace needs, Steen (2003) regarded mathematical thinking as "an essential component of virtually every competency. Reasoning, making decisions, solving problems, managing resources, interpreting information, understanding systems, applying technology – all these and more build on quantitative and mathematical acumen" (p. 56). When the information to be dealt with is quantitative in nature and keen quantitative discernment is required, statistics and statistical thinking play an important role in the mathematical curriculum.

This chapter aims to emphasise the necessity of complementing statistical thinking and mathematical thinking in school and generating didactic strategies allowing statistics and mathematics to evolve together, in a harmonious way.

2   Why Statistics is Taught in School Mathematics

To be part of a modern society in a competent and critical way requires citizens to know and to interpret collective/social phenomena in a broad sense, and understand the variability, dispersion, and heterogeneity which cause uncertainty in interpreting, in making decisions, and in facing risks. To pose and solve problems in everyday life may require data collection and the ability to analyse the data in order to get information to be interpreted and used in suitable ways.

However, in reality citizens will seldom have the opportunity to control all stages of the statistical process of inquiry, particularly when they have at their disposal only data collected, organised and interpreted by others to address others' aims. In this case, statistical competences and thinking become more and more important as they encourage caution before using those data in a superficial way. That is why modern citizens require both basic knowledge of statistics and statistical concepts, and also statistical thinking.

The role of data, statistics and probability in school curriculum has been recognised in the Organisation for Economic Co-operation and Development

(OECD) study titled Programme for International Student Assessment (PISA). Mathematical literacy is one of the three domains assessed by PISA in order to measure how well young adults, at the age of 15, are prepared to meet the challenges of today's knowledge society. According to the project:

> Mathematical literacy is an individual's capacity to identify and understand the role that mathematics plays in the world, to make well-founded judgements and to use and engage with mathematics in ways that meet the needs of that individual's life as a constructive, concerned and reflective citizen. (OECD, 2006, p. 72)

For PISA assessment purposes the mathematical content that a person might utilise in solving a problem has been organised by four overarching ideas: *space and shape*, *change and relationships*, *quantity* and *uncertainty* (OECD, 2006). The three first ideas form the heart of any mathematics curriculum, but it is not the same for the fourth. The recognition from OECD that dealing with uncertainty is essential in everyday life is obviously of primary importance in promoting the teaching of statistics and elements of probability theory in school mathematics.

3   Teaching Statistics in the Mathematics Classroom

Statistics is appearing more and more in school curricula; in some countries, statistics has recently even entered the curricula of elementary schools. The situation in various countries is described in the first chapter of this book. Statistics in schools is linked to mathematics so mathematics teachers are responsible for its implementation.

Curriculum developers suggest a data-oriented approach to teaching statistics (Moore, 1997; Burrill & Camden, 2005) where students should formulate research questions; design investigations; collect data using observations, surveys, and experiments; describe and compare data sets; and propose and justify conclusions and predictions based on data.

The GAISE project, for example, has developed useful guidelines for statistics education (Franklin et al., 2005). However, as discussed in the Joint ICMI/IASE Study Conference (Batanero, Burrill, Reading & Rossman, 2008), these recommendations are seldom followed and doing statistics too often becomes synonymous with doing computations and following protocols. Consequently, students finishing high school understand very little statistics and are usually unable to utilise it in a critical way.

The problem is that the teachers generally have no preparation for teaching statistics, little knowledge about statistics and almost never any training in statistics education. They need a framework for understanding statistics, so that they can understand where their students are coming from and where they are going (Ottaviani, Peck, Pfannkuch, & Rossman, 2005). Although there has been a lot of progress in the implementation of statistics in the school curriculum, statistics education for future teachers is almost non-existent.

The situation is serious for elementary teachers who have little or no experience in this field, and often demonstrate little interest in mathematics although they have to teach it. The situation is not much better for secondary teachers. Their mathematical knowledge is more important but in some ways, particularly if mathematics is seen in a formalistic view, this may even hinder their grasp of statistics. Most trainee secondary teachers will follow a course in statistics but very few teacher training programmes include the didactic of statistics. In fact, mathematic educators often casually admit their lack of qualification in the subject.

In addition to gaps in teachers' statistical knowledge, negative attitude and beliefs towards statistics complicate the situation. "Negative attitudes are linked to perceived difficulty, lack of knowledge and overly formal learning experience" (Estrada & Batanero, 2008, p. 5). Meletiou (2003) argued that beliefs about the nature of mathematics affect instructional approaches and curricula in statistics, and act as a barrier to the kind of instruction that would provide students with the skills necessary to recognise and intelligently deal with uncertainty and variability. Although the teaching of mathematics has undergone many changes and proposes a constructivist approach, long-held beliefs and attitudes of teachers are difficult to change. Statistical concepts linked to context should be approached as social constructs, following the way suggested by the data-oriented approach. In reality, concepts are too often presented to students without any links to the real-world context or at the most within artificial examples and using a traditional and procedural approach that in many cases meet students' and parents' expectations.

Obviously, knowing the theory of statistics is not enough to teach it. Teachers must have the opportunity to develop their own statistical thinking. The education of pre-service and in-service teachers has to be taken seriously. According to Batanero (2008), initial and continuing teacher training courses for mathematics teachers need to be redesigned completely. Future teachers must experience the same activities proposed for students and experience the same difficulties, but obviously teacher knowledge needs to be broader and deeper than that of the students they are teaching. In fact, many teachers have no experience with data analysis and do not understand the role of variability and the idea of distribution, which are key concepts for the development of statistical thinking.

Today, teacher training is mostly under the auspices of mathematics educators. However, statisticians involved with statistics education, and statistics educators must cooperate and be involved in developing resources for teachers including high-quality teaching materials that, promoting the issue of teaching statistics, could help motivate students to learn mathematics. To achieve these goals it is fundamentally necessary to describe characteristics of statistics and to identify differences and similarities between mathematical thinking and statistical thinking.

4   Statistics and Statistical Thinking

If statistics is different from mathematics, what is statistics? As is the case for any science, to define statistics is difficult. In recent years it has been recognised that:

> Statistics has developed from two disciplines: The mathematical study of probabilities and chance events and the scientific attempt to draw conclusions from data in the face of inevitable error and imprecision. Modern statistics does not simply apply mathematical results to determine the properties of particular statistical methods; it includes a concern for discerning, describing, and confirming patterns and relationships in data. (Thisted & Velleman, 1992, p. 41)

In fact, "statistics makes a heavy and essential use of mathematics, yet has its own territory to explore and its own core concepts to guide the exploration" (Cobb & Moore, 1997, p. 814). It "is a subject whose goal is to solve real-world problems" (Moore & Cobb, 2000, p. 617). Statistics may be considered both as a discipline in itself and as a technique: "A special technique suitable for the quantitative investigation of mass or collective phenomena, those phenomena (…) whose measurement requires a collection of observations" (Gini, 1966, p. 17).

The process of statistical investigation begins with some study questions providing a basis for the design used to produce data, it goes on with the collection of the data, their exploration and description, and eventually formal inductive inference is required if conclusions are needed about the population or process from which the data were drawn. The interpretation of the results coming from the data is the crucial point where statistics comes in touch again with the questions that started all the process. Only at this point does it become evident whether both the statistical methods used and the statistical reasoning followed were effective in solving the problems giving rise to the study. The investigative cycle: from problem, to data (collected, analysed and reported), to problem forms the core of statistical thinking. In this vision of statistics, there are concepts – such as centre and variability – and measures of concepts – such as arithmetic mean, median, mode and standard deviation, interquartile range, range – and not just numbers and formulae.

To debate the differences between statistics and mathematics is important for statistics educators who need "to carefully define the unique characteristics of statistics and in particular the distinction between statistical literacy, reasoning and thinking" (Garfield & Ben-Zvi, 2007, p. 380). Each of these three capabilities can be differentiated according to the level of statistical tools and concepts people understand and the connections people are able to make among them. The focus in this chapter is on statistical thinking.

To simplify the comparison with mathematical thinking, this chapter uses the definition of statistical thinking proposed by Scheaffer (2003): "Data analysis and statistical thinking … develop knowledge, beliefs, dispositions, habits of mind, communication capabilities and problem solving skills that people need to engage effectively in quantitative situations arising in life and work" (pp. 146–147), particularly in those situations involving processes and their variation.

An active group of educators, psychologists, and statisticians have studied and examined the change in the importance given by the statistics instruction when evolving from the statistical techniques, formulas, computations and procedures towards conceptual understanding of statistics (Garfield & Ben-Zvi, 2004), and have also made connections between the research results and practical suggestions for teachers (Garfield & Ben-Zvi, 2008).

5 Differences Between Mathematical Thinking and Statistical Thinking in School

In some ways mathematical thinking and statistical thinking may appear contrary, but when we underline their differences, we will see that they may support each other. Where mathematics exploits deductive reasoning, statistics uses more inductive reasoning. While mathematics promotes abstraction, statistics insists on interpretation in context. Variation and measurement are dealt with differently in the two disciplines. In summary, reasoning in mathematics and statistics is different. A more comprehensive picture of the situation can be found in Rossman, Chance, and Medina (2006) and Scheaffer (2006).

Although, more and more mathematics educators encourage a constructivist approach for mathematics in the classroom, teaching too often is dominated by presenting deterministic procedures even if most curricula propose a broader view of mathematics. "One question has one answer". Traditional teaching is all too often focused on developing procedures to solve closed problems. Even in the so-called open-ended problems, the solutions are often predetermined. This misleads students who look for "what the teacher wants". Mathematics is about logical and deductive reasoning, modelling, optimising, and proving results that come logically from axioms and definitions. Although not all mathematics teaching in schools follows this line, it is too often procedural, allocating more space to calculation than to understanding. However, more and more mathematics educators and researchers are rejecting the traditional approach and proposing that learning mathematics should develop the ability to create mathematical models of real phenomena, pose hypotheses and verify them using mathematical tools (Sierpinska & Kilpatrick, 1998). In statistics, the same question with the same data may lead to different ways of analysing and different solutions that are equally defendable. This requires inductive reasoning, working with randomness, dealing with counterintuitive results, drawing uncertain conclusions, and interpreting results.

Mathematics and statistics are different in the ways that they use numbers. Mathematics mostly deals with numbers, their operations, generalisations and "abstractions", while for statistics numbers are "data linked to a context", which is essential to statistical reasoning as well as to mathematical modelling. When doing statistics, one must know the nature of data, and where and how they are produced, to be able to go on with the analysis and to draw some conclusions. Mathematics, on the contrary, may rely on context for motivation in the classroom, or as a source of research problems, but its goal is abstracting, finding patterns and generalising. The context has to be put to one side to grasp the model or the structure. To synthesise: "In data analysis the emphasis is on answering real questions rather than trying to fit those questions into established theories" (Scheaffer, 2003, p. 145).

Variability and variation are found in mathematics and in statistics but with a different sense. In the mathematics classroom students study the dependence of one variable on the other, and the form of the link between the variables. In statistics variability, that is the propensity of the observations for one data set to change, is a fundamental idea supporting the concept of distribution. Looking at averages without taking account of variability (spread) is useless and will not lead to the understanding of the distribution, thus missing the whole pattern.

Furthermore mathematics and statistics have a different approach to measurement. In mathematics, measurement goes with spatial configurations, and their transformation, and abstraction. For example, at secondary school in a geometry problem there is no need for rulers to show that two sides of a triangle have equal length; the equality of the length can be deduced from hypotheses, definitions, and theorems. Although a figure may help understanding or finding the proof, its measures do not need to be accurate and can be assessed approximately. Because statistics is mostly about understanding, measuring and describing the real world, taking valid measurements is crucial. In any investigation, the study question has to be well formulated and the data have to be accurate.

6 Advantages of Doing Statistics and Mathematics Together in School

Despite the differences between statistical thinking and mathematical thinking, there are certain advantages in studying statistics in the mathematics classroom. First, statistics can stimulate motivation and develop problem-solving abilities such as posing questions, analysing, representing and communicating quantitative information. If well chosen and close to students' interests, context, which is essential to statistics, often has a positive effect on students' motivation and involvement also in mathematics. According to Kranendonk (2006), students playing with real data that makes sense connect with these data, and they get curious and often go beyond what they were asked to do. Finding a new interest can modify a negative attitude towards mathematics. Using context also agrees with new curricula in mathematics that advocate problem solving imbedded in real-world situations.

Second, much of statistics involves posing questions and finding ways to answer them. A problem leading to the collection of data and analysis, even if elementary, will enrich a child's mathematical thinking. At the beginning, there is no need for complex mathematics, but only for the ability to classify and group. The ability to formulate a question and to be critical about it can be practised even by kindergarten children (Schwartz, 2006) and will transfer to the study of mathematics; is it not said that, in problem solving, when the question is posed the problem is almost solved?

Third, statistical analysis is not a linear process. After collecting and grouping data, analysis comes next. Comparing groups, looking at the characteristics of the distribution, identifying clusters, outliers, examining the differences in the medians, means, modes and measures of spread may suggest a "rerouting" in the analysis of data. To go back and forth to find a solution or a proof can also be very helpful in mathematics, but in the classroom it is unusual and not often shown. Instead, the result is traditionally exposed in a straightforward manner without revealing the trials and errors preceding the optimal solution shown on the board, as it would come from some magical inspiration leaving the students helpless.

Finally, the construction of representations is essential in the study of data. Not only does the representation have to be adequate and complete, but it helps to visualise statistical distributions and give evidence to relationships among variables. Different representations may also lead to a different grasp of the distribution. This is useful also in mathematics where a graphical representation is a necessary prerequisite to modelling, even if "the standard mathematical models ignore data production" (Cobb & Moore, 1997, p. 807).

The competence to communicate mathematical results is nowadays part of the mathematics curricula recommendations. In statistics, interpreting and communicating the results to answer the original question follows statistical analysis. It requires convincing with "numerical" arguments placed in their context and is completed with a discussion of the various possibilities investigated, thus assuming (more or less consciously) variability and probability. Again, the development of this competency may benefit both disciplines.

The points underpinned above are about conceptual understanding and thinking both of statistics and of mathematics. Besides, it should not be forgotten that when going through various statistical procedures, a lot of mathematics is applied. From elementary arithmetic (especially proportional reasoning) to advanced functions (e.g., the least squares method), many examples of mathematical concepts and tools are employed while doing statistics and mathematical learning can only profit from this use (Gattuso, 2006, 2008). Teachers must be aware of the benefits of making statistics part of their mathematics teaching but at the same time be familiar with the specificity of each discipline.

7   Implications for Teacher Training

An important key to the development of statistics teaching is teacher training. Well-prepared teachers will willingly include statistics in their teaching. With adequate training, teachers will be more confident and they will be able to encourage students to speculate and explore phenomena, create their own data representations, make and test their own conjectures, use appropriate technological tools, and spend time on discussion and reflection instead of limiting the students to the practice of procedural skills and execution of calculations.

Teachers surely need to acquire statistical knowledge and develop their statistical thinking, but they also need training in the didactic of statistics to be able to follow their students' learning and reasoning and be able to spontaneously take advantage of classroom situations to promote student learning. The didactic of statistics will introduce teachers to misconceptions, difficulties, and common errors involved in learning statistics and will propose ways to handle them, thus allowing teachers to develop the self-assurance needed to teach adequately.

Also, it is important to pay attention to the teachers' concerns about leaving out some mathematical content by assuring and showing them that, while doing statistics, they are really doing a lot of mathematics. It is also necessary to match mathematical concepts to their applications in statistics so that one supports the development of the other (Dunkels, 1990). Statistics can contribute to the learning of mathematics by introducing mathematical concepts in realistic and motivating contexts. Measurement of phenomena (such as bullying, free time, fertility, poverty), proportional reasoning and percentages, graphical displays, averages, data modelling, and inductive reasoning are all points of contact and tension between statistics and mathematics in school (Biehler, 2008). Research is necessary to understand how to transform a possible uneasy junction of such different disciplines into a fruitful one. This may require statistics educators to work side by side with mathematics educators, respecting each other and showing how concepts and knowledge of the two disciplines may evolve together in the classroom in a harmonious way (Ottaviani, 2008).

During their training in statistics teachers should also be exposed to the use of technological tools. Technology, in fact, can assist students in "doing" and "seeing" statistics and in reflecting on data. Different kinds of statistical tools exist. Some are useful to visualise data and to analyse it in a simple way, some are more suitable for developing an understanding of data and data exploration, and others are more useful for understanding concepts connected to probability distributions. Besides this, the Internet offers a large set of downloadable data to support exploratory data analysis and to assist in understanding variability (Garfield & Ben-Zvi, 2004). By navigating the Internet it is possible to find resources for teachers to use in classrooms or improve their knowledge of statistics and resources for those training the teachers. In particular, the International Statistical Literacy Project (ISLP), under the umbrella of the International Association for Statistical Education provides an online repository of national and international activities to disseminate statistical thinking (www.stat.auckland.ac.nz/~iase/islp/). Through this, teaching statistics in school mathematics has the added bonus of students acquiring a greater familiarity with technological instruments used in everyday life.

Teacher training should also include discussion on assessment methods. Mathematics teachers are used to utilising multiple choice, "right or wrong" answer or short answer questions, thus focusing on accuracy of computation, correct application of formulas or correctness of graphs and charts. These kinds of questions are not useful when statistical thinking is involved. To get information about students' statistical reasoning processes requires the teachers to identify assessment methods that can reveal student understanding of basic statistical concepts such as variability, visual representation of data, centre and spread (Gal & Garfield, 1997). The importance of assessment is evident when we notice that teachers are more and more motivated to do a better job with statistics as long as assessing the achievement of statistical curriculum is required.

The support of statistics educators and practising statisticians for mathematics teachers is essential to help them cope with their new role as statistics teachers.

8 Conclusion

The "marriage" of statistics and mathematics in schools is difficult particularly due to the school teachers' general lack of statistical knowledge that makes it hard for them to develop their own statistical thinking. In fact, this chapter shows that mathematics and statistics are different, at least, in the way that reasoning takes place, in the way they use numbers, in the way that variability and variation are taken into account, and in their approach to measurement. However, there are good reasons for mathematics school teachers to teach statistics in their classes, such as while students are doing statistics they are really doing a lot of mathematics, and students with a negative attitude towards mathematics can find a new interest. For statistics to be taught in an adequate way in school mathematics will take a long time. A lot of research activity needs to be carried out by statistics educators in collaboration with mathematics education in order to offer mathematics teachers appropriate instructional resources and strategies. There is no doubt that to have statistical thinking diffused in society is fundamental so that both pre-service and in-service mathematics teachers have to receive high-quality training in statistics.

References

Batanero, C. (2008, September). *Training school teachers to teach statistics: an international challenge*. Paper presented in the Premier Colloque Francophone International sur l'Enseignement de la Statistique. Lyon, France: Société Française de Statistique.

Batanero, C., Burrill, G., Reading, C., & Rossman, A. (Eds.) (2008). *Joint ICMI/IASE Study: Teaching Statistics in School Mathematics. Challenges for Teaching and Teacher Education. Proceedings of the ICMI Study 18 and 2008 IASE Round Table Conference*. Monterrey, Mexico: International Commission on Mathematical Instruction and International Association for Statistical Education. Online: www.stat.auckland.ac.nz/~iase/publications

Biehler, R. (2008). From statistical literacy to fundamental ideas in mathematics: How can we bridge the tension in order to support teachers of statistics. In C. Batanero, G. Burrill, C. Reading, & A. Rossman (2008).

Burrill, G., & Camden, M. (Eds.) (2005). *Curricular Development in Statistics Education: International Association for Statistical Education 2004 Roundtable*. Voorburg, The Netherlands: International Statistical Institute. Online: www.stat.auckland.ac.nz/~iase/publications

Cobb, G. W., & Moore, D. S. (1997). Mathematics, statistics, and teaching. *The American Mathematical Monthly, 104*, 801–823.

Dunkels, A. (1990). Examples from the in-service classroom (age group 7–12). In A. Hawkins (Ed.), *Training teachers to teach statistics: Proceedings of the International Statistical Institute Round Table Conference* (pp. 102–109). Voorburg, The Netherlands: International Statistical Institute.

Estrada, A., & Batanero, C. (2008). Explaining teachers' attitudes towards statistics. In C. Batanero, G. Burrill, C. Reading, & A. Rossman (2008).

Franklin, C., Kader, G., Mewborn, D., Moreno, J., Peck, R., Perry, M., & Scheaffer, R. L. (2005). *Guidelines for assessment and instruction in statistics education (GAISE) report: A preK-12 curriculum framework.* Alexandria, VA: American Statistical Association. Online: www. amstat.org/Education/gaise/

Gal, I., & Garfield, J. B. (1997). *The assessment challenge in statistics education.* Amsterdam: IOS Press and International Statistical Institute.

Garfield, J. B., & Ben-Zvi, D. (2004). Research on statistical literacy, reasoning and thinking: Issues, challenges and implications. In D. Ben-Zvi & J. B. Garfield (Eds.), *The challenge of developing statistical literacy, reasoning and thinking* (pp. 397–409). Dordrecht, The Netherlands: Kluwer.

Garfield, J., & Ben-Zvi, D. (2007). How students learn statistics revisited: A current review of research on teaching and learning statistics. *International Statistical Review, 75*(3), 372–396.

Garfield, J., & Ben-Zvi, D. (2008). *Developing students' statistical reasoning: Connecting research and teaching practice.* Dordrecht, The Netherlands: Springer.

Gattuso, L. (2006). Statistics and mathematics: Is it possible to create fruitful links? In A. Rossman & B. Chance (Eds.), *Proceedings of the 7th International Conference on Teaching Statistics.* Salvador, Bahia, Brazil: International Statistical Institute and International Association for Statistical Education. Online: www.stat.auckland.ac.nz/~iase/publications

Gattuso, L. (2008). Mathematics in a statistical context? In C. Batanero, G. Burrill, C. Reading, & A. Rossman (2008).

Gini, C. (1966). *Statistical methods.* Roma, Italy: Università degli Studi di Roma.

Kranendonk, H. (2006). A statistical study of generations. In G. Burrill (Ed.), *Thinking and reasoning with data and chance, 68th NCTM yearbook (2006)* (pp. 103–116). Reston, VA: National Council of Teachers of Mathematics.

Meletiou, M. (2003). On the formalist view of mathematics: Impact on statistics instruction and learning. In A. Mariotti (Ed.), *Proceedings of the Third European Conference in Mathematics Education.* Bellaria, Italy: European Research in Mathematics Education Society. Online: www.dm.unipi.it/~didattica/CERME3/proceedings

Moore, D. S. (1997). New pedagogy and new content: The case of statistics. *International Statistical Review, 65*(2), 123–137.

Moore, D. S., & Cobb, G. (2000). Statistics and mathematics: Tension and cooperation. *The American Mathematical Monthly, 107*, 615–630.

National Council of Teachers of Mathematics (1989). *Curriculum and evaluation standards for school mathematics.* Reston, VA: Author.

Organisation for Economic Co-operation and Development (2006). *Assessing scientific, reading and mathematical literacy: A framework for PISA 2006.* Online: www.oecd.org/dataoecd/63/35/37464175.pdf

Ottaviani, M. G. (2008). Statistica e matematica a scuola: due discipline e un solo insegnamento. Confronto culturale e opportunità interdisciplinare (Statistics and mathematics at school: Two disciplines in one subject. Cultural comparisons and interdisciplinary potentialities). *Induzioni, 36*, 17–38.

Ottaviani, M. G., Peck, R., Pfannkuch, M., & Rossman, A. (2005). Working group report on teacher preparation for statistics education. In G. Burrill & M. Camden (Eds.), *Curricular Development in Statistics Education: International Association for Statistical Education 2004 Roundtable.* Voorburg, The Netherlands: International Statistical Institute. Online: www.stat.auckland.ac.nz/~iase/publications

Rossman, A., Chance, B., & Medina, E. (2006). Some important comparisons between statistics and mathematics, and why teachers should care. In G. Burrill (Ed.), *Thinking and reasoning with data and chance, 68th NCTM Yearbook (2006)* (pp. 323–334). Reston, VA: National Council of Teachers of Mathematics.

Scheaffer, R. L. (2003). Statistics and quantitative literacy. In B. L. Madison & L. A. Steen (Eds.), *Quantitative literacy: Why numeracy matters for schools and colleges* (pp. 145–152). Princeton, NJ: National Council on Education and the Disciplines. Online: www.maa.org/ql/qltoc.html

Scheaffer, R. L. (2006). Statistics and mathematics: On making a happy marriage. In G. Burrill (Ed.), *Thinking and reasoning with data and chance: 68th NCTM Yearbook (2006)* (pp. 309–321). Reston, VA: National Council of Teachers of Mathematics.

Schwartz, S. L. (2006). Graphing with four-year-olds: Exploring the possibilities through staff development. In G. Burrill (Ed.), *Thinking and reasoning with data and chance, 68th NCTM Yearbook (2006)* (pp. 5–17). Reston, VA: National Council of Teachers of Mathematics.

Sierpinska, A., & Kilpatrick, J. (Eds.). (1998). *Mathematics education as a research domain: A search for identity. An ICMI Study.* Dordrecht, The Netherlands: Kluwer.

Steen, L. A. (2003). Data, shapes, symbols: Achieving balance in school mathematics. In B. L. Madison & L. A. Steen (Eds.), *Quantitative literacy: Why numeracy matters for schools and colleges* (pp. 53–74). Princeton, NJ: National Council on Education and the Disciplines. Online: www.maa.org/ql/qltoc.html

Stein, S. (2001). *Equipped for the future, content standards.* Washington, DC: National Institute for Literacy.

Thisted, R. A., & Velleman, P. F. (1992). Computers and modern statistics. In D. C. Hoaglin & D. S. Moore (Eds.), *Perspectives on contemporary statistics* (MAA notes 21, pp. 41–53). Washington, DC: Mathematical Association of America.

United Nations Educational, Scientific and Cultural Organisation. (1997). *UNESCO's sponsorship.* Resolution 29 C/DR126. Online: wmy2000.math.jussieu.fr/unesco.html

Chapter 16
Assessment of Learning, for Learning, and as Learning in Statistics Education

Joan Garfield and Christine Franklin

Abstract Assessing student learning of statistics poses unique challenges to mathematics teachers at the elementary and secondary level. This chapter describes some guiding principles for developing or selecting assessment items, building on general pillars of good assessment practice as well as important features of the discipline of statistics. The chapter concludes with some specific recommendations regarding the improvement of assessment of student learning of statistics.

1 Introduction

Assessment plays an important role in teaching and learning statistics. This topic has received much attention in the statistics education community (see Gal & Garfield, 1997; Chance, 2004; Phillips & Weldon, 2007). Ideas about effective uses of assessment in the context of teaching and learning statistics have been greatly influenced by the mathematics education community (e.g., Romberg, 1992; Mathematical Sciences Education Board [MSEB], 1993; National Council of Teachers of Mathematics [NCTM], 1995) and the educational measurement community (Pellegrino, Chudowsky, & Glaser, 2001). This chapter outlines some issues and challenges regarding assessment of student learning of statistics, as well as offering some guiding principles regarding the preparation of statistics teachers. Particular attention is paid to the unique aspects of assessing ideas of data

J. Garfield (✉)
Educational Psychology, University of Minnesota, 178 Ed Sciences Bldg,
56 East River Road, Minneapolis, MN 55455, USA
e-mail: jbg@umn.edu

C. Franklin
Department of Statistics, University of Georgia, 204 Statistics Bldg, 101 Cedar Street,
GA 30602, Athens
e-mail: chris@stat.uga.edu

C. Batanero, G. Burrill, and C. Reading (eds.), *Teaching Statistics in School*   133
Mathematics-Challenges for Teaching and Teacher Education: A Joint ICMI/IASE Study,
DOI 10.1007/978-94-007-1131-0_16, © Springer Science+Business Media B.V. 2011

understanding and exploration, as opposed to assessing other topics in the K-12 mathematics curriculum.

2   Purposes of Educational Assessment

Assessment of student learning provides data that may be used for different purposes, such as informing students of their progress and achievement, informing teachers of the effect of their teaching, and providing evidence of student achievement of desired student learning outcomes. Three broad purposes of assessment are described in Pellegrino et al. (2001): to assist learning, to measure individual achievement, and to evaluate programmes. This report cautions that assessment results are only estimates of what a person knows and can do and that every assessment, regardless of its purpose, rests on three pillars: (a) model of how students represent knowledge and develop competence in the subject domain; (b) tasks or situations that allow one to observe student performance; and (c) an interpretation method for drawing inferences from the performance evidence thus obtained.

These three foundational pillars – labelled *cognition*, *observation*, and *interpretation* – comprise an "assessment triangle" that underlies all assessment and must be explicitly connected and designed as a coordinated whole (Pellegrino et al., 2001).

Connected to all three pillars is the *purpose* of assessment. While traditional descriptions of assessment distinguish between Formative (assessments used to provide formative feedback to improve student learning) and Summative (assessments used to provide a summative indication of student achievement), more recent publications examine the purposes and use of student assessment as falling into three categories: Assessment *of* learning, Assessment *for* learning, and Assessment *as* learning (see Earl & Katz, 2006). Statistics teachers have traditionally used summative assessment to provide information *of* student learning, while using some types of formative assessments as agents *for* student learning, that is, to provide feedback to students to help them better learn statistics. The use of assessment *as* learning, which could encompass both summative and formative methods, situates the student at the integral junction between learning and assessment. In this unique purpose of assessment, students engage in new learning by monitoring and adapting their own understanding via the assessment process. Examples in a statistics course could include having students create or invent a unique model as part of a problem-solving activity that has them reflect and make sense of their own knowledge throughout the creation process (e.g., Lesh, Hoover, Hole, Kelly, & Post, 2000), or in an authentic task such as completing a statistical investigation or project (e.g., Holmes, 1997; Starkings, 1997). Assessment of, for, and as learning each serve different, yet interrelated purposes in helping students learn statistics. It is important for classroom teachers and writers of curriculum and assessments to consider each purpose so that their assessment process can provide meaningful information about, and evaluation of, student learning.

3   Assessing Statistical Learning

Regardless of the purposes for which an assessment is to be used, it is important to think carefully about the three pillars of assessment (cognition, observation, and interpretation) in order to develop or select assessment items of the highest quality that are appropriate for a given purpose.

3.1   *Cognition: Models of Learning and Desired Learning Outcomes*

The cognition pillar requires a model of how students represent knowledge and develop competence in statistics. National and local curriculum standards provide listings of desired content for student assessments (e.g., NCTM, 2000; Franklin & Garfield, 2006; Ministry of Education New Zealand, 2007). However, in designing or selecting assessment items, the level of cognitive outcome to be assessed must also be considered. The most well-known framework for delineating the cognitive outcomes of student learning is Bloom's (1956) Taxonomy. Many assessments have been created based on those levels of cognitive outcomes (knowledge, comprehension, application, analysis, synthesis, and evaluation).

More recently, researchers in statistics education (e.g., delMas, 2002; Garfield & Ben-Zvi, 2008; Garfield & delMas, 2010) suggested the following categorisation of cognitive statistical learning outcomes:

- *Statistical literacy*, understanding and using the basic language and tools of statistics
- *Statistical reasoning*, reasoning with statistical ideas and making sense of statistical information
- *Statistical thinking*, recognising the importance of examining and trying to explain variability and knowing where the data came from, as well as connecting data analysis to the larger context of a statistical investigation.

delMas (2002) distinguished between these related outcomes by suggesting types of words used in the assessment of each outcome, while Garfield and Ben-Zvi (2008) provided examples of assessment items for each type of outcome. By considering these types of outcomes teachers may be able to produce more balanced assessment items than ones that rely primarily on literacy-type items. Alternative frameworks for describing statistical learning outcomes, such as the Structure of Observed Learning Outcome (SOLO) taxonomy, may be found in Jones, Langrall, Mooney, and Thornton (2004).

Distinguishing between types of learning outcomes also leads educators to reflect on the content and instructional methods they are using. Wiggins and McTighe (1998) stressed the importance of carefully delineating desired learning goals in designing assessment. They outlined the need for educators to design assessments backward from the task, asking at each step of the way: "What is the evidence I need of students' understanding? Will this assessment get at it?" This is especially important in distinguishing between desired learning goals that are mathematically based versus those that are statistically based.

Consideration of desired content learning outcomes can focus attention on important differences in the assessment of statistical learning versus mathematical learning. This is a crucial point to make because statistics is part of the mathematics curriculum, and is frequently taught by educators trained in mathematics. There are several important readings to demonstrate to educators the differences between the disciplines of mathematics and statistics (Cobb & Moore, 1997; Franklin et al., 2005; Rossman, Chance, & Medina, 2006).

One of the differences often seen between mathematicians and statisticians is how they view and assess data analysis. An item labelled "data analysis" on a standardised test is often, in fact, a question assessing some form of mathematical computation or reasoning. Some illustrative examples are provided in the following sections. Often, educators trained in mathematics view data analysis as simply the computational aspects of the analysis – that is, finding numerical summaries or creating graphical representations. However, statisticians view data analysis as involving the *process* of formulating a scientific question that can be answered with data, designing a plan to collect the data, analysing the data with appropriate graphical and numerical summaries, and interpreting the results as they relate to the original question of interest (see Franklin & Garfield, 2006).

In assessing student learning in statistics, it is important to:

- *Include real data and real problem contexts*: In statistics, it is important to use real data that are provided in the context of a statistical investigation of interest (Groth, 2010) rather than "story problems" that are artificially constructed and do not represent real problems.
- *Include recognising and understanding the concept of variability*: In statistics, recognising and examining variability is equally (if not more) important than the trend or pattern in the data.
- *Include opportunities to select methods of graphing and analysing data*: In statistics, the choice of how data will be analysed is equally (if not more) important than the actual computations and calculations that are used to carry out the procedure.
- *Maintain a balance between items assessing understanding probability concepts and understanding statistics concepts.*
- *When appropriate, require students to provide interpretations of data analysis as well as justifications for their analyses and conclusions.*

This section concludes with a set of guiding principles regarding consideration of the role of cognition in designing and selecting assessment items.

1. It is important to construct an assessment blueprint, a table outlining the important learning goals that are to be assessed in a particular course along with how they will be assessed.

2. The most important learning goals should be identified and used to drive the selection or creation of assessment items. "What is tested is what gets taught. Tests must measure what is most important" (National Research Council, 1989, p. 69).
3. Assessment should balance procedural proficiency, conceptual understanding, and the use of contexts of statistical investigations.
4. The assessment should reflect the values of the discipline of statistics, for example, emphasis on data and data exploration, rather than on mathematical computations.

3.2   Observation: Assessment Methods and Practical Issues Involved in Collecting Evidence of Student Learning of Statistics

The observation pillar involves specifying the tasks or situations that allow one to observe student understanding and proficiency with statistical content. There are many ways to gather evidence of student learning, such as quizzes, exams, homework assignments, student projects, and informal observations or communications with students (see Garfield & Ben-Zvi, 2008, for more details on various assessment methods). Attention in this section is focused on the development and use of items that could be used in homework, quizzes, or examinations. The main goal is to distinguish between the use of these items to assess student learning of statistics (in particular, data analysis) from student assessment of other learning outcomes in elementary and mathematics classes.

There are many factors to consider regarding the evidence that is observed in an assessment item or task. For example, the way assessment tasks are constructed, the way scores are given or rubrics applied (as detailed in the next section on Interpretation), and the quality of both task and response. It is important that the item or task meets the intended goals for assessment and that the item be scored in a way that it reveals useful and accurate information to the students and teacher. Students learn to value what they know will be assessed (Garfield, 1995) so it is important for teachers to assess not only what they value, but also what is valued by the discipline. Assessment of statistical learning needs to measure authentic concepts and skills in data analysis, rather than mathematical skills.

3.2.1   Examples of Statistics Assessment Items

Scheaffer (2006) discussed examples of sample assessment items (see Figs. 16.1 and 16.2) which were classified as "data analysis" questions.

An examination of these items suggests that they do not actually assess statistical learning or students' ability to analyse data. The item in Fig. 16.1 requires the student to apply the algorithm, or formula, for finding the mean and the relationship of the

The average weight of 50 prize-winning tomatoes is
2.36 pounds. What is the combined weight, in pounds,
of these 50 tomatoes?

> A. 0.0472
> B. 11.8
> C. 52.36
> D. 59
> E. 118

Source: NAEP Sample Questions; nces.ed.gov/nationsreportcard/

Fig. 16.1 National Assessment of Educational Progress (NAEP) item example 1

The table below shows the daily attendance at two movie theaters for 5 days
and the mean (average) and the median attendance.

	Theater A	Theater B
Day 1	100	72
Day 2	87	97
Day 3	90	70
Day 4	10	71
Day 5	91	100
Mean	75.6	82
Median	90	72

(1) Which statistic, the mean or the median, would you use to describe the typical daily attendance for the 5 days at Theater A? Justify your answer.

(2) Which statistic, the mean or the median, would you use to describe the typical daily attendance for the 5 days at Theater B? Justify your answer.

Source: NAEP Sample Questions; nces.ed.gov/nationsreportcard/

Fig. 16.2 National Assessment of Educational Progress (NAEP) item example 2

mean value to the total sum of all 50 observations. This example is simply a mathematical computational problem. The mean is used in statistics; however, this problem required no statistical reasoning on the part of the student relating the answer to the context of the problem. In fact, the item did not even include the unit of measurement (pounds) in the responses to select from. It is difficult to learn something meaningful about a student's understanding of statistics based on their response to this item, whether correct or incorrect. The item is not aligned with important learning goals and a model of student learning as described by the Cognition Pillar.

The item in Fig. 16.2 goes beyond asking the student to complete a routine computation, challenging students to notice that Theater A has an outlier of 10, while Theater B has no outliers and to reason about which measure of centre is most appropriate. However, the context of the problem is irrelevant in answering the two questions.

An excellent source of assessment items that ask students to answer a statistical question, often with realistic data available, by analysing the data and justifying their conclusions is the *United States College Board AP Central Website* (AP Statistics Exam, *AP Central*). Free response questions, open-ended problems to be solved by students, are available at this site, along with detailed scoring rubrics, student sample papers, and commentary for the sample student papers. Allowing students to explore real (or realistic) statistics questions and giving them a chance to present and explain their conclusions to classmates and others, followed by a discussion, positions this task to become an assessment *for* learning as well as *of* learning.

3.2.2   Principles for Selecting, Modifying, and Creating Assessment Items of Statistics Learning

Writing, selecting, and revising statistical assessment items and tasks are very challenging for all educators and writers of high-stakes tests, and particularly difficult for educators trained in mathematics. The following principles are offered as a guide for the choice of good assessment items:

1. Assessment tasks should be situated in a context for which there is a good explanation of what question is being asked and why data were collected.
2. Assessment tasks should use real or realistic data (such as rounded numbers appropriate for the level of the students).
3. Assessment items asking for computations should have a context – the item should show more than computed answers and the computed answer should have a meaningful purpose.
4. Forced-choice items should have meaningful distractors that reveal common errors in learning/reasoning or misconceptions. These may be identified through the examination of how students learn statistics (cognitive model).
5. Free-response items should be used to allow students to create, explain, and communicate their understanding. The Advanced Placement (AP) Examination (Roberts, Scheaffer, & Watkins, 1999) is a model for developing good open-ended tasks and scoring rubrics, developing the item and rubric side-by-side.
6. Decisions about whether to use forced-choice or free-response items need to be informed by the purpose of the assessment.
7. When appropriate for the level of the student, an assessment plan should include opportunities for students to plan, conduct, and describe a statistical investigation.

3.3   Interpretation: Using Assessment Data to Make Inferences About What Students Have Learned

Once assessment data has been gathered, the interpretation and use of that evidence are important to consider. An interpretation method is needed for drawing inferences from the performance evidence obtained from assessment tasks. Often, this interpretation is more of an intuitive process than a statistical one.

Assessment results may be used to provide feedback to students about the quality of the student learning, may be used by teachers to identify gaps in student learning, may be used to document development and progress in learning, and may be used to assign summative grades. The same assessment items might be used for formative or summative purposes. In addition, aggregated assessment data may also provide evaluation information regarding curriculum and teaching.

One aspect of interpretation of assessment results involves the scoring of open-ended tasks and projects. While some assessments are scored using a holistic approach, other tasks have more detailed scoring rubrics, which take more time to construct and apply. The College Board Advanced Placement Statistics programme has provided an exemplary model (College Board, n.d.-a, b) for the development of scoring rubrics and the training of readers to effectively and efficiently use scoring rubrics for open-ended problems. In the process of developing and applying rubrics, there is much collaboration among the faculty raters which leads to improved rubrics and assessments. At a more local level, most classroom teachers tend to develop and apply their own scoring rubrics without such collaboration. We recommend a modified version of the AP method, where teachers and writers of high-stakes tests will share their rubrics with colleagues and discuss ways to improve and modify them to obtain better reliability as well as information about student learning. However, the scoring method needs to be linked to the purpose of assessment as well as to the learning model so that the results provide useful information on the nature of student learning. Some principles to guide the interpretation of assessment results are provided:

1. Be cautious about making inferences about student learning based on assessments. The quality, nature, and purpose of the assessment should guide the interpretation of results. For example, scores on a poorly constructed or ambiguous task should not be used to draw conclusions about student learning.
2. Select or create assessments that include a balance of tasks used to provide appropriate data for gathering formative and summative data, that is, assessment *for* learning items as well as assessment *of* learning items.
3. Consider using items embedded in classroom activities, or even out-of-class projects, as assessment *for* learning as well as data used for interpretation of important learning outcomes.
4. When developing scoring rubrics, seek the collaboration of colleagues to provide feedback and improvements. This collaboration is also suggested for reflection on assessment results.
5. Use assessment as a way of learning more about the discipline itself (statistics) as well as about student achievement. Rich and open-ended assessment tasks may build not only the student knowledge but also the teacher content knowledge.

4   Addressing Issues in Assessment

Assessment does not occur in isolation of teachers, classrooms, resources, and demands of school districts. There are many factors that affect the assessment of student learning. In addition to the constraints or demands of curriculum, there are three important issues to address: the role of technology, the power of high-stakes tests, and the role of teacher preparation in statistics and assessment.

4.1 Role of Technology

Technology is an important tool in exploring data, performing statistical analyses, and in helping students visualise abstract concepts. The NCTM Principles and Standards for School Mathematics (2000, p. 24) stated that "technology is essential in teaching and learning mathematics; it influences the mathematics that is taught and enhances students' learning". Using technology in the statistics classroom can allow more time for the student to reason statistically by avoiding tedious computations or graphical constructions. Simulations can allow students to visualise difficult relationships, for example, how measures of centre are or are not affected by outliers, as well as important theorems, for example, the Central Limit Theorem.

The following recommendations explain how technology can enhance and improve assessment of student learning of statistics:

- *Align student assessment with the technology used in student learning.* If technology is an integral part of the statistics curriculum and the way students learn, then students should be assessed appropriately with this technology, for example, software, Web applet, or calculators, in a manner consistent with how the technology was used in the curriculum.
- *Provide resources that show use of technology even if the actual technology, such as computers or calculators, is not accessible to the student.* Output, for example, from computer software or calculators, can be provided for students, allowing them to practise and learn how to interpret statistical analysis of data. Most important is providing statistical output, allowing the student to develop and communicate appropriate conclusions in a statistical context. There is a crucial need for more of these resources and examples of how assessment items can be developed using these technology resources.

4.2 Power of High-Stakes Tests

Many schools, districts, and countries require students to take tests that are used for purposes that lead to outcomes such as funding or international comparisons. These "high-stakes" tests often lead teachers to "teach to the test" that may not match the curriculum as defined by national, state, or local standards. Unfortunately, these tests may include few if any questions on statistics, and tend to focus primarily on computation rather than on data analysis skills, statistical reasoning, and statistical thinking. There is a need for such tests to include more authentic items that assess statistical reasoning, statistical concepts, and data analyses. It is vital that statisticians become involved with the writing of high-stakes test. As mentioned earlier, the high-stakes AP Statistics exam (AP Statistics Exam, *AP Central*) is unique in that it does assess the process of data analysis. Designing the questions for this examination is a deliberate, time-intensive process. The AP Statistics committee (composed of college statisticians and high school master teachers) made the decision in the early years of the examination to integrate technology into the examination by giving students access to a graphing calculator during the examination and including statistical output with some of the questions. This allows the committee to design questions that require students to conceptualise the analysis of the data instead of spending the test time computing (Roberts et al., 1999).

4.3 Role of Teacher Preparation in Statistics and Assessment

Due to the inclusion of data analysis and probability in the curriculum and standards, there is an increased expectation of teachers regarding the teaching and assessing of probability and statistics. Franklin and Mewborn (2006) discussed the importance of building a nucleus of teachers who can effectively teach the data analysis required in the new curricula by improving the pre-service and in-service teacher preparation programmes. *The Mathematical Education of Teachers (MET)* report (Conference Board of the Mathematical Sciences, 2001) noted that teachers should gain "both technical and conceptual knowledge" (p. 34) of the statistics and probability content that appears in the curriculum for their students and that secondary teachers, in particular, need to "appreciate and understand the major themes of statistics" (p. 44). The *MET* report also emphasises the necessity for teacher education to be the shared responsibility of mathematical scientists and education faculty. Franklin and Mewborn (2006) suggested that this collaboration be expanded to include statisticians and statistics departments. This collaboration in teacher preparation must also extend to preparing teachers in the area of desired assessment in statistics where statisticians help mathematicians and mathematical educators understand how to design items that assess statistical learning and the ability to analyse data. Ball (2003) and Ferrini-Mundy and Findell (2010) promoted the development of statistics courses for teachers where content, pedagogy, and assessment issues are an integrated part of the course curriculum. Ideally, these courses would be taught in collaboration with statisticians. Given that statistics is a relatively recent addition to the mathematics curriculum, there is a critical need for adequate and accessible resources for these teachers so that they may learn to design good items for assessing statistical learning outcomes. This is a critical area of research in statistics education.

5 Summary and Recommendations

In this chapter some general issues regarding assessment of student learning of statistics were outlined. Guiding principles were offered for each of the three pillars of an assessment triangle: cognition, observation, and interpretation. These pillars are used to consider the overall quality and use of an assessment. These three pillars also relate to assessments used for different purposes. A good assessment item that has been designed with the three pillars in mind could be used as an assessment *of* learning (e.g., an item in a final exam); an assessment *for* learning (an item used as a review or homework exercise in order to provide formative feedback for students); or an assessment *as* learning (e.g., an item used to structure a small group activity where students are developing new learning or understanding through their discussion of this item).

Educators were encouraged to think more broadly about the purposes of assessment beyond evaluation and grading, to use assessment *for* learning and *as* part of the learning process. Distinctions between assessments of statistical learning as opposed to assessing mathematical learning were offered, and educators and writers of high-stakes assessments were encouraged to use assessments that are appropriate and authentic to the discipline of statistics.

For assessment to influence learning, curriculum, and teaching in the most positive ways, the following recommendations are offered:

1. In-service and pre-service statistics educators need to learn, as part of their preparation, appropriate methods of assessing student learning. This includes general principles and techniques of student assessment as well as unique issues regarding assessment of statistical learning outcomes.
2. High-quality and accessible resources should be utilised by teachers of statistics as well as test developers.
3. Collaborative activities that involve development and evaluation of students' assessments should be encouraged amongst educators.

References

Ball, D. (2003). *What mathematical knowledge is needed for teaching mathematics?* Washington, DC: Secretary's Summit on Mathematics, U.S. Department of Education.

Bloom, B. S. (Ed.). (1956). *Taxonomy of educational objectives: The classification of educational goals: Handbook I, cognitive domain.* New York: Longmans, Green.

Chance, B. (Ed.). (2004). *Proceedings of the ARTIST Roundtable Conference on Assessment in Statistics.* Online: www.rossmanchance.com/artist/Proctoc.html

Cobb, G. W., & Moore, D. (1997). Mathematics, statistics, and teaching. *American Mathematical Monthly, 104,* 801–823.

College Board. (n.d.-a). AP statistics course home page. Online: apcentral.collegeboard.com/apc/public/courses/teachers_corner/2151.html

College Board. (n.d.-b). AP statistics exam. Online: apcentral.collegeboard.com/apc/members/exam/exam_questions/8357.html

Conference Board of the Mathematical Sciences. (2001). *The mathematical education of teachers.* Providence, RI/Washington, DC: American Mathematical Society and Mathematical Association of America.

delMas, R. C. (2002). Statistical literacy, reasoning, and learning: A commentary. *Journal of Statistics Education, 10*(3). Online: www.amstat.org/publications/jse

Earl, L., & Katz, S. (2006). *Rethinking classroom assessment with purpose in mind: Assessment for learning, assessment as learning and assessment of learning.* Winnipeg, Canada: Minister of Education, Citizenship and Youth.

Ferrini-Mundy, J., & Findell, B. (2010). The mathematical education of prospective teachers of secondary school mathematics: Old assumptions, new challenges. In *CUPM Discussion papers about mathematics and the mathematical sciences in 2010: What should students know?* (pp. 31–41). Washington, DC: The Committee on the Undergraduate Program in Mathematics, Mathematics Association of America.

Franklin, C., & Garfield, J. (2006). The GAISE Project: Developing statistics education guidelines for pre K-12 and college courses. In G. Burrill (Ed.), *Thinking and reasoning with data and chance: 2006 NCTM Yearbook* (pp. 345–375). Reston, VA: National Council of Teachers of Mathematics.

Franklin, C., Kader, G., Mewborn, D., Moreno, J., Peck, R., Perry, M., & Scheaffer, R. (2005). *Guidelines and assessment for instruction in statistics education (GAISE) report: A pre-K-12 curriculum framework.* Alexandria, VA: American Statistical Association.

Franklin, C., & Mewborn, D. (2006). The statistical education of grades pre-K-12 teachers: A shared responsibility. In G. Burrill (Ed.), *Thinking and reasoning with data and chance: 2006 NCTM Yearbook* (pp. 335–344). Reston, VA: National Council of Teachers of Mathematics.

Gal, I., & Garfield, J. (Eds.). (1997). *The assessment challenge in statistics education.* Amsterdam: IOS Press.

Garfield, J. (1995). How students learn statistics. *International Statistical Review, 63,* 25–34.

Garfield, J., & delMas, R. (2010). A website that provides resources for assessing students' statistical literacy, reasoning, and thinking. *Teaching Statistics, 3*(1), 2–7.

Garfield, J. B., & Ben-Zvi, D. (2008). *Developing students' statistical reasoning: Connecting research and teaching practice.* New York: Springer.

Groth, R. E. (2010). Three perspectives on the central objects of study in PreK-8 statistics. In B. J. Reys & R. E. Reys (Eds.), *Mathematics curriculum issues, trends, and further directions: 2010 NCTM Yearbook* (pp. 157–170). Reston, VA: National Council of Teachers of Mathematics.

Holmes, P. (1997). Assessing project work by external examiners. In I. Gal & J. Garfield (Eds.), *The assessment challenge in statistics education* (pp. 153–164). Amsterdam: IOS Press and International Statistical Institute.

Jones, G. A., Langrall, C. W., Mooney, E. S., & Thornton, C. A. (2004). Models of development in statistical reasoning. In D. Ben-Zvi & J. Garfield (Eds.), *The challenge of developing statistical literacy, reasoning, and thinking* (pp. 97–117). Dordrecht, The Netherlands: Kluwer Academic Publishers.

Lesh, R., Hoover, M., Hole, B., Kelly, A., & Post, T. (2000). Principles for developing thought-revealing activities for students and teachers. In A. Kelly & R. Lesh (Eds.), *Handbook of research design in mathematics and science education* (pp. 591–646). Mahwah, NJ: Lawrence Erlbaum.

Mathematical Sciences Education Board. (1993). *Measuring what counts: A conceptual guide for mathematics assessment.* Washington, DC: National Academy Press.

Ministry of Education New Zealand. (2007). *The New Zealand curriculum.* Wellington, New Zealand: Learning Media Limited.

National Council of Teachers of Mathematics. (1995). *Assessment standards for school mathematics.* Reston, VA: Author.

National Council of Teachers of Mathematics. (2000). *Principles and standards for school mathematics.* Reston, VA: Author.

National Research Council. (1989). *Everybody counts: A report to the nation on the future of mathematics education.* Washington, DC: National Academy Press.

Pellegrino, J. W., Chudowsky, N., & Glaser, R. (Eds.). (2001). *Knowing what students know: The science and design of educational assessment.* Washington, DC: National Academy Press.

Phillips, B., & Weldon, L. (Eds.). (2007). Assessing student learning in statistics. *Proceedings of the IASE/ISI Satellite Conference on Statistical Education.* Voorburg, The Netherlands: International Association for Statistical Education. Online: www.stat.auckland.ac.nz/~iase/publications

Roberts, R., Scheaffer, R., & Watkins, A. (1999). Advanced placement statistics: Past, present, and future. *The American Statistician, 53*, 307–320.

Romberg, T. (1992). *Mathematics assessment and evaluation: Imperatives for mathematics educators (SUNY series, reform in mathematics education)*. Albany, NY: State University of New York Press.

Rossman, A., Chance, B., & Medina, E. (2006). Some important comparisons between statistics and mathematics, and why teachers should care. In G. F. Burrill (Ed.), *Thinking and reasoning about data and chance, 68th NCTM Yearbook (2006)* (pp. 323–334). Reston, VA: National Council of Teachers of Mathematics.

Scheaffer, R. (2006). Statistics and mathematics: On making a happy marriage. In G. Burrill (Ed.), *Thinking and reasoning with data and chance, 68th NCTM Yearbook (2006)* (pp. 309–321). Reston, VA: National Council of Teachers of Mathematics.

Starkings, S. (1997). Assessing student projects. In I. Gal & J. B. Garfield (Eds.), *The assessment challenge in statistics education* (pp. 139–152). Amsterdam: IOS Press.

Wiggins, G., & McTighe, J. (1998). *Understanding by design*. Alexandria, VA: Association for Supervision and Curriculum Development.

Part III
Teachers' Beliefs, Attitudes and Knowledge

Carmen Batanero

Among the different topics of the Study identified by the International Programme Committee, Topic 2 was focused on *Teachers' attitudes, knowledge, conceptions and beliefs in relation to statistics education.* Discussions in the related working group made clear that we face a challenge to increase and improve the quality of research related to the preparation of teachers to teach statistics, since research in this topic is very scarce, in spite of the attention that the education of teachers has received in other areas of mathematics education.

Focusing on this Topic, this part presents what we have achieved in this Joint ICMI/IASE Study. It is aimed at providing a synthesis of the relevant research related to the topic, as well as suggesting future research directions and recommendations to train teachers. There are 12 chapters in the part that are arranged into the following three themes: (a) teachers' beliefs and attitudes; (b) teachers' statistical knowledge; and (c) teachers' knowledge to teach statistics.

Since the study of teachers' attitudes, beliefs and principles forms part of the process of understanding how teachers conceptualise their work, the first three chapters in this part dealt with the emotional component in teacher education and the impact of this component on teachers' instructional decisions.

In the first chapter of this part, Pierce and Chick analyse teachers' beliefs about statistics education, including the relationship between statistics and mathematics, the place of statistics in the curriculum, what statistics content is important for students to learn, and how students learn statistics. These beliefs can be thought of as lenses through which a person looks when interpreting the world, derived from teachers' previous experiences with the topic, which usually influence the way they teach statistics and the way the students learn statistics. Suggestions for further research are also proposed.

In the next chapter, Estrada, Batanero, and Lancaster clarify the differences among attitudes, emotions, and beliefs, then describe the main components of teachers'

C. Batanero (✉)
Departamento de Didáctica de la Matemática, Universidad de Granada,
Facultad de Ciencias de la Educación, Campus de Cartuja, 18071 Granada, Spain
e-mail: batanero@ugr.es

attitudes towards statistics and review some instruments for measuring these attitudes. The authors summarise research on teacher' attitudes towards statistics, and describe the variables that affect these attitudes. They suggest that professional development which engages teachers in a direct exploration of their beliefs may provide the opportunity for changing attitudes. Some implications for training teachers in statistics are finally discussed.

On the basis of the cognitive constructivist framework, educational research has begun to look at teachers' implicit and explicit theories of teaching and learning as determinants of teaching practices and student learning. Eichler highlights three relevant and complementary points that depend on statistics teachers' thinking and influence their decisions in the classroom: (a) teachers' planning for teaching, (b) the relationship between teachers' planning and their classroom practice, and (c) the relationships among teachers' classroom practices and students' learning. For each of these issues, the author provides an overview of the relevant research from mathematics and statistics education and, finally, some results taken from his own research on these issues.

The next group of chapters in this part (Chaps. 20–25) summarise research related to teachers' statistical knowledge in specific topics: graphs, variation, distribution, sampling and inference, and correlation.

Statistical graphs are essential for exploring, analysing and communicating data, and are tools for transnumeration, a basic component in statistical reasoning and thinking. In their chapter, González, Espinel, and Ainley firstly explore the levels and components in graphical competence, and then summarise existing research on teachers' graphical competence. They finish their chapter with some implications about how graphing may be taught and the abilities about this topic needed in teachers, and learners.

Since it has been a part of the curriculum for a long time, students' understanding of averages has been one of widest areas explored in statistics education; however research focused on teachers' understanding of averages is still very scarce. Jaccobe and Carvalho firstly analyse some studies related to school students' understanding of averages and then focus on teachers' understanding and professional knowledge about averages. The authors conclude that there is not much difference in the knowledge of students and teachers, since research shows reliance upon procedural algorithms and a general lack of conceptual understanding by both students and teachers. Consequently they suggest that the way to impact on students' understanding is by addressing teachers' statistical preparation.

Several authors have suggested that variation is at the heart of statistics; however, this is not a simple concept as it is linked to other statistical ideas, such as uncertainty, change, variable, distribution or outlier. Moreover, understanding of variation is a pre-requisite for other important ideas such as distribution, probability or inference. This concept and the related literature are analysed by Sánchez, Borim, and Coutinho, who discuss teachers' understanding of informal and formal expression of variation, and their ability to deal with variation in comparing groups and in random situations. Some studies on students' understanding of variation that focussed on these same topics are also reviewed.

Knowledge of distribution is also supported in the understanding of key concepts such as centre, spread, and shape. At the same time, this knowledge is needed in other complex statistical ideas, such as sampling distribution, statistical confidence or statistical significance. Using the metaphor of a "web of statistical knowledge" as a set of interrelated ideas needed to reason statistically, Reading and Canada offer a deep analysis of distribution as an important point in this web of knowledge. The authors summarise research studies that have investigated the knowledge development of teachers as regards the idea of distribution, both while training and while teaching. They finish with some recommendations for teacher learning and future research into teachers' knowledge of distribution.

Statistical inference is a main tool in research and management; however the application and interpretation of formal inference procedures, such as significance tests or confidence intervals is often incorrect. Until very recently, these topics were only studied at University level; however, in the past few years, ideas of statistical inference are being increasingly included in the high school curriculum in many countries. Harradine, Batanero, and Rossman analyse in their chapter the basic components of statistical inference and summarise the most relevant research related to understanding of formal inference, part of which has dealt with teachers. Implications from their survey include the need to develop multiple meanings of sample in students and the possibility of teaching informal inference procedures, before a formal study of the topic.

The last concepts analysed in this part are correlation and regression that expand to random situations the concept of functional dependence. Engel and Sedlmeier suggest that people's reasoning about association between statistical variables is conditioned by their experience with deterministic mathematical functions. In the study of regression, different mathematical functions are used to model the data. Understanding regression and correlation requires, however, apart from basic knowledge about functions, an appreciation of the role of variation. The authors revise some common errors and fallacies related to the concepts of correlation and regression and provide some recommendations to overcome these difficulties.

All teaching requires teachers to have knowledge, of both the content to be covered, and of effective ways to teach it. The last three chapters in this part (Burgess; Godino et al.; Callingham & Watson) discuss the pedagogical content knowledge that teachers need to teach statistics.

Isolated understandings of fundamental statistical ideas do not guarantee that teachers are successful when dealing in the classroom with statistical investigations and projects. Burgess in his chapter presents a theoretical model of the knowledge that teachers need to successfully implement the teaching of statistics through projects and investigations, as recommended in the new curricula. A summary of research related to teacher knowledge of statistics investigations and teacher pedagogical knowledge to teach statistics through investigations are used to describe components in this knowledge and offer examples of situations where teachers can use each of these components.

In addition to Burgess, other authors have offered different theoretical analyses of the knowledge that teachers need to successfully manage the complexity of teaching mathematics or statistics. Pedagogical content knowledge in the sense proposed by Shulman is a widely accepted approach to conceptualising teachers' content-specific belief systems about students' learning and appropriate ways of teaching. Following Shulman's research many authors have analysed the knowledge put in play in effective teaching. Godino et al. summarise and compare part of this research related to the education of mathematics teachers, as well as a few models that statistics educators have offered to describe the knowledge of teachers. Based on these analyses, the authors offer a new framework where different facets and levels of knowledge that should be taken into account when educating mathematics and statistics teachers are considered.

As with other topics, efforts to increase teachers' pedagogical knowledge to teach statistics should be based on previous evaluation of this knowledge. Callingham and Watson summarise the scarce research related to building adequate instruments to measure teachers' statistical pedagogical knowledge. They also present some findings from a large-scale Australian study that is directed at preparing instruments to assess the teachers' knowledge and discuss the implications for future research.

In summary, although each chapter in this part deals with a different topic, each of them contributes a picture of research efforts related to teacher's attitudes, beliefs and knowledge in statistics. We hope this survey will contribute to improving the preparation of teachers' educators and will make them conscious of the efforts needed to prepare mathematics teachers to teach statistics. At the same time, we hope the different research questions included in each chapter may attract researchers towards this priority area of research.

Chapter 17
Teachers' Beliefs About Statistics Education

Robyn Pierce and Helen Chick

Abstract Beliefs have long been known to affect teaching and learning. In statistics education, little research has been conducted on the nature of teachers' beliefs, despite the likely impact these beliefs have on teachers' activities. This chapter first considers content-focused beliefs about statistics, its relationship with mathematics, and its place in the curriculum, before addressing beliefs associated with teaching and learning statistics. Influences on beliefs and the impact of beliefs on teaching are considered, and suggestions for further research are proposed.

1 Introduction

Teachers' beliefs influence the actions of teachers conducting statistics lessons. In teaching about measures of central tendency, for example, teachers' approaches will be influenced by beliefs about whether students need to practice computing the mean, whether students should see statistics as associated with real-world situations, whether technology might help students learn, and whether it is important that students learn how to choose appropriate measures. The study of students' and teachers' beliefs relating to mathematics education has a long and extensive history; the story for statistics education is sparser and comparatively short, especially for teachers' beliefs.

This chapter uses Philipp's definition of "beliefs" (2007), which derives from and clarifies the term's uses in the literature. He defines beliefs as "psychologically held understandings, premises, or propositions about the world that are thought to be true" (p. 259). They are regarded as cognitive (so are "known" in some sense);

R. Pierce (✉) and H. Chick
Melbourne Graduate School of Education,
University of Melbourne, VIC, Australia 3010
e-mail: r.pierce@unimelb.edu.au; h.chick@unimelb.edu.au

C. Batanero, G. Burrill, and C. Reading (eds.), *Teaching Statistics in School*
Mathematics-Challenges for Teaching and Teacher Education: A Joint ICMI/IASE Study,
DOI 10.1007/978-94-007-1131-0_17, © Springer Science+Business Media B.V. 2011

he uses the metaphor of "lenses" through which we view the world (p. 258). Beliefs may be held with varying degrees of conviction, and may seem inconsistent or contradictory from an observer's point of view (p. 260). As a result, beliefs are not amenable to measurement using scales.

Philipp contrasts beliefs with attitudes, which are associated with emotions. Attitudes are "manners of acting, feeling or thinking that show one's disposition or opinion" (p. 259). Unlike beliefs, attitudes are commonly assessed using various scales. In a separate chapter, teachers' attitudes towards statistics education are described and discussed (Estrada, Batanero, & Lancaster, this book). A person's beliefs will affect but not determine their attitudes.

The importance of *students'* beliefs regarding statistics has been asserted for many years. Gal, Ginsburg, and Schau (1997, p. 38) highlight that, for students, beliefs influence (a) the teaching/learning process and (b) students' relationship with statistics beyond the classroom. The first point applies equally well to *teachers'* beliefs; the second, as highlighted by Estrada and Batanero (2008), is influenced by teachers' beliefs. The study of teachers' beliefs in statistics education is thus essential.

1.1   Contextual Issues

There are three background issues that must be raised. The first is the scope of "statistics". What counts as "statistics" in the school curriculum varies widely, from simple data representation at the primary (elementary) level, to beginning inference at the secondary level. There are some who believe that pre-secondary data representation work should not be called statistics at all.

Second, statistics, as a discipline, has only recently entered the curriculum in a substantial way. While some countries have had statistics as part of the high school curriculum for 40 years (see, e.g., Parsian & Rejali, 2008), only in the last 20 years has it received a major push (see, e.g., National Council of Teachers of Mathematics, 1989; Australian Education Council, 1991). In some countries the inclusion of statistics dates only to the past decade (e.g., Ainley & Monteiro, 2008; Newton, Dietiker, & Horvath, 2008; Opolot-Okurut, Mwanamoiza, & Opyene-Eluk, 2008; Wessels, 2008).

Finally, teachers have varied life and academic experiences. Some have studied statistics formally and others have not. For those who have studied statistics, their views as a teacher may reflect the views they held as a student. If teachers' encounters with statistics have been within other disciplines or in everyday situations then this experience may influence their belief framework. Even teachers who have studied statistics may have varied beliefs because of the relative emphases on theoretical statistics, applied statistics, and statistics education issues within their course. The three factors – the scope of statistics, the recency and place of statistics in the school curriculum, and teachers' backgrounds – must be considered when discussing beliefs.

1.2   Overview of This Chapter

With this as background, there are particular domains where beliefs are significant for teachers and school statistics teaching. In 1997, Gal et al. proposed some key areas for investigation, such as what teachers believe about statistics itself, the relationship between mathematics and statistics, the place of statistics in the curriculum, what statistics is important for students to learn, and how students learn statistics. The early sections of this chapter examine these questions, and present some results and speculations. Shaughnessy (2007, p. 1001), however, points out that – despite the years since the questions were proposed and a reiterated call by Batanero, Garfield, Ottaviani, and Truran (2000) – very little work has been done. The surveys by McLeod (1992), on students' beliefs in mathematics, and by Thompson (1992) and Philipp (2007) on teachers' beliefs, give insights into possible issues, but statistics education is absent from their considerations. There were few papers on the topic presented at the Joint ICMI/IASE Study conference in 2008 (Chick & Pierce, 2008; Eichler, 2008; Sedlmeier & Wassner, 2008), and what little has been done involves case studies and/or small or convenience samples. Results about teachers' beliefs in mathematics education and tertiary students' beliefs in statistics education may supplement what is known about teachers and statistics education. Other sections will consider influences on and impacts of beliefs, and belief change. The chapter concludes by suggesting areas needing critical attention.

2   Teachers' Beliefs About Statistics: Discipline and Curriculum Issues

Teachers' beliefs about statistics education involve their beliefs about statistics itself and its place in the curriculum. Do teachers' beliefs about statistics match the views of statisticians and statistics educators? In asking this, it is necessary to identify the views of the latter group, since their perceptions about statistics education may suggest certain "desirable beliefs". A strong theme at the Joint ICMI/IASE Study conference was that teachers must see that statistics is not defined by procedural computations but rather by investigative processes in the context of societal activity (Gattuso & Ottaviani, this book). Pfannkuch (2008) expressed concern that with statistical graphs, for example, schools emphasise construction techniques rather than the thinking needed for data-based decision-making. This highlights a possible mismatch between teachers' beliefs about statistics and how statistics educators view it.

Over a decade ago Cobb and Moore (1997, p. 801) also drew attention to features of the discipline of statistics, and asserted that "Statistics requires a different *kind* of thinking, because *data are not just numbers, they are numbers with context*" (emphasis in original). Wild and Pfannkuch (1999) highlighted ways in which statistical thinking is different from mathematical thinking, having investigative cycles, distinctive types of thinking, interrogative cycles, and characteristic dispositions. This underpins Pfannkuch's (2008) discussion of the implication of these for teaching: "To be a teacher of statistics is to realise that one is not teaching a branch of mathematics but … a discipline that has its own independent intellectual method" and that "statistical thinking or reasoning or literacy needs to be recognised as a key educational goal for all students" (p. 5). These views provide a background to an examination of the beliefs about statistics held by teachers themselves, as opposed to statistics educators.

2.1 Beliefs About Statistics

Teachers' beliefs about statistics itself will influence their attitude towards teaching statistics and their practice, and will depend on their own experiences with statistics. Primary school teachers seldom will have studied tertiary statistics, so their beliefs may reflect those of the secondary school students they once were. Secondary school teachers, in contrast, probably have studied at least one tertiary statistics subject. With few studies on teachers' actual beliefs, some information may be extrapolated from students' beliefs, on the assumption that those beliefs leave a legacy when such students become teachers.

For example, teachers who have studied tertiary statistics may have beliefs matching the views of 20 Australian university students in Reid and Petocz's (2002) phenomenographic study. From interviews with students taking a first course in statistics (typically descriptive statistics, probability and inference) or a third year course in statistics (regression analysis), six conceptions of statistics emerged. These were: Statistics is (1) individual numerical activities, (2) using individual statistical techniques, (3) a collection of statistical techniques, (4) the analysis and interpretation of data, (5) a way of understanding real life using different statistical models, and/or (6) an inclusive tool used to make sense of the world and develop personal meanings. Conception 1 suggests a belief that statistics is a particularly mathematical activity; Conception 6 recognises that statistics involves ways of thinking and sense-making, reflecting Pfannkuch's (2008) views. The intervening conceptions omit aspects of the more sophisticated ones. Such conceptions or beliefs are likely to influence a teacher's approach to teaching statistics.

In a study of primary teachers, Begg and Edwards (1999) collected views related to statistics from 22 practising and 12 pre-service teachers. When asked about the usefulness of statistics several themes coinciding with Conceptions 5 and 6 (above) emerged, including that statistics helps us make sense of our world; plan for the future; summarise information; and compare, organise, and predict. However, these teachers also felt that statistics can be "easily manipulated to support any view, be it wrong or right" (p. 2). Despite this perception, however, teachers generally disagreed with the statement "statistics are fairly worthless because people who have contrasting views on a certain issue can each use the same statistical finding to support their view" (p. 2).

Such a mixture of beliefs was also found by Chick and Pierce (2008). Their data from 27 pre-service primary teachers employed a statistics attitudes and beliefs survey using the SCAS instrument reported by Garfield (1996; see also Gal et al., 1997). This group of teachers had not studied tertiary statistics. They did not hold strong beliefs about statistics or its value, although there was strong agreement with "To be an intelligent consumer, it is necessary to know something about statistics". On the other hand, a majority agreed that "When buying a new car, asking a few friends about problems they have had with their cars is preferable to consulting an owner satisfaction survey in a consumer magazine", suggesting a belief that personal opinions have more value than statistical reports.

2.2   *Beliefs About the Relationship Between Mathematics and Statistics*

In most countries, in both primary and secondary schools, the same teacher is responsible for teaching mathematics *and* statistics. Teachers' and pre-service teachers' beliefs about the relationship between mathematics and statistics at the school level vary. Statistics typically is included within the mathematics curriculum. Anecdotal evidence suggests that primary teachers, for example, may not think of themselves as teaching "statistics" but rather applied number work. Begg and Edwards (1999) found that most of their practising and pre-service primary teachers believed that statistics was part of mathematics, but thought that a good understanding of mathematics was not necessary in order to grasp basic statistical concepts. Many claimed "statistics gives students who might have had a 'bad' experience with maths another chance" (p. 2). On the other hand, the pre-service primary teachers surveyed by Chick and Pierce (2008) were split in their responses to the statement "You must be good at mathematics to understand basic statistical concepts".

Although the teachers in Begg and Edwards' (1999) study recognised the cross-curricular nature of statistics, they taught it as part of mathematics. Those who valued statistics did so because, in their view, statistics gives meaning to mathematics and they believed that students find statistics motivating and fun. The majority of practising teachers saw teaching statistics as the same as teaching mathematics, while those in the pre-service group were not as sure of this, but still viewed them similarly: "it's part of maths; we know it's a maths thing" (p. 5).

Finally, little work has been done on whether teachers hold different beliefs about how mathematical and statistical activities are conducted. Gal and Ginsburg (1994) recommended considering Schoenfeld's list (1992, p. 359) of typical student beliefs about the nature of mathematics and mathematical activity. Schoenfeld suggests that students come to believe that mathematics problems have a single right answer and one method of solution, that mathematics is a solitary activity, and that problems have quick solutions. Whether or not *teachers* believe this about mathematics is another question; more salient for this chapter is whether these beliefs are held by teachers or students for *statistics*. There may be differences;

certainly lesson plans produced by pre-service primary teachers (Chick & Pierce, 2008) seemed to reflect a belief that group work – rather than working alone – is appropriate for learning statistics.

2.3   Beliefs About the Place of Statistics in the Curriculum

Statistical literacy and quantitative data analysis are required across school disciplines and outside the classroom (Watson, 2006). Although statistics may be taught as part of mathematics, statistical literacy is needed, for example, to understand articles in the media, record or interpret results in science, monitor performance of sporting teams, or to quantify social problems. When considering the place of statistics in the curriculum Begg and Edwards' (1999) respondents all thought that studying statistics was important for primary school children but only a quarter thought it was "really important" or "one of the most important areas" (p. 6). Of Chick and Pierce's (2008) pre-service primary teachers less than a quarter believed primary school students did not spend enough time on statistics/data. Sedlmeier and Wassner (2008) surveyed 40 secondary mathematics teachers in German high schools (Gymnasium) about the importance of statistics in daily life as compared with other areas of mathematics. Just over half said it was of higher importance (and very few said statistics was less important than other topics); however, when asked if statistics should be "given more hours per week even if this meant other mathematics topics got less" few teachers agreed.

Across this limited number of studies, the majority of teachers surveyed believed that understanding statistics is important for everyday life. However it is not known where they believe this teaching is best placed: under the umbrella of mathematics or in the context of other disciplines. Statistics seems to be "accepted" as part of mathematics, yet this sits uneasily with the idea, discussed earlier, that it is separate. In addition we know nothing of the views of secondary teachers who specialise in teaching disciplines other than mathematics and statistics. It is not clear whether they believe that statistics is integral to their curriculum area, or if they see it as the responsibility of mathematics teachers.

3   Teachers' Beliefs About the Teaching and Learning of Statistics

3.1   Beliefs About What Statistics Is Important for Students to Learn

The view, sometimes externally imposed by curriculum or policy, that an informed citizen in today's world needs a basic understanding of statistics impacts on individual teacher's beliefs. Believing that "to be an intelligent consumer it is necessary to know something about statistics" – a view held by most respondents in the studies of Begg and Edwards (1999) and Chick and Pierce (2008) – reflects this common perception. Such beliefs may influence what statistics teachers believe students should learn.

In the Begg and Edwards study (1999) teachers expressed a belief that statistics has utilitarian value for functioning in all areas of life. Nearly all these teachers mentioned teaching graphing and data collection. These are essentially procedural skills. Beyond this, teachers indicated that they believed graphs were valuable for communication although far fewer referred to graphs as data exploration tools. Watson (2001) profiled 43 primary and secondary teachers with respect to teaching chance and data, and noted a typical response from a primary teacher: "Children live in a world where data is flowing so fast that they must be able to comprehend what is going on" (p. 314). The secondary teachers believed it was important to teach graph interpretation and construction, central measures, spread of data, practical applications, and probability and how it is used in society. Primary teachers were more likely to suggest student surveys, focusing on interests and hobbies.

These few examples show a trend towards an increased emphasis on statistical thinking and literacy, although it is unclear if deep statistical reasoning, as called for by Pfannkuch (2008), is being fostered.

3.2   Beliefs About Teaching and Learning Statistics

Beliefs about teaching and learning statistics will naturally be linked to the age of the students involved and to teachers' views about teaching in general and teaching mathematics in particular. Learning in context, with discussion as an important class activity, is believed by statistics education researchers to be fundamental. This view is not held by all teachers, however. Eichler (2007; see also Eichler, this book) developed case studies of 13 German upper secondary mathematics teachers, focusing in part on their "individual curricula", meaning what teachers planned to do. Eichler's analysis developed four categories reflecting the teachers' beliefs: traditionalists, application preparers, everyday life preparers, and structuralists. *Traditionalists* emphasised mathematical theory and were less concerned about applications; they believed students should gain algorithmic skills. *Application preparers* taught mathematical theory and algorithms so students could use this theory to solve real-world problems. *Everyday life preparers* taught through applications to develop abilities to address real stochastic problems. Finally, *structuralists* examined applications but as a starting point for exemplifying mathematical theory and abstract systems.

Traditionalists and structuralists appear to hold views about statistics and its teaching at odds with Shaughnessy (2007, p. 1002), who emphasised context and the view that statistics is fundamentally different from mathematics. In terms of the list of conceptions found by Reid and Petocz (2002) such teachers might hold Conceptions 1, 2, or 3 but are less likely to hold Conceptions 4, 5, or 6.

The mathematics teachers in Sedlmeier and Wassner's (2008) study rated the following strategies highly for good instruction in statistics: relating content taught to daily issues, discussing different problem solutions, and making connections with other (non-mathematical) topics. Such responses would tend to suggest that these teachers fit Eichler's (2007) "everyday life preparers" and "application preparers" categories. In contrast, Sedlmeier and Wassner also claimed many teachers were neither keen to base their instruction on students' own data collection nor emphasise students' interests. This was particularly true of older teachers; younger teachers, in contrast, believed more strongly in making connections to daily life, using relevant examples, and conducting real experiments.

4   Influences on and Impacts of Teachers' Beliefs

4.1   Influences on Beliefs

Many factors are likely to influence teachers' beliefs about statistics education, although this, too, has been studied very little. Begg and Edwards (1999) found that teachers' beliefs were related to their prior experiences, with evidence that beliefs about statistics being process-oriented reflected personal learning experiences. This suggests likely differences between primary and secondary teachers based on the number, depth, and nature of statistics courses experienced. More specifically, Carvalho (2008) suggests that teachers may find it hard to implement interactive, experiential, and practical statistics learning experiences without experiencing these activities themselves.

Beliefs about teaching statistics may also be influenced by beliefs about statistics itself (including its relationship to mathematics) and about teaching more generally. For example, if teachers teach mathematics in a decontextualised way then they may use a similar approach in statistics, perhaps practising procedures first before giving "application" examples with a weak context as window-dressing. Beliefs may also be influenced by the extent to which teachers see the value and use of statistics, including its real-world relevance. Similarly, mathematics teachers who do not strongly value group work, but who feel pressure to conduct it, may believe that statistics lessons afford this opportunity more than other topics.

Given the potential mismatches between teachers' beliefs and those of statistics educators, as discussed earlier, further work is needed on how to modify the beliefs of those who, for example, perceive statistics primarily as context-free algorithms. Among limited research in this area there is, at least, some evidence of positive change from professional development. Frierson, Friel, Brerenson, Bright, and Tremblay (1993) asked teachers in a professional development programme about statistics concepts they believed were appropriate for Grade 3. Prior to the professional development programme the teachers nominated "isolated content" such as graphing, probability, or organising data, but following the programme their views had shifted and they were advocating "conceptual ideas such as formulating questions, or interpreting data" (p. 42). More work is needed to determine what professional development experiences offer the greatest potential for change.

4.2 Impacts of Beliefs

The connection between beliefs and actions is a key reason for investigating beliefs, and has been part of the mathematics education literature. Chick and Pierce (2008) examined the lesson plans of 27 pre-service primary teachers asked to teach statistical concepts from a given data set. Common features of the lessons included the intention to encourage class discussions, and to have students share their findings, but with little emphasis on teaching students to engage with and interpret the data. Despite limited content and pedagogical content knowledge, which hampered their ability to convey statistical ideas, these pre-service teachers' belief in "group work" and "class discussion" appears to recognise that an interactive approach would best serve the purpose of engaging the students with statistical ideas.

The case studies of Eichler (2008) also provide a rare direct examination of the impacts of beliefs. He explored connections between teachers' beliefs, expressed in their individual curriculum intentions, and their enacted curriculum in the classroom. He found strong links between the two (see also Eichler, this book). One teacher espoused the importance of real statistical problems and actually used them to develop statistical methods in class. Another emphasised a theoretical foundation for statistics and used more routine tasks and traditional methods. Eichler also investigated connections between the teachers' beliefs and actions, and the beliefs about statistics expressed by their students after completing the course. In one case a teacher allowed students to make up their own problems or gave examples in real contexts but with unrealistic data, and these strategies resulted in students believing that statistics had no relevance in their lives.

Implicit in Eichler's study, and in the discussion of the impact of beliefs, is the causality connection. This has not, however, been explored in detail, so it is not yet known if particular beliefs about statistics education are likely to imply particular outcomes, or vice versa.

5 Implications for Research

In 1994 Gal and Ginsburg highlighted the importance of researching the role of beliefs and attitudes in statistics education. They discussed some of the typical tests used at that time and called for more qualitative approaches. This has not been heeded, apart from the relatively small body of work discussed here. Future studies of teachers might provide richer data, however, if questions and interviews are better framed to target the statistics education issues raised here.

Although statistics is present in the curricula of many countries, and teachers believe that understanding statistics is important for educated citizens, there is considerable diversity in how it is treated and what other beliefs teachers hold about it. As seen, this variability is influenced by past experiences, learning or using statistics, and beliefs about what it means to do and understand statistics. This affects beliefs about what aspects of statistics should be taught in schools and how. More general beliefs about mathematics and teaching also have an effect.

The questions of Gal et al. (1997) have thus been answered only incompletely and require further research. Larger, more systematic studies to enhance past small-scale studies are needed. Finally, the following directions for research are recommended:

- Investigate teachers' beliefs about statistics itself and how this is influenced by the teachers' backgrounds. Explore teachers' beliefs about the relationship between mathematics and statistics.
- Investigate the beliefs about statistics education held by non-mathematics/ statistics teachers whose subject areas require statistical literacy.
- Identify what key features of statistical thinking teachers think they should develop in their students. Identify what teachers believe to be barriers and enablers for teaching statistical thinking, not just procedural routines.
- Evaluate professional development activities to establish which activities lead to changes in beliefs.
- Ascertain the extent to which the local situation (culture, history and curriculum) affects teachers' beliefs.
- Identify the interactions amongst beliefs, technology use, and statistics learning.
- Explore the impact of teachers' beliefs on actual classroom practices.

Information gathered from such research would help us better understand the factors that influence classroom practice in statistics education.

References

Ainley, J., & Monteiro, C. (2008). Comparing curricular approaches for statistics in primary school in England and Brazil: A focus on graphing. In C. Batanero, G. Burrill, C. Reading, & A. Rossman (2008).

Australian Education Council. (1991). *A national statement on mathematics for Australian schools*. Melbourne: Curriculum Corporation.

Batanero, C., Burrill, G., Reading, C., & Rossman, A. (Eds.). (2008). *Joint ICMI/IASE Study: Teaching Statistics in School Mathematics. Challenges for Teaching and Teacher Education. Proceedings of the ICMI Study 18 and 2008 IASE Round Table Conference.* Monterrey, Mexico: International Commission on Mathematical Instruction and International Association for Statistical Education. Online: www.stat.auckland.ac.nz/~iase/publications

Batanero, C., Garfield, J. B., Ottaviani, M. G., & Truran, J. (2000). Research in statistical education: Some priority questions. *Statistics Education Research Newsletter, 1*(2), 2–6. Online: www. stat.auckland.ac.nz/~iase/serj/

Begg, A., & Edwards, R. (1999). Teachers' ideas about teaching statistics. *Proceedings of the 1999 Combined Conference of the Australian Association for Research in Education and the*

New Zealand Association for Research in Education. Melbourne: Australian Association for Research in Education & New Zealand Association for Research in Education. Online: www. aare.edu.au/99pap/beg99082.htm

Carvalho, C. (2008). Collaborative work in statistics classes: Why do it? In C. Batanero, G. Burrill, C. Reading, & A. Rossman (2008).

Chick, H. L., & Pierce, R. U. (2008). Teaching statistics at the primary school level: Beliefs, affordances, and pedagogical content knowledge. In C. Batanero, G. Burrill, C. Reading, & A. Rossman (2008).

Cobb, G. W., & Moore, D. S. (1997). Mathematics, statistics, and teaching. *American Mathematical Monthly, 104*, 801–823.

Eichler, A. (2007). Individual curricula: Teachers' beliefs concerning stochastics instruction. *International Electronic Journal of Mathematics Education, 2*(3). Online: www.iejme.com/

Eichler, A. (2008). Teachers' classroom practice and students' learning. In C. Batanero, G. Burrill, C. Reading, & A. Rossman (2008).

Estrada, A., & Batanero, C. (2008). Explaining teachers' attitudes towards statistics. In C. Batanero, G. Burrill, C. Reading, & A. Rossman (2008).

Frierson, D., Friel, S., Brerenson, S., Bright, G., & Tremblay, C. (1993). Teach-stat: A professional development program for elementary teachers (grades K-6) in North Carolina, U.S.A. In L. Brunelli & G. Cicchitelli (Eds.), *Proceedings of the First Scientific Meeting of the International Association for Statistical Education.* Perugia, Italy: IASE. Online: www.stat. auckland.ac.nz/~iase/publications

Gal, I., & Ginsburg, L. (1994). The role of beliefs and attitudes in learning statistics: Towards an assessment framework. *Journal of Statistics Education, 2*(2). Online: www.amstat.org/ publications/jse/

Gal, I., Ginsburg, L., & Schau, C. (1997). Monitoring attitudes and beliefs in statistics education. In I. Gal & J. B. Garfield (Eds.), *The assessment challenge in statistics education* (pp. 37–51). Amsterdam: IOS Press and International Statistical Institute.

Garfield, J. (1996). Assessing student learning in the context of evaluating a chance course. *Communications in Statistics: Theory and Methods, 25*, 2863–2873.

McLeod, D. B. (1992). Research on affect in mathematics education: A reconceptualisation. In D. A. Grouws (Ed.), *Handbook of research on mathematics teaching and learning* (pp. 575–596). New York: Macmillan.

National Council of Teachers of Mathematics. (1989). *Curriculum and evaluation standards for school mathematics.* Reston, VA: Author.

Newton, J., Dietiker, L., & Horvath, A. (2008). Statistics: A look across K-8 state standards. In C. Batanero, G. Burrill, C. Reading, & A. Rossman (2008).

Opolot-Okurut, C., Mwanamoiza, M., & Opyene-Eluk, P. (2008). The current teaching of statistics in schools in Uganda. In C. Batanero, G. Burrill, C. Reading, & A. Rossman (2008).

Parsian, A., & Rejali, A. (2008). A report on preparing mathematics teachers to teach statistics in high school. In C. Batanero, G. Burrill, C. Reading, & A. Rossman (2008).

Pfannkuch, M. (2008). Training teachers to develop statistical thinking. In C. Batanero, G. Burrill, C. Reading, & A. Rossman (2008).

Philipp, R. A. (2007). Mathematics teachers' beliefs and affect. In F. K. Lester (Ed.), *Second handbook of research on mathematics teaching and learning* (pp. 257–315). Charlotte, NC: Information Age Publishing.

Reid, A., & Petocz, P. (2002). Students' conceptions of statistics: A phenomenographic study. *Journal of Statistics Education, 10*(2). Online: www.amstat.org/publications/jse/

Schoenfeld, A. H. (1992). Learning to think mathematically: Problem solving, metacognition, and sense making in mathematics. In D. A. Grouws (Ed.), *Handbook of research on mathematics teaching and learning* (pp. 334–370). New York: Macmillan.

Sedlmeier, P., & Wassner, C. (2008). German mathematics teachers' views on statistics education. In C. Batanero, G. Burrill, C. Reading, & A. Rossman (2008).

Shaughnessy, J. M. (2007). Research on statistics learning and reasoning. In F. K. Lester (Ed.), *Second handbook of research on mathematics teaching and learning* (pp. 957–1009). Charlotte, NC: Information Age Publishing.

Thompson, A. (1992). Teachers' beliefs and conceptions: A synthesis of research. In D. A. Grouws (Ed.), *Handbook of research on mathematics teaching and learning* (pp. 127–146). New York: Macmillan.

Watson, J. M. (2001). Profiling teachers' competence and confidence to teach particular mathematics topics: The case of chance and data. *Journal of Mathematics Teacher Education, 4*, 305–337.

Watson, J. M. (2006). *Statistical literacy at school*. Mahwah, NJ: Lawrence Erlbaum.

Wessels, H. (2008). Statistics in the South African school curriculum: Content, assessment and teacher training. In C. Batanero, G. Burrill, C. Reading, & A. Rossman (2008).

Wild, C., & Pfannkuch, M. (1999). Statistical thinking in empirical enquiry. *International Statistical Review, 67*(3), 223–248.

Chapter 18
Teachers' Attitudes Towards Statistics

Assumpta Estrada, Carmen Batanero, and Stephen Lancaster

Abstract Teachers' attitudes towards statistics play a significant role in assuring success in implementing any new statistical curriculum. In this chapter, attitudes and their component factors are conceptualised, and the primary instruments available to assess attitudes are reviewed. Following this, the research on teacher attitudes towards statistics is summarised. Finally, some implications for training teachers in statistics are discussed.

1 Introduction

Teacher training in statistics is generally focused on improving the cognitive aspects of instruction with little attention paid to the emotional component of learning. However, this latter factor "can impede learning of statistics, or hinder the extent to which students will develop useful statistical intuitions and apply what they have learned outside the classroom" (Gal & Ginsburg, 1994, p. 1).

The focus of this chapter is attitudes towards statistics, which might influence a person's statistical behaviour inside and outside the classroom and their willingness to attend statistics courses in the future (Pajares, 1996; Bandura, 1997; Gal, Ginsburg, & Schau, 1997). This is particularly relevant in the preparation of teachers, since

A. Estrada (✉)
Departament de Matemàtica, Facultat de Ciències de l'Educació, University of Lleida,
Avinguda Estudi General 4, 25001 Lleida, Spain
e-mail: aestrada@matematica.udl.es

C. Batanero
Departamento de Didáctica de la Matemática, Universidad de Granada,
Facultad de Ciencias de la Educación, Campus de Cartuja, 18071 Granada, Spain
e-mail: batanero@ugr.es

S. Lancaster
Department of Mathematics, California State University, Fullerton,
800 N State College Blvd, Fullerton, CA 92832, USA
e-mail: slancaster@fullerton.edu

C. Batanero, G. Burrill, and C. Reading (eds.), *Teaching Statistics in School Mathematics-Challenges for Teaching and Teacher Education: A Joint ICMI/IASE Study*, DOI 10.1007/978-94-007-1131-0_18, © Springer Science+Business Media B.V. 2011

positive attitudes towards statistics would help them to understand that statistics is useful in their students' professional and personal lives, and that their students can be trained to understand and use statistics (Schau, 2003). In addition, statistics teachers would be more likely to transmit to their students a positive view of statistics and an appreciation for the potential uses of statistics in future personal and professional life (Gal & Ginsburg, 1994).

This chapter clarifies the differences among attitudes, emotions, and beliefs, and then describes components of teacher attitudes and instruments measuring these attitudes. This is followed by a summary of the scarce research related to teachers' attitudes towards statistics that tries to identify affective factors that teachers associate with both statistics learning and the importance of continuing professional development in statistics. Finally, some recommendations about how teacher training may attempt to improve teacher's attitude towards statistics are included.

2   Attitudes Towards Statistics

2.1   Conceptualising Attitudes

McLeod (1992) analysed the mathematics education affective domain and discriminated between emotions, attitudes, and beliefs. These are distinguished by the stability of the affective responses that they represent, the degree to which cognition plays a role in the response, and the time that they take to develop:

- Emotions are feelings or states of consciousness, distinguished from cognition (Philipp, 2007). They involve positive (e.g., satisfaction) and negative (e.g., panic) responses triggered by one's immediate experiences, for example, while studying statistics. Emotions are transient and hard to measure but can be intense and serve as a source for development of attitudes (Gal et al., 1997). They may involve little cognitive appraisal and may change rapidly (McLeod, 1992).
- Beliefs are psychologically held premises or propositions about the world that are thought to be true (Philipp, 2007). They are largely cognitive in nature, and are developed over a relatively long period of time (McLeod, 1992). Beliefs include ideas about statistics, about oneself as a learner of statistics, and about the social context of learning that together provide a context for statistics experiences (Gal et al., 1997; Chick & Pierce, this book; Eichler, this book).
- Attitudes can be viewed as "affective responses that involve negative or positive feelings of moderate intensity" (McLeod, 1992, p. 581). More recently Philipp (2007) described attitudes as manners of acting, feeling, or thinking that show a person's disposition or opinion towards a topic. They are relatively stable, resistant to change, and comprise a larger cognitive component and less emotional intensity than emotions. They develop as repeated positive or negative emotional responses and are automatised over time (Gal et al., 1997).

Consequently, the sequence of beliefs, attitudes, and emotions represents increasing levels of affect and intensity of response, from cold beliefs related to liking or disliking mathematics, to strong emotional reactions such as feeling frustration when not being able to solve non-routine problems. The sequence also represents decreasing levels of cognitive involvement and response stability (McLeod, 1992).

In summary, attitudes collectively form an important psychological construct that is often used to understand and predict an individual's reaction to an object and how behaviour can be influenced (Fishbein & Ajzen, 1975). Attitudes are expressed along a positive-negative continuum, such as like-dislike, pleasant-unpleasant, and may represent, for example, feelings towards a teacher, a topic, or an activity. They change more slowly than emotions because they are more cognitively based (Philipp, 2007). Attitudes towards a topic derive from positive or negative experiences over time devoted to learning. Students may have had such experiences – in the case of statistics – at school or in informal learning out of school. Students may have a fuzzy understanding of what statistics might be about, or be unaware about life domains where statistics may be used, and this lack of knowledge can affect their attitudes. In other cases, students transfer their negative feelings towards mathematics into statistics (Gal & Ginsburg, 1994). All of these considerations are applicable to prospective teachers or in-service teachers with no specific training in statistics.

2.2   Components in Attitudes

As suggested earlier, several authors conceptualise attitudes as a multidimensional concept. Wise (1985), for example, distinguished between attitudes towards a course in basic statistics that the students were taking (course subscale) and attitudes towards the use of statistics in their field of study in the future (field subscale). Three of the generally accepted components of the term "attitude" (Aiken, 1980; Auzmendi, 1991; Olson & Zanna, 1993; Gómez-Chacón, 2000) are: (a) *Affective*: feelings about the object in question, (b) *Cognitive*: the person's self-perception as regards the object, and (c) *Behavioural*: the person's inclination to act towards the attitude object in a particular way. Schau, Stevens, Dauphine, and del Vecchio (1995) assumed four dimensions: (a) *Affect*: feelings concerning statistics; (b) *Cognitive competence*: perception of self-competence, knowledge, and intellectual skills when applied to statistics; (c) *Value:* appreciation of the usefulness, relevance, and worth of statistics in personal and professional life; and (d) *Difficulty*: perceived difficulty of statistics as a subject.

Depending on the above established sub-constructs, over the last two decades a large number of tools to measure attitudes towards statistics have been developed. Below we describe only the most widely used instruments.

3   Measuring Attitudes Towards Statistics

Since attitude is a psychological construct it cannot be directly measured and the use of only a single statement or question to assess attitude will not provide reliable responses. The most common approach is to use a Likert-type survey that requires individuals to decide on their level of agreement with a number of statements related to the different components of attitudes. Responses are generally on a five-point scale (i.e., strongly agree, agree, neither agree nor disagree, disagree, strongly disagree). For example, the sentence "I enjoy taking statistics courses" is related to the affect component and strong agreement suggests a positive attitude in this component. At the same time the sentence "I am under stress in the statistics class" is related to a negative attitude, also in the affect component. Scores in items that are presented in negative form should be reversed, before the scale and component scores are formed by adding the scores in the different items. Three of the most widely used scales measuring attitudes towards statistics are described below:

- *Statistics Attitude Survey* (SAS; Roberts & Bilderback, 1980; Roberts & Saxe, 1982) – the first instrument to measure attitude towards statistics. This scale was designed to be one-dimensional, with 33 Likert-type items, each of them with five response possibilities ranging from strongly disagree to strongly agree.
- *Attitudes Toward Statistics* (ATS; Wise, 1985). The ATS is a 29-item, Likert-type scale with five response possibilities consisting of two subscales. The *Field* (20 items) and *Course* (9 items) subscales respectively aim to measure attitudes towards the particular statistics course in which students are enrolled and the use of statistics in their fields of study.
- *Survey of Attitudes Towards Statistics* (SATS; Schau et al., 1995; Cashin & Elmore, 2005). The scale consists of 28 items measuring four components of students' attitudes towards statistics: *Affect* (six items measuring feelings concerning statistics), *Cognitive competence* (six items assessing perception of self competence, knowledge, and intellectual skills when applied to statistics), *Value* (nine items that concern appreciation of the usefulness and worth of statistics in personal and professional life) and *Difficulty* (seven items measuring the perceived difficulty of statistics, as a subject).

These and other instruments have been extensively used to measure attitudes towards statistics in undergraduate students (see Carmona, 2004, for a survey). There is, however, little research concerning teachers' attitudes towards statistics, and most of it has dealt only with prospective teachers. In the next section, we summarise research measuring teachers' attitudes towards statistics and then analyse research that identifies variables affecting these attitudes.

4   Teacher Attitudes Towards Statistics

The scarce research describing teachers' attitudes towards statistics deals with three different themes: (a) measuring teachers' global attitudes towards statistics (Begg & Edwards, 1999; Estrada, 2002; Estrada, Batanero, Fortuny, & Diaz, 2005; Chick & Pierce, 2008) and comparing these attitudes with those of undergraduate students in other fields (Onwuegbuzie, 1998, 2003); (b) focussing on a specific part of teachers' *cognitive competence*, namely, their *attitudes towards their role as continuing learners of statistics* (Lancaster, 2007) and (c) analysing teachers' attitudes in relation to the teaching of statistics (Begg & Edwards, 1999; Watson, 2001).

4.1   Teacher Global Attitudes Towards Statistics

Begg and Edwards (1999) used interviews, an unspecified survey scale and concept maps to study attitudes in a sample of 22 in-service and 12 prospective primary teachers. Results showed negative attitudes in the affective component, with the teachers expressing feelings like "fear", "horror", or "lack of interest". As regards value, some teachers considered statistics important, since "It helps us make sense of our world" or "It helps us compare and organise things, shows trends, and enables us to predict". Despite the teachers' concern about how statistics can be manipulated, they generally disagreed with the statement "Statistics is fairly worthless". Most of the teachers thought that a good understanding of mathematics was not needed to grasp basic statistical concepts. They did not consider themselves good with statistics, although they were confident about their ability to read and understand statistical terms and graphs in the media.

Estrada (2002) measured the attitudes towards statistics in 66 in-service and 74 prospective primary school teachers. She developed her own scale that contained 25 items taken from SAS, ATS and another scale (Auzmendi, 1991) built in Spain. In her scale, Estrada complemented the three classical attitude components (affect, cognition, behaviour), with another three possible components: (a) *Social:* perception of the value of statistics in society; (b) *Educational:* interest in learning and teaching statistics; and (c) *Instrumental:* perceptions of the use of statistics in other areas. Results of this study showed neutral attitudes towards statistics in both groups with better scores in items measuring the instrumental role of statistics (e.g., "I understand better the results of elections when they are presented through statistical graphs") and the educational value of the topic (e.g., "You should learn statistics in school"). Lower scores corresponded to items measuring confidence in statistics (e.g., "Reality can be manipulated with statistics") and affect (e.g., "I enjoy taking statistics courses").

Later Estrada et al. (2005) gave the SATS instrument to a sample of 367 prospective primary school teachers in Spain. Results showed moderately positive attitudes on cognitive competence items (e.g., "I can learn statistics") and value items (e.g., "Statistics is worthless") having the highest scores. Correlations were found between the subscales Affect and Cognitive Competence, and Affect and Value. Consequently liking or disliking statistics was related in these teachers to their perception of self-capacity to learn statistics and to the value given to statistics.

Chick and Pierce (2008) gave 10 items taken from SATS to 27 prospective primary school teachers. Results showed that the teachers had neutral attitudes. As regards the value of statistics, even if they agreed with the item "To be an intelligent consumer, it is necessary to know something about statistics"; a majority also agreed that "When buying a new car, asking a few friends is preferable to consulting an owner satisfaction survey in a consumer magazine".

Onwuegbuzie (1998, 2003) used ATS to compare prospective teachers' attitudes towards statistics with those of graduate students enrolled in other courses and found that teachers in his sample had fewer positive attitudes towards statistics than did other graduate students.

4.2   Teachers' Attitudes Towards Their Role as Continuing Learners of Statistics

Continuing professional development is an important strategy to remedy the lack of teachers' statistical content and pedagogical content knowledge (Hill & Ball, 2004) and therefore, it is important to motivate teachers to participate in professional development in statistics (Gould & Peck, 2004).

Lancaster (2007) investigated cognitive competence as continuing students of statistics in a sample of 56 prospective primary school teachers in the United States of America that had received a course with statistical content. The questionnaire, a Likert-type survey with five response possibilities for each question, posed six questions such as "Would you be interested in participating in a workshop or other professional development that focuses on improving your knowledge of statistics for the grade level you wish to teach?" and "Do you agree that attending a workshop or other professional development that focuses on improving your knowledge of statistics would improve your ability to teach statistics at your desired grade level?" Results indicated that the teachers' attitudes towards statistics affected their willingness to participate in activities related to statistics in the future.

The prospective teachers who were still in the early part of their mathematical training, and had an experienced instructor with a reputation as a motivational teacher, had higher cognitive competence for statistics than did prospective teachers who were in the late part of their mathematical training and were learning statistical concepts in a class with a novice instructor. Though the numbers involved with this comparison were relatively small, these results underscore the possibility that instructor competence and teaching style may affect student attitudes and beliefs towards statistics.

4.3   Teacher Attitudes Towards Their Ability to Teach Statistics

Two studies have focused on teacher attitudes towards statistics as a subject to teach. In their research, Begg and Edwards (1999) analysed the teachers' attitudes towards teaching statistics and found that about 75% of the practising teachers in their sample felt secure when teaching the topic. These teachers were confident in their statistical abilities and showed low interest in continued professional development in statistics.

Watson (2001) designed and administered a multi-faceted survey to 15 primary school teachers and 28 secondary school teachers in Australia with the aim of assessing professional development needs for teachers arising from changes in the mathematics curriculum. The survey included Likert-type questions, open-ended questions, and the option of an interview. A part of the survey measured the teacher confidence to teach specific statistical concepts. Teacher confidence was highest for "graphical representations" and lowest for "odds". Teachers were asked what type of professional development they would prefer. Four out of every ten indicated school-based sessions while two out of every ten indicated independent readings or a University course. Of those who provided responses to this question, many believed that there was a need for more professional development opportunities.

In summary, research described in this section shows poor or neutral attitudes in prospective teacher. This might be explained by the fact that prospective teachers' attitudes depend on their previous experiences with statistics, which is often restricted to studying statistics in school or studying statistics with a very formal mathematical approach. Moreover, if a teacher has low cognitive competence, then he/she may be less likely to be motivated to participate in continuing professional development in statistics. Results related to teachers' confidence to teach statistics are scarce and inconclusive, so more research is needed in this area.

5   Variables That Affect Teacher Attitudes Towards Statistics

Research that has tried to identify factors affecting the attitudes of teachers focuses mainly on three themes: (a) the relationship between attitudes and statistical knowledge (Nasser, 2004; Estrada et al., 2005; Estrada & Batanero, 2008) (b) relating prospective teachers' attitudes to other affective variables (Lancaster, 2007, 2008); and (c) assessing differences in attitudes in prospective and practising teachers (Estrada et al., 2005; Estrada & Batanero, 2008).

5.1   Relating Prospective Teachers' Attitudes and Statistical Knowledge

Nasser (2004) used SATS to examine the relationships among attitudes towards statistics, anxiety, mathematical aptitude, and statistics achievement of 167 prospective teachers in Egypt taking part in an introductory statistics course. Teachers' achievement in statistics was assessed using ten open-ended questions including descriptive statistics (frequency tables, central tendency, variation, distributions, and association) and inferential statistics (estimation, hypothesis tests). Nasser found a small positive effect of attitudes towards statistics on achievement in statistics.

Estrada et al. (2005) also analysed the relationship between prospective teachers' attitudes and their statistical knowledge as assessed by nine open-ended items taken from the *Statistical Reasoning Assessment* (*SRA*) questionnaire (Garfield, 2003). The particular SRA items used in this study assessed understanding of the main statistics content in the Spanish primary school curriculum: reasoning about data, graphs, average and spread, uncertainty and sampling bias. The authors found a significant and worrying percentage in the sample of prospective teachers who did not understand some of the elementary statistical concepts they will have to teach to their future students. There was a significant correlation between attitudes and the number of statistics courses previously taken by the participants, as well as between attitudes and performance on SRA items. Detailed analyses of scores showed that attitudes improved consistently with the number of courses and knowledge of statistics.

In order to understand prospective teachers' attitudes and misconceptions better, Estrada and Batanero (2008) carried out a complementary study of a new sample of prospective teachers ($n = 121$) who were only given the ten SATS items that had yielded lower scores in the Estrada et al. (2005) study. These participants were asked to first complete the ten items and then justify their responses. A qualitative analysis of their open justifications served to classify the main reasons for positive and negative scorings in the Estrada and Batanero (2008) study. The main explanations given for positive attitudes included: (a) considering statistics as an easy topic, e.g., "pretty logical and simple"; (b) satisfactory learning experiences, e.g., "the teacher explained it well"; (c) novelty of the topic, e.g., "I like topics that are non-routine"; (d) perception of the usefulness of statistics for a teacher, e.g., "you have to know about statistics to be able to teach it to children"; and (e) the formative value of statistics, e.g., "essential in many different kinds of work". The main reasons for negative scoring were: (a) lack of previous knowledge or training, e.g., "I only studied statistics at primary school and I hardly remember anything"; (b) difficulty with statistical reasoning, e.g., "you need too much logical thinking"; (c) content too formal, e.g., "symbols and equations are too strange"; (d) considering that statistics is not valued in society, e.g., "statistical knowledge is not required when you look for a job"; and (e) lack of knowledge of applications, e.g., "I found no possible applications of statistics".

Onwuegbuzie (1998, 2003) also examined attitudes and knowledge among teachers. He concluded that attitudes towards statistics have a strong relation to achievements on statistics; however, the impact on reasoning abilities in statistics is not so high because statistical reasoning abilities are more strongly tied to mathematics outcomes than to statistics outcomes.

In summary, the research reported above related attitudes to statistics achievement. Positive attitudes increase when students have good learning experiences and perceive value for their own professional work or for their students' education. Negative attitudes are linked to perceived difficulty, lack of knowledge and overly formal content.

5.2   Relating Prospective Teachers' Attitudes to Other Affective Variables

Lancaster (2007) also investigated the effect of several variables on prospective teacher attitudes towards their role as continuing students of statistics, examining: (a) attitudes towards statistics measured by the ATS instrument; (b) current self-efficacy to use statistics that has been learned; (c) current self-efficacy to learn statistics in the future and (d) knowledge of basic statistical concepts, measured by the ARTIST scales (delMas, R, Garfield, Ooms, & Chance, 2006). Current statistics self-efficacy was defined by Finney and Schraw (2003) as confidence in one's ability to solve specific tasks related to statistics, and self-efficacy to learn statistics was defined as confidence in one's ability to learn the skills necessary to solve specific tasks related to statistics. The results of the study showed no correlations between the affective variables, a result also found by Tempelaar (2003) in his study with undergraduates.

Current self efficacy to learn statistics in the future predicted the teachers' beliefs that continuing professional development in statistics would benefit them in their classroom teaching. Furthermore, current self-efficacy to use statistics that has been learned combined with attitudes towards the current course served to predict current self-efficacy to learn statistics in the future. These results confirm suggestions by Finney and Schraw (2003) that, for research into statistics attitudes, specific affective measures are more likely than general affective measures to provide meaningful results.

5.3   Comparing Prospective and In-Service Teachers' Attitudes Towards Statistics

In her study, Estrada (2002) compared the attitudes towards statistics of in-service primary school teachers and prospective primary school teachers and related these attitudes with gender, number of previous statistics courses, specialty (topic in which the prospective teachers were specialising or topic the teachers taught), and number of years of teaching experience in mathematics (for in-service teachers). The results of Estrada favoured the prospective teachers group, that is, attitudes seemed to deteriorate with the actual practice of teaching. When analysing the items in which there were significant differences by group, in-service teachers were more critical of the use of statistics in the media. Because they found statistics to be more useful for everyday life and gave it more value for the education of citizens, the prospective teachers tended to assume that they would include statistics in their teaching and found it easier to understand as well as more interesting than in-service teachers did. The number of previous mathematics courses with a statistics component taken had a significant effect, with attitudes improving as this number increased. There was no difference by gender and only a small difference regarding the specialty in which the prospective teachers were majoring. In the case of in-service teachers attitudes became less positive as teachers got older. The analysis of the specific items suggests that senior teachers had a greater tendency to suppress statistics when possible and found statistics more difficult than younger teachers. Finally, the results also suggested that teachers who did not use statistics (or used it very little) in their professional lives (e.g., in assessment or to compare performances of different groups) tended to have poorer attitudes towards statistics.

6 Implications for Research into Training Teachers in Statistics

Students learn statistics more effectively in settings where collaboration is encouraged, and where progressive teaching methods such as discovery learning and problem solving are the focus. To successfully implement such strategies, teachers must possess the necessary attitude structures as well as good knowledge of mathematical and didactic aspects of school learning of statistics (Wilson & Cooney, 2002). The first step in achieving these aims is to continue with research aimed at assessing teachers' attitudes towards statistics and finding possible explanatory variables. This research is particularly needed as regards secondary school teachers, where no research is currently available.

Secondly, reported research also suggested that a main influence on teachers' attitudes is their previous knowledge of statistics as well as good learning experiences (non-abstract, with examples of applications in everyday and professional life). Attention is drawn from these results to the need to find more methods to improve teachers' knowledge and provide them with examples that statistics is valuable and that basic statistics can be taught in an attractive and easy way. Teacher educators are then responsible for creating an emotionally and cognitively supportive environment in statistics training, where prospective teachers explore different statistical methods, gain confidence in their own ability to learn, and learn to value the role of statistics in modern society.

Acknowledgements Research supported by the project EDU2010-14947, MICIIN-FEDER.

References

Aiken, L. R. (1980). Attitudes measurement research. In D. A. Payne (Ed.), *Recent developments in affective measurement* (pp. 1–24). San Francisco, CA: Jossey-Bass.
Auzmendi, E. (1991, April). *Factors related to statistics: A study with a Spanish sample.* Paper presented at the Annual Meeting of the American Educational Research Association, Chicago, IL.

Bandura, A. (1997). *Self-efficacy: The exercise of control.* New York: Freeman.

Batanero, C., Burrill, G., Reading, C., & Rossman, A. (Eds.). (2008). *Joint ICMI/IASE Study: Teaching Statistics in School Mathematics. Challenges for Teaching and Teacher Education. Proceedings of the ICMI Study 18 and 2008 IASE Round Table Conference.* Monterrey, Mexico: International Commission on Mathematical Instruction and International Association for Statistical Education. Online: www.stat.auckland.ac.nz/~iase/publications

Begg, A., & Edwards, R. (1999). Teachers' ideas about teaching statistics. *Proceedings of the 1999 Combined Conference of the Australian Association for Research in Education and the New Zealand Association for Research in Education.* Melbourne: AARE & NZARE. Online: www.aare.edu.au/99pap/beg99082.htm

Carmona, J. (2004). Una revisión de las evidencias de fiabilidad y validez de los cuestionarios de actitudes y ansiedad hacia la estadística (Revising the reliability and validity evidence of attitudes and anxiety towards statistics questionnaires). *Statistics Education Research Journal, 1*(1), 5–28. Online: www.stat.auckland.ac.nz/~iase/serj/

Cashin, F., & Elmore, P. (2005). The Survey of Attitudes Toward Statistics scale: A construct validity study. *Educational and Psychological Measurement, 65,* 509–524.

Chick, H. L., & Pierce, R. U. (2008). Teaching statistics at the primary school level: Beliefs, affordances, and pedagogical content knowledge. In C. Batanero, G. Burrill, C. Reading, & A. Rossman (2008).

delMas, R., Garfield, J. B., Ooms, A., & Chance, B. (2006). The ARTIST scales. Online: https://app.gen.umn.edu/artist/

Estrada, A. (2002). *Análisis de las actitudes y conocimientos estadísticos elementales en la formación del profesorado (Analysing attitudes and elementary statistical knowledge in the training of teachers).* Unpublished Ph.D. dissertation, Universidad Autónoma de Barcelona, Spain.

Estrada, A., & Batanero, C. (2008). Explaining teachers' attitudes towards statistics. In C. Batanero, G. Burrill, C. Reading, & A. Rossman (2008).

Estrada, A., Batanero, C., Fortuny, J. M., & Diaz, C. (2005). A structural study of future teachers' attitudes towards statistics. In M. Bosch (Ed.), *Proceedings of the Fourth Congress of the European Society for Research in Mathematics Education* (pp. 508–517). Barcelona: IQS Fundemi. Online: ermeweb.free.fr/CERME4/

Finney, S. J., & Schraw, G. (2003). Self-efficacy beliefs in college statistics courses. *Contemporary Educational Psychology, 28,* 161–186.

Fishbein, M., & Ajzen, I. (1975). *Belief, attitude, intention and behaviour: An introduction to theory and research.* London: Addison-Wesley.

Gal, I., & Ginsburg, L. (1994). The role of beliefs and attitudes in learning statistics: Towards an assessment framework. *Journal of Statistics Education, 2*(2). Online: www.amstat.org/publications/jse/

Gal, I., Ginsburg, L., & Schau, C. (1997). Monitoring attitudes and beliefs in statistics education. In I. Gal & J. B. Garfield (Eds.), *The assessment challenge in statistics education* (pp. 37–51). Voorburg, The Netherlands: IOS Press.

Garfield, J. B. (2003). Assessing statistical reasoning. *Statistics Education Research Journal, 2*(1), 22–38. Online: www.stat.auckland.ac.nz/~iase/serj/

Gómez-Chacón, I. (2000). Affective influences in the knowledge of mathematics. *Educational Studies in Mathematics, 43*(2), 149–168.

Gould, R., & Peck, R. (2004, July). *Preparing secondary teachers to teach statistics.* Paper presented at the 10th International Congress on Mathematics Education, Copenhagen.

Hill, H., & Ball, D. L. (2004). Learning mathematics for teaching: Results from California's mathematics professional development institutes. *Journal for Research in Mathematics Education, 35*(5), 330–351.

Lancaster, S., (2007). *Preservice teachers and statistics: Interrelationships between content confidence, knowledge, and attitudes; pedagogical beliefs; classroom practices; and teacher interest in professional development in statistics.* Unpublished Ph.D. dissertation, University of Oklahoma, Norman.

Lancaster, S. (2008). A study of preservice teachers' attitudes toward their role as students of statistics and implications for future professional development in statistics. In C. Batanero, G. Burrill, C. Reading, & A. Rossman (2008).

McLeod, D. B. (1992). Research on affect in mathematics education: A reconceptualization. In D. A. Grows (Ed.), *Handbook of research on mathematics teaching and learning* (pp. 575–596). New York: Macmillan and National Council of Teachers of Mathematics.

Nasser, F. M. (2004). Structural model of the effects of cognitive and affective factors on the achievement of Arabic-speaking prospective teachers in introductory statistics. *Journal of Statistics Education, 12*(1). Online: www.amstat.org/publications/jse/

Olson, J. M., & Zanna, M. P. (1993). Attitude and attitude change. *Annual Review of Psychology, 44*, 117–154.

Onwuegbuzie, A. J. (1998). Teachers' attitudes toward statistics. *Psychological Reports, 83*, 1008–1010.

Onwuegbuzie, A. J. (2003). Modelling statistics achievement among graduate students. *Educational and Psychological Measurement, 63*, 1020–1038.

Pajares, F. (1996). Self-efficacy beliefs in academic settings. *Review of Educational Research, 66*(4), 543–578.

Philipp, R. A. (2007). Mathematics teachers' beliefs and affects. In F. Lester (Ed.), *Second handbook of research on mathematics teaching and learning* (pp. 257–315). Charlotte, NC: Information Age Publishing and National Council of Teachers of Mathematics.

Roberts, D. M., & Bilderback, E. W. (1980). Reliability and validity of a statistics attitude survey. *Educational and Psychological Measurement, 40*, 235–238.

Roberts, D. M., & Saxe, J. E. (1982). Validity of a statistics attitude survey: A follow-up study. *Educational and Psychological Measurement, 42*, 907–912.

Schau, C. (2003, August). *Students' attitudes: The other important outcome in statistics education.* Paper presented at the Joint Statistical Meeting of the American Statistical Association, San Francisco, CA.

Schau, C., Stevens, J., Dauphine, T., & del Vecchio, A. (1995). The development and validation of the survey of attitudes towards statistics. *Educational and Psychological Measurement, 55*(5), 868–875.

Tempelaar, D. (2003). Statistical reasoning and its relationship to attitudes towards statistics and achievement. *Proceedings of the International Statistical Institute 54th Session*, Berlin, Germany. Online: www.stat.auckland.ac.nz/~iase/publications

Watson, J. M. (2001). Profiling teachers' competence and confidence to teach particular mathematics topics: The case of chance and data. *Journal of Mathematics Teacher Education, 4*(4), 305–337.

Wilson, M. S., & Cooney, T. J. (2002). Mathematics teacher change and development: The role of beliefs. In G. C. Leder, E. Pehkonen, & G. Torner (Eds.), *Beliefs: A hidden variable in mathematics education?* (Vol. 31, pp. 127–148). London: Kluwer Academic Publishers.

Wise, S. L. (1985). The development and validation of a scale measuring attitudes toward statistics. *Educational and Psychological Measurement, 45*, 401–405.

Chapter 19
Statistics Teachers and Classroom Practices

Andreas Eichler

Abstract Why do statistics teachers teach certain topics, how do they teach these topics, and to what extent does the teaching affect students' learning? In this chapter, a theoretical framework combining a curriculum model with the construct of beliefs will be provided to analyse previous research concerning teachers' instructional planning, their classroom practices, and the impact of these practices on their students' learning. Each section includes a brief discussion of research results referring to mathematics education in general and statistics education in particular, and exemplifying results from research that address the three questions posed.

1 Introduction

> How teachers make sense of their professional world [...], and how teachers' understanding of teaching, learning, children, and the subject matter informs their everyday practice are important questions that necessitate an investigation of the cognitive and affective aspects of teachers' professional lives (Calderhead, 1996, p. 709).

Like Calderhead, many researchers in mathematics education recognise the importance of teachers' sense-making in their professional work for the following reasons:

- Teachers' thinking about mathematics and the teaching and learning of mathematics have a high impact on their instructional practice (Philipp, 2007); and
- Teachers' instructional practice, which is considerably determined by teachers' thinking about their professional world, has a high impact on students' learning and beliefs concerning mathematics (Hiebert & Grouws, 2007).

A. Eichler (✉)
Institute for Mathematics Education, University of Education Freiburg, Kunzenweg 21,
79117 Freiburg, Germany
e-mail: andreas.eichler@ph-freiburg.de

C. Batanero, G. Burrill, and C. Reading (eds.), *Teaching Statistics in School*
Mathematics-Challenges for Teaching and Teacher Education: A Joint ICMI/IASE Study,
DOI 10.1007/978-94-007-1131-0_19, © Springer Science+Business Media B.V. 2011

Despite the importance of research that focuses on mathematics teachers' thinking, research in this field is sparse for the teaching of statistics and probability (Jones, Langrall, & Mooney, 2007; Shaughnessy, 2007). This is despite the fact that teachers' thinking has been declared a crucial research topic in statistics education (Batanero, Garfield, Ottaviani, & Truran, 2000; Shaughnessy, 2007). This chapter will highlight three specific issues associated with statistics teachers' thinking and actions in their professional world: (a) Statistics teachers' planning of statistics teaching; (b) the relationship between statistics teachers' planning and their classroom practice; and (c) the relationships among statistics teachers' classroom practices and students' learning. The first section of the chapter provides a theoretical framework for describing the three issues listed above. Using this framework the issues will be examined in separate sections, each including a brief overview of the relevant research from mathematics education in general, a discussion of research approaches in statistics education and, finally, some results taken from research that directly addressed these issues. Implications associated with changing statistics teachers' classroom practices will be described in the last section.

2   A Theoretical Framework

Research addressing why statistics teachers teach certain topics, how they teach these topics, and the extent to which students learn can be described by using the enlarged model of the curriculum proposed by Stein, Remillard, and Smith (2007, p.322) (see Fig. 19.1).

2.1   Four Phases of the Curriculum

The *written curriculum* involves both instructional content and teaching objectives – or, standards – often prescribed by national governments. The teachers' interpretation of the written curriculum – that is, the individual teacher's transformation of the written curriculum – is called the *intended curriculum*. The interactions of a teacher,

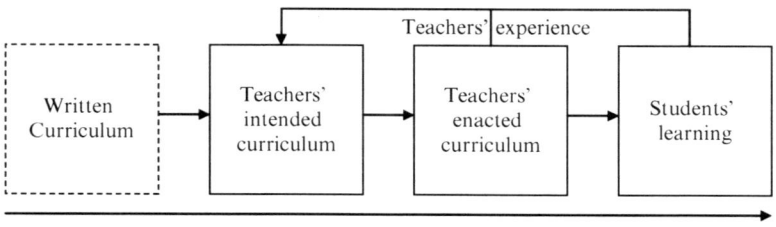

Process of transformation

Fig. 19.1   Four phases of the curriculum according to Stein et al. (2007)

his or her students, and the instructional content "bring the curriculum to life and, in the process, create something different than what could exist […] in the teacher's mind" (Stein et al., 2007, p. 321). This transformation of the intended curriculum is called the *enacted curriculum.* Finally, the students transform the content addressed in the enacted curriculum into their own personal subjective knowledge and develop their own beliefs about the content. This is the *students' learning.*

These phases are not static. A teacher's own experiences with his or her classroom practice (the *enacted curriculum*) as well as his or her awareness of the beliefs and knowledge attained by the students (the *students' learning*) in turn have an impact on the teacher's intended curriculum (Hofer, 1986), so that it actually develops over time. In this chapter, the focus is mainly on the three latter phases of the curriculum model, namely, teachers' intended curricula, their enacted curricula, and their students' learning.

2.2   Belief Systems in the Perspective of the Curriculum Model

The process of curriculum transformation, as shown in Fig. 19.1, is affected by teachers' beliefs. The term *beliefs* is understood as an individual's personal conviction concerning a specific subject, which shapes an individual's ways of both receiving information about a subject and acting in a specific situation (Pajares, 1992; Thompson, 1992; Furinghetti & Pehkonen, 2002). Beliefs and knowledge are both components of an individual's conviction, and so are inextricably intertwined (Pajares, 1992). For this reason, the term "beliefs" predominantly will be used (in contrast to "knowledge"). Further, an individual's internal organisation of beliefs is called a *belief system* (Thompson, 1992). Belief systems might include contradictory clusters of beliefs (in contrast to "objective" systems of knowledge), and might include beliefs that have different degrees of importance (centrality) for an individual (Thompson, 1992).

A teacher's *intended curriculum* is represented by a belief system including all the beliefs that a teacher takes into account when planning (in his or her view) appropriate classroom practices. Hence, intended curricula might include beliefs about specific content, teaching goals linked to this content, the best way to teach mathematics or statistics, and the way students learn mathematics or statistics.

A teacher's *enacted curriculum* involves the observable part of the teacher's intended curriculum, transformed by the interaction of the teacher, the students involved and the content within the classroom practice. Finally, *students' learning* is represented by students' belief systems concerning mathematics or statistics that are strongly determined by the classroom practice. These belief systems are understood to involve the students' statistical knowledge (Broers, 2006) and the students' beliefs about the benefit of statistics for society and students' own lives (Eichler, 2008a).

Using the theoretical framework discussed above, the following sections will discuss mathematics teachers' classroom practice and, in particular, statistics teachers' classroom practice, starting with teachers' intended curricula.

3   Teachers' Intended Curricula

3.1   Mathematics Teachers' Intended Curricula

The first step in researching teachers' classroom practice is to investigate what teachers intend to do – that is, the teachers' intended curricula. This involves an examination of belief clusters concerning mathematics, the teaching and learning of mathematics, and the curriculum (Pajares, 1992). One purpose of this research is to describe and classify belief systems that represent the intended curricula of groups of teachers. A well-known classification of Thompson (1992) distinguishes among beliefs about mathematics as "(a) a dynamic, problem-driven discipline; (b) a static, unified body of knowledge; or (c) a bag of tools" (Philipp, 2007, p. 260). Grigutsch, Raatz, and Törner (1998) add to these (d) the application view, to distinguish teachers who emphasise applied mathematics that has relevance for solving real-world problems, in contrast to pure and abstract mathematics or a tool kit of rules and formulae.

In respect to teachers' beliefs concerning teaching and learning mathematics, Thompson (1992, p.136) distinguishes two main teaching styles: a constructivist "learner focused view", and a "content focused view" that, more recently, Staub and Stern (2002) called a teacher's cognitive constructivist orientation and a teacher's traditional orientation (so-called direct transmission), respectively. However, most of the increasing body of research on teachers' beliefs does not consider that teachers' beliefs may vary across different mathematical areas (Franke, Kazemi, & Battey, 2007). For this reason, it is crucial to examine the specific beliefs of statistics teachers referring to the teaching of statistics.

3.2   Statistics Teachers' Intended Curricula

It is obvious that issues concerning what statistics teachers are *able* to teach appropriately (see Chaps. 20–26 of this volume) and what teachers are *required* to teach (see Chaps. 10–16 of this volume) are important directions for research. In this section, however, the focus is on what content ordinary statistics teachers intend to teach and what instructional goals they have for their every-day classroom practice.

Also, it seems obvious that teachers who do not accept statistics as worthwhile or enjoyable are less likely to incorporate it in their own teaching. Research investigating teachers' general beliefs about statistics, however, shows a high acceptance of statistics (e.g., Gattuso & Pannone, 2002; Chick & Pierce, 2008). Further research related to these attitudes and beliefs towards statistics is reviewed by Estrada, Lancaster, and Batanero, and also by Chick and Pierce (in this volume). Given this high acceptance, it is appropriate to ask what aspects of statistics teachers choose to incorporate in their teaching.

Turning to more specific research results about beliefs towards statistics teaching, Begg and Edwards (1999), investigating 34 Australian primary teachers (using interviews, questionnaires, and concept maps), reported data collection, graphs, data interpretation, and probability as topics predominant in the teachers' intended curriculum. Watson's research (2001), involving 43 Australian primary and secondary teachers (interviews and written reports), yielded similar results regarding primary teachers, and a focus on data analysis and probability among secondary teachers. One important result beyond the specific instructional content in both studies was that the teachers' intended curricula seemed to fit the written curricula. The same result arose from a survey of 110 German secondary teachers by Eichler (2008b). In contrast to Australia, however, the German teachers placed a heavy emphasis on probability that seems to be common for Europe (e.g., Broers, 2006), a little emphasis on inference, and there was an absence of data analysis.

Concerning instructional goals, Watson (2001, p. 313) identified four significant factors relating to "the teachers themselves, the students, the content and school issues" that provided individual reasons for teachers to teach statistics. For example, some of these reasons are the "relevance of statistics to the real world", "the use of technology" or, loosely, "motivation" (Watson, 2001, p. 313).

In his qualitative interview study with 13 upper secondary mathematics teachers, Eichler (2007, 2008a) described four types of teachers' intended curricula for teachers teaching similar content. Of the four types, *traditionalists* and *everyday-life preparers* represent the extremes of teachers' intended curricula. The main objective of the *traditionalists* is to establish a theoretical basis for statistics involving algorithmic skills and insights into the abstract structure of mathematics, but they neglect applications. In contrast, the *everyday-life preparers* intend to develop statistical methods in a process, the results of which will be both the students' ability to cope with real stochastic problems and the students' ability to criticise decision-making processes in real life. These differences will be clarified using some excerpts from the interviews with two teachers involved in the research: Mr. D (an everyday-life preparer) and Mr. J (a traditionalist). In discussing his goals for teaching statistics Mr. D argued:

> Mr. D: And that's what I am trying to illustrate …, that you will of course get quite far with relative frequency, but that if you have similar situations afterwards, such as elections or opinion polls, you will … need to develop the use of confidence intervals. This means showing them (students), as well, that mathematics really has applications … that there are quite often problems which you can solve with maths. … Students should be enabled to better categorise mathematical models which determine our economic condition.

Whereas Mr. D emphasised the goal of having his students cope with real problems, Mr. J valued the role of context considerably less:

> Mr. J: Personally, concerning statistics, I emphasise the mathematical background involving, for instance, set theory. Other teachers think the students do not need a broad background, but must understand how to apply statistical methods in real situations. This is for me a step away from mathematics, only pure application.

These quotations illustrate the *central* objectives of the two teachers. However, both teachers also mentioned peripheral objectives. For example, Mr. D also referred to the formal mathematical aspects that could be explored in statistics, and even Mr. J mentioned that applying statistics was a goal in his teaching practice but a peripheral one. A striking result in this research was that traditionalists tend to neglect the relevance of the *role of the context* that Shaughnessy (2007) mentioned as one of the main aspects of teaching statistics. Moreover, in a quantitative survey with 110 teachers Eichler (2008b) showed that underemphasising the role of context in statistics education is common in German secondary high schools. About 70% of the teachers in this survey predominantly agreed with the objectives of the traditionalists, while only about 30% of the teachers agreed with the objectives of the everyday-life preparers.

4   Relationships Between Teachers' Intended and Enacted Curricula

4.1   Mathematics Teachers' Intended and Enacted Curricula

The results of research into the relationship between teachers' intended curricula and teachers' enacted curricula are ambivalent. Some researchers found inconsistencies between these two aspects, while other researchers noted consistency (Thompson, 1992; Philipp, 2007). The differences between teachers' beliefs and their instructional practice are explained by the experience of the observed teachers (Artzt & Armour-Thomas, 1999), the specific situation of different classrooms (Hiebert & Grouws, 2007), and the inconsistency of peripheral beliefs, in spite of the consistency of central beliefs (Putnam & Borko, 2000). Further, several studies revealed that the classroom practices of different teachers differ considerably even if they address the same tasks (Stein et al., 2007).

4.2   Statistics Teachers' Intended and Enacted Curricula

Although the research of Chick and Pierce (2008) did not include an observation of the (pre-service) teachers' enacted curricula, it yielded a noticeable result concerning a phenomenon that one task yield considerably different classroom practices. Thus, although the 27 prospective teachers involved in the qualitative research were asked to plan a lesson on the basis of the same data, their lesson plans showed a variety of approaches and topics. This highlights that the same data or even the same task could yield different classroom practices.

Further, Burgess (2008) reports the classroom practice of two teachers (grade 5/6 and grade 7). Using a two-dimensional framework concerning teachers' knowledge and five aspects of statistical thinking (from Wild & Pfannkuch, 1999), he found substantial differences between the practices of the two teachers in their ability to take advantage of the learning opportunities of a task given by the researcher.

The qualitative study of Paparistodemou, Potari, and Pitta (2006) involved the planning of several lessons by 23 prospective teachers and their resulting classroom practice. In this research, the case of Macy showed heavy differences between her appropriate planning of a lesson and her inappropriate teaching practice that lacked central aspects of her planning.

Pfannkuch (2006) reported a case study of one teacher whose teaching was focusing on comparing two data sets. The intervention study involved instructional planning by the teacher supported by the researcher, and the videotaped observation of 15 lessons. Analysis of the observations yielded "elements of reasoning" (Pfannkuch, 2006, p. 33) that were based on the collaborative planning of the lessons, but also elements that primarily arose during the classroom practice of the teacher.

In contrast to the four studies mentioned above, the case studies of Eichler (2007, 2008a) provided a direct investigation of the impact of ordinary teachers' intended curricula on their enacted curricula. His observation of four teachers' classroom practice lasting about half of one year provided strong evidence that the teachers pursue their main objectives (Eichler, 2008a) or, rather, their central beliefs (Putnam & Borko, 2000). For example, the observation of the two teachers discussed earlier, Mr. D and Mr. J, yielded relevant differences in teaching styles. Mr. D's students predominantly worked on realistic problems comprising real data sets, and new statistical concepts often evolved from previous problem solutions. Mr. J's lessons, in contrast, involved teacher-directed explanations of new statistical concepts followed by student work on routine tasks. He seldom used real data sets or realistic problems but preferred traditional tasks involving dice, cards or urns.

These observations provide evidence that both teachers enacted their central instructional goals, whereas they seemed to neglect their peripheral goals. For Mr. D, this meant emphasising formal aspects of statistics, and, in the case of Mr. J, emphasising the role of context. Again, the role of context seems to emerge as the main difference between the two teachers.

5 Relationships Among Teachers' Intended and Enacted Curricula, and Their Students' Learning

5.1 *Mathematics Teachers' Intended and Enacted Curricula in Relation to Their Students' Learning*

The relationship between teachers' classroom practices and their students' learning is probably the most crucial, but also the most challenging question in mathematics education. Although there has been considerable research effort in this field, Hiebert and Grouws (2007, p.373) stated that "theories that specify the ways in which the key components of teaching fit together to form an interactive, dynamic system for achieving particular learning goals have not been sufficiently developed". However, there exist some research results that give, for example, evidence that:

- Different teachers affect patterns in students' learning (Hiebert & Grouws, 2007).
- Emphasising the connections between mathematical concepts and procedures, and using cognitively demanding tasks could increase students' conceptual knowledge (Hiebert & Grouws, 2007).
- Using a constructivist teaching approach promotes students' learning (Franke et al., 2007).

5.2   Statistics Teachers' Intended and Enacted Curricula in Relation to Their Students' Learning

Castro (1998) investigated the impact of a curriculum defined by the researcher and taught by the regular teachers of six high school classes. The curriculum prescribed the same syllabus of instruction, but different teaching methods for each of two sets of three classes. As suggested by the previous section, the research showed a significantly higher performance in skills and probability reasoning for the students of the three classrooms where the teachers taught with a constructivist orientation, in contrast to three classrooms where the teachers taught using an expository teaching style.

The research of Pfannkuch and Horring (2005) and Pfannkuch (2006) focused on the development of students' statistical reasoning based on lessons planned collaboratively by the teacher and researcher and involving the comparison of two data sets. The analysis of videotaped lessons and student questionnaires provided evidence that the intended emphasis on the statistical investigation process oriented the students' beliefs towards statistical analysis (Pfannkuch & Horring, 2005). Moreover, the analysis also showed a direct connection between the students' inability to draw conclusions when comparing two data sets and the missed opportunities of the teacher to communicate ways of drawing such conclusions.

The case studies of Eichler (2008a) highlighted possible relationships among four teachers' intended curricula, their classroom practice and achievement of five of their students who were interviewed after the courses about their statistical knowledge and their beliefs concerning statistics. Although the students of the four teachers showed similar capacities to explain statistical concepts and to draw connections among different statistical concepts, there was a direct impact on the students' beliefs about the relevance of statistics from the teachers' differing emphases on real problems, real data sets and the role of context (Eichler, 2008a):

- The students of Mr. D (everyday-life preparer) predominantly gave meaningful explanations of statistical concepts and were able to mention connections among statistical concepts, but seldom used formal explanations. The students believed that statistics is highly relevant for society. To explain this relevance, the students used various realistic situations that they had mostly examined in school. However, the students believed that statistics would have little relevance for their own life.

- The students of Mr. J predominantly gave formal explanations of the statistical concepts that were often vague and lacked connections among statistical concepts. All the students assigned statistics little relevance for society using situations solely from school to explain the possible relevance of statistics. Most of these situations concerned games of chance. None of the students gave statistics relevance for their own life.

The quantitative survey of Eichler (2008b, 2009) involving 110 teachers and 323 students supported the pattern mentioned above. The teachers who showed a strong emphasis on statistical applications (everyday-life preparers) significantly promoted their students' beliefs concerning the relevance of statistics. In contrast, the teachers who showed a preference for a traditional curriculum seemed to influence their students' lack of appreciation for statistics.

6 Implications for Teaching and Research

This overview of research into statistics teacher's practices provides an insight into the teaching and learning of statistics in ordinary classrooms. Combining the curriculum model and the results from both research into mathematics teaching and statistics teaching, some important results emerge.

Firstly, teachers assigned to teach statistics tend to meet the recommendations of the written curriculum with regard to the instructional content. Hence, referring to content, written curricula appear to be attended to in statistics teaching. Nevertheless, the research of Eichler (2008a, 2008b) concerning German teachers found that, although teachers may intend to teach similar content, they differ considerably concerning the objectives linked to this content. The differences in the lesson plans of Australian prospective teachers (Chick & Pierce, 2008) concerning the same data set as well as the differences in the classroom practice of two New Zealand teachers using the same task (Burgess, 2008) might also arise from different instructional objectives of the teachers.

Further, teachers' intended curricula appear associated with teachers' enacted curricula. This is particularly the case regarding teachers' central objectives for teaching statistics and, hence, the teachers' intentions appear to be relevant to classroom practice. Within the teachers' intentions and classroom practice, the role of context seems to play a significant role in explaining differences among teachers.

Finally, obtaining evidence concerning the impact of classroom practice on the students' learning remains the most challenging aspect of research related to statistics teachers' beliefs. Although the existing research yields patterns in students' learning influenced by individual teachers (see, for example, the work of Eichler, 2008b, 2009 reported earlier), there is currently only weak evidence concerning this impact of teaching on students' learning.

Franke et al. (2007) and psychological research related to teachers' actions (e.g., Hofer, 1986) suggest it is the nature of teachers' thinking, and, in particular, the system of instructional goals the teachers hold, that determines the teachers' intended curricula, teachers' classroom practice, and, finally, students' learning. Accepting this statement, understanding statistics teachers' thinking and their instructional goals are thus key factors for achieving changes in statistics teachers' instructional practice.

Although the review of Franke et al. (2007) gave evidence that changing teachers' classroom practice is possible, research also highlights many obstacles to changing mathematics teachers' beliefs. These obstacles, in particular, seem to exist with respect to the central beliefs that teachers have formed in their professional lives according to their experiences with classroom practice and their students' learning (e.g., Philipp, 2007).

One of the most striking results of the overview of research described in this chapter is the minor status of research on statistics teachers' intended and enacted curricula and their influence on students' learning. If we accept that a potentially successful way to change teachers' central beliefs is through teachers' assimilation of new ideas in contrast to accommodation (Pajares, 1992) it seems worthwhile to increase the research addressing the understanding of statistics teachers' central beliefs, and to understand the relationships among teachers' central beliefs, their classroom practice, and students' learning.

References

Artzt, A. F., & Armour-Thomas, E. (1999). A cognitive model for examining teachers' instructional practice in mathematics: A guide for facilitating teacher reflection. *Educational Studies in Mathematics, 40*, 211–235.

Batanero, C., Burrill, G., Reading, C., & Rossman, A. (Eds.). (2008). *Joint ICMI/IASE Study: Teaching Statistics in School Mathematics. Challenges for Teaching and Teacher Education. Proceedings of the ICMI Study 18 and 2008 IASE Round Table Conference.* Monterrey, Mexico: International Commission on Mathematical Instruction and International Association for Statistical Education. Online: www.stat.auckland.ac.nz/~iase/publications

Batanero, C., Garfield, J. B., Ottaviani, M. G., & Truran, J. (2000). Research in statistical education: some priority questions. *Statistics Education Research Newsletter, 1*(2), 2–6. Online: www.stat.auckland.ac.nz/~iase/serj/

Begg, A., & Edwards, R. (1999). Teachers' ideas about teaching statistics. *Proceedings of the 1999 Combined Conference of the Australian Association for Research in Education and the New Zealand Association for Research in Education.* Melbourne: Australian Association for Research in Education & New Zealand Association for Research in Education. Online: www.aare.edu.au/99pap/

Broers, N. J. (2006). Learning goals: The primacy of statistical knowledge. In A. Rossman & B. Chance (Eds.), *Proceedings of the Seventh International Conference on Teaching Statistics.* Salvador, Bahia, Brazil: International Statistical Institute and International Association for Statistical Education. Online: www.stat.auckland.ac.nz/~iase/publications

Burgess, T. (2008). Teacher knowledge for teaching statistics through investigations. In C. Batanero, G. Burrill, C. Reading, & A. Rossman (2008).

Calderhead, J. (1996). Teachers: Beliefs and knowledge. In D. C. Berliner (Ed.), *Handbook of education* (pp. 709–725). New York: Macmillan.

Castro, C. S. (1998). Teaching probability for conceptual change. *Educational Studies in Mathematics, 35*, 233–254.

Chick, H. L., & Pierce, R. U. (2008). Teaching statistics at the primary school level: Beliefs, affordances, and pedagogical content knowledge. In C. Batanero, G. Burrill, C. Reading, & A. Rossman (2008).

Eichler, A. (2007). Individual curricula – teachers' beliefs concerning stochastics instruction. *International Electronical Journal of Mathematics Education 2*(3). Online: http://www.iejme.com/

Eichler, A. (2008a). Teachers' classroom practice and students' learning. In C. Batanero, G. Burrill, C. Reading, & A. Rossman (2008).

Eichler, A. (2008b). Statistics teaching in German secondary high schools. In C. Batanero, G. Burrill, C. Reading, & A. Rossman (2008).

Eichler, A. (2009). Teachers' teaching of stochastics and students' learning. In M. Tzekaki, M. Kaldrimidou, & H. Sakonidis (Eds.), *Proceedings of the 33th Conference of the International Group for the Psychology of Mathematics Education* (Vol. 2, pp. 385–392). Thessaloniki, Greece: Psychology of Mathematics Education.

Franke, M. L., Kazemi, E., & Battey, D. (2007). Understanding teaching and classroom practice in mathematics. In F. K. Lester (Ed.), *Second handbook of research on mathematics teaching and learning* (pp. 225–256). Charlotte, NC: Information Age Publishing.

Furinghetti, F., & Pehkonen, E. (2002). Rethinking characterizations of beliefs. In G. C. Leder, E. Pehkonen, & G. Törner (Eds.), *Beliefs: A hidden variable in mathematics education?* (pp. 39–58). Dordrecht: Kluwer.

Gattuso, L., & Pannone, M. A. (2002). Teachers' training in a statistics teaching experiment. In B. Phillips (Ed.), *Proceedings of the Sixth International Conference on Teaching Statistics.* Cape Town, South Africa: International Statistical Institute and International Association for Statistics Education. Online: www.stat.auckland.ac.nz/~iase/publications

Grigutsch, S., Raatz, U., & Törner, G. (1998). Einstellungen gegenüber Mathematik bei Mathematiklehrern (Teachers' beliefs concerning mathematics). *Journal für Mathematikdidaktik, 19*(1), 3–45.

Hiebert, J., & Grouws, D. A. (2007). The effect of classroom mathematics teaching on students' learning. In F. K. Lester (Ed.), *Second handbook of research on mathematics teaching and learning* (pp. 371–404). Charlotte, NC: Information Age Publishing.

Hofer, M. (1986). *Sozialpsychologie des erzieherischen Handelns (Social psychology of educational action).* Göttingen, Germany: Hogrefe.

Jones, G. A., Langrall, C. W., & Mooney, E. S. (2007). Research in probability. In F. K. Lester (Ed.), *Second handbook of research on mathematics teaching and learning* (pp. 909–956). Charlotte, NC: Information Age Publishing.

Pajares, F. M. (1992). Teachers' beliefs and educational research: Cleaning up a messy construct. *Review of Educational Research, 62*(3), 307–332.

Paparistodemou, E., Potari, D., & Pitta, D. (2006). Prospective teachers' awareness of young children's stochastic activities. In A. Rossman & B. Chance (Eds.), *Proceedings of the Seventh International Conference on Teaching Statistics.* Salvador, Bahia, Brazil: International Statistical Institute and International Association for Statistical Education. Online: www.stat.auckland.ac.nz/~iase/publications

Pfannkuch, M. (2006). Comparing box plot distributions: A teacher's reasoning. *Statistics Education Research Journal, 5*(2), 27–45. Online: www.stat.auckland.ac.nz/~iase/publications

Pfannkuch, M., & Horring, M. (2005). Developing statistical thinking in a secondary school: A collaborative curriculum development. In G. Burrill & M. Camden (Eds.), *Curricular Development in Statistics Education International Association for Statistical Education 2004 Roundtable* (pp. 163–173). Voorburg, The Netherlands: International Statistical Institute.

Philipp, R. A. (2007). Mathematics teachers' beliefs and affect. In F. K. Lester (Ed.), *Second handbook of research on mathematics teaching and learning* (pp. 257–315). Charlotte, NC: Information Age Publishing.

Putnam, R. T., & Borko, H. (2000). What do new views of knowledge and thinking have to say about research on teacher learning? *Educational Researcher, 29*(1), 4–15.

Shaughnessy, M. (2007). Research on statistics learning and reasoning. In F. K. Lester (Ed.), *Second handbook of research on mathematics teaching and learning* (pp. 957–1010). Charlotte, NC: Information Age Publishing.

Staub, F., & Stern, E. (2002). The nature of teacher's pedagogical content beliefs matters for students' achievement gains: Quasi-experimental evidence from elementary mathematics. *Journal of Educational Psychology, 94*(2), 344–355.

Stein, M. K., Remillard, J., & Smith, M. S. (2007). How curriculum influences student learning. In F. K. Lester (Ed.), *Second handbook of research on mathematics teaching and learning* (pp. 319–369). Charlotte, NC: Information Age Publishing.

Thompson, A. G. (1992). Teachers' beliefs and conceptions: A synthesis of the research. In D. A. Grouws (Ed.), *Handbook of research on mathematics teaching and learning* (pp. 127–146). New York: Macmillan.

Watson, J. (2001). Profiling teachers' competence and confidence to teach particular mathematics topics: The case of chance and data. *Journal of Mathematics Teacher Education, 4*, 305–337.

Wild, C., & Pfannkuch, M. (1999). Statistical thinking in empirical enquiry. *International Statistical Review, 67*(3), 223–248.

Chapter 20
Teachers' Graphical Competence

M. Teresa González, M. Candelaria Espinel, and Janet Ainley

Abstract Statistical graphs have an important role in our society because they are present in many fields of life. Competence in understanding and working with graphs is therefore a key feature of statistical literacy. In this chapter, what is meant by *graphical competence* is analysed, based on discussion of the findings of a range of research projects. Then research on teachers' content, and pedagogical content knowledge about graphs is summarised. The chapter concludes with some pedagogical implications for teacher education and some recommendations for future research.

1 Introduction

Graphs originated as tools to present data visually in ways that are easy to understand and to analyse. The first use of graphical representation is generally credited to the Scottish engineer William Playfair who, in 1789, designed the first bar charts to help keep track of his business accounts (Wainer & Spence, 2005). In 1865, Florence Nightingale pioneered the use of statistical graphs showing the causes of mortality during the Crimean war as a means of drawing the attention of the authorities to the conditions in her hospital. She is credited with inventing the pie chart, another example of how graphical representations were developed

M.T. González (✉)
Department of Didactics of Mathematics and Didactics of Experimental Sciences,
University of Salamanca, Salamanca, Spain
e-mail: maite@usal.es

M.C. Espinel
Department of Mathematical Analysis, University of La Laguna, La Laguna, Spain
e-mail: mespinel@ull.es

J. Ainley
School of Education, University of Leicester, Leicester, United Kingdom
e-mail: janet.ainley@le.ac.uk

C. Batanero, G. Burrill, and C. Reading (eds.), *Teaching Statistics in School Mathematics-Challenges for Teaching and Teacher Education: A Joint ICMI/IASE Study*, DOI 10.1007/978-94-007-1131-0_20, © Springer Science+Business Media B.V. 2011

and used to identify and communicate important messages about specific contexts. Other statistical representations have been invented more recently, for example, steam and leaf plots and box-plots were introduced by John Tukey in 1976 as part of his development of Exploratory Data Analysis (EDA) (Tufte, 2001).

Graphs are used primarily in two ways: to communicate information and as tools to analyse data. The majority of graphs that appear in the media are used to communicate information and contain a statistical summary of the original data. These representations are mainly line and bar graphs (or their variants), pie charts and pictograms, and generally show data in the form of percentages or proportions. By contrast, in *professional* activity, graphs are both part of the language of communication in many professions, and tools by means of which processes, relationships, and numerical results are presented and analysed to allow an appreciation of patterns that are hidden in the data. *Data Visualisation* is an approach to data analysis where the goal is to reveal some aspects of the data that might not be perceived, appreciated, or absorbed by other means. Graphs may be seen as essential for the exploration, analysis, and presentation of numerical data and are tools for transnumeration, a basic component in statistical reasoning (Wild & Pfannkuch, 1999).

Instruction about statistical graphs features prominently as one of the objectives of statistics curricula in many countries. In primary and secondary education the range of graphical representations introduced to students generally includes histograms, bar charts, pie charts, scatter graphs, stem and leaf diagrams, and frequency polygons (Shaughnessy, 2007). However, didactical research suggests that the emphasis of teaching is often put on the construction of such graphs, with little attention to their interpretation. The argument proposed in this chapter is that there is a need for pedagogical approaches that enable students to develop the *graphical competence* necessary to use statistical graphs effectively.

The aim of this chapter is to explore what might be meant by graphical competence, to review existing research on the graphical competence of teachers, and to suggest implications for teacher education and for future research.

2   Graphical Competence

A range of research focuses on the knowledge students have about graphs, the levels of graphical understanding they display, and the difficulties they demonstrate (Curcio, 1989; Wainer, 1992; Friel, Curcio, & Bright, 2001; Aoyama, 2007). Drawing on these studies, researchers have offered different analyses of what is involved in the competent use of graphs. Friel et al. (2001) summarised previous research concerning students reading and interpreting graphs to produce a definition of *graph comprehension* as the readers' ability to derive meaning from graphs created by others or by themselves. They described the development of graph comprehension as gradual, through the repeated construction and use of a variety of graphs in problem contexts that require the learner to make sense of data.

To analyse students' graph comprehension researchers have characterised different levels in the critical understanding of graphs that vary from a complete inability to make sense of the graph, through reading isolated elements or being able to compare elements, to the ability to predict or extrapolate data that are not included in the graph. One of the most widely used models identifies four stages (Curcio, 1989; Friel et al., 2001), which might be seen as both a hierarchy of competence, and a framework for posing questions in a pedagogical context. The four stages are:

- *Reading the data*: focused on extracting data from the graph;
- *Reading between the data*: characterised by finding relationships between data;
- *Reading beyond the data*: requiring extrapolation and identification of relationships in order to make predictions or generalisations; and
- *Reading behind the data*: looking for possible causes of variation and relationships among variables in the data.

In addition to reading graphs, creating and interpreting statistical graphs are essential elements in the acquisition of statistical literacy, which Gal (2002) describes as the union of two related competences: (a) interpreting and critically evaluating statistically based information from a wide range of sources; and (b) formulating and communicating a reasoned opinion on such information. Statistical literacy also involves realising that different graphs allow different views of the data. Thus the choice of an appropriate graph in relation to both the situation and the data to be represented is crucial. In addition it is important to take into account the possible biases that may be voluntarily or involuntarily transferred to the graph, especially when choosing the scale. Wu (2004), drawing on a study involving 907 13–15 year-old students, summarises these competences in the form of four essential skills:

- *Reading graphs*: to extract data directly from one or more graphs and to generate information by calculating with data explicitly shown in one or more graphs.
- *Interpreting graphs*: to formulate opinions about one or more graphs.
- *Building graphs*: to present and edit data in graphic form.
- *Evaluating graphs*: to evaluate the accuracy and effectiveness of a graph.

Although these competences are defined separately, they are related to each other. There has been an increasing interest amongst researchers in the need for a critical ability when reading graphs, both in their increasingly frequent appearance in news and advertising media, and in professional contexts (Monteiro & Ainley, 2007). Analysing deeper aspects of students' graphical interpretation, Aoyama (2007) has established the following hierarchy of graph comprehension based on the statistical literacy framework, constructed by Watson and Callingham (2003) that describes what a student can or cannot do in each level:

1. *Idiosyncratic level*: Students cannot correctly read the graph or see tendencies.
2. *Basic graph reading level*: Students can read values on graphs and see tendencies, but they cannot use the graph features to explain the contextual meaning.

3. *Rational/literal level*: Students can read particular values on graphs and see tendencies; they can explain contextual meanings literally using the graph features, but they cannot suggest any alternative interpretations.
4. *Critical level*: Students can read graphs, understand the context, and evaluate the information reliability, but they are unable to suggest alternative hypotheses.
5. *Hypothesising and modelling level*: Students can read graphs, accept and evaluate the information, and can suggest their own explanatory hypotheses or models.

As a summary of these attempts to identify the elements that contribute to statistical literacy in relation to graphing, the following definition of graphical *competence* as the union of three different capacities will be used:

• The ability to extract data from different sorts of graphs and to interpret meanings from them by reading between, beyond, and behind the data displayed to form hypotheses about the phenomena represented in the graph;
• The capacity to select and create appropriate graphs for specific situations, with or without the support of technology; and
• The ability to critically evaluate graphs and to distinguish the relative strengths and limitations of particular graphical representations, recognising that creating a graph involves an interpretation of the original data.

If the teaching of statistics at school level is to enable students to develop *graphical competence,* then it is clear that the graphical competence of teachers, and their understanding of the pedagogy relating to this, is important. In the following section research in this field is described.

3   Research on Teachers' Graphical Competence

Relatively few studies focus on teachers' knowledge and conceptions about statistical graphs, and most of these are related to pre-service teachers. The limited study of teachers' knowledge was recognised by Batanero, Garfield, Ottaviani, and Truran (2001), who emphasised the need for teachers to develop specific knowledge about teaching statistics, in the form of pedagogical content knowledge (PCK) as defined by Shulman (1986). (See related chapters by Godino et al., and Callingham & Watson, in this book.)

Research dealing with two basic components of professional knowledge (Shulman, 1986), teachers' statistical knowledge and teachers' pedagogical content knowledge, is summarised below. The first section is focussed on teachers' graphical competence, that is, teachers' knowledge about statistical graphs and the difficulties they have in reading, constructing, interpreting, and evaluating graphs. The next section is related to teachers' pedagogical content knowledge, including how graphing may be taught and the conceptions teachers, and of course learners, may have about this topic.

3.1   Teachers' Statistical Knowledge

Research focused on statistical content knowledge has shown that, in general, pre-service teachers have a low level of statistical knowledge and that specifically their graphical competence is limited.

Bruno and Espinel (2009), in a study conducted with 29 pre-service primary teachers in Spain, found that they experienced a large number of conceptual and procedural difficulties with the construction of histograms and frequency polygons. These included separating histogram rectangles, inadequate labelling of real numbers on the axes, not considering zero frequency intervals or not completing the frequency polygon. Moreover, when the same pre-service teachers had to assess the graphs constructed by other students and identify the mistakes made, the errors they had made in constructing their own graphs became evident. In another study by Espinel, Bruno, and Plasencia (2008) with 190 pre-service primary teachers, the authors found that the teachers had great difficulties with the interpretation of statistical graphs. Participants did not take into account the distribution as a whole, focused instead on some specific aspects such as the average or the outliers and were unable to associate descriptions of different variables with the appropriate distribution graph.

Monteiro and Ainley (2006, 2007) analysed the competence of 218 pre-service primary teachers from Brazil and England when interpreting statistical graphs published in print media. They used a questionnaire in which these teachers had to ask questions about the graphs presented and give their opinion about the messages the graph was conveying. In this study the authors found evidence of participants displaying elements of critical sense in the interpretation of graphs, that is, "mobilising and balancing statistical skills, with contextual knowledge and experience" in order to think critically about various aspects of the data represented in the graphs (Ainley & Monteiro, 2008, p. 1). In their interpretation of the media graphs pre-service teachers drew not only on technical knowledge about graphs but also on other resources such as opinions or feelings about the data, as well as knowledge about the context. While this study suggests a higher level of graphical competence than that found by Espinel et al. (2008), the statistical knowledge demanded by the graphs used in the study (essentially bar graphs) was relatively low.

To explore the statistical and mathematical knowledge of 30 pre-service primary teachers in New Zealand, Burgess (2002) analysed their reports about a multivariate data set. The different elements used by the teachers were first tabulated and then searched to see if there was any relationship between these factors and the statements made in their reports. Some of the teachers made graphs in their reports but appeared to do so as an end point in the work, without knowing when and why a particular kind of graph should be used instead of some other kind. Many participants in this study were not able to make generalisations about the data. These teachers seemed to think it was sufficient to construct a graph in order for the task to be completed. It appears that they had learned statistical content as isolated skills and therefore were not able to integrate this knowledge with the problem context. They showed a low level of graphical competence when applying these skills in the research process.

This difficulty in seeing graphs as tools for establishing conclusions has been also shown in several research studies. Batanero, Arteaga, and Ruiz (2010) conducted a study with 93 Spanish pre-service primary teachers based on a statistical project in which they had to consider the results of a coin that is thrown 20 times, perform the experiment, and then select the data to compare the results of simulated and real sequences. The students were provided with a sheet of the results from the whole group, and they had to make a report on the group's intuitions about randomness. The researchers classified the graphs students included in their reports into four groups according to their semiotic complexity. For each of these four groups the students' ability to interpret the graphs to reach some conclusion about the research questions was assessed, and misunderstandings evidenced by the selection and construction of the graphs were analysed. In general, the study found that although the pre-service teachers could interpret some graphs correctly, the difficulty of the interpretation increased according to the semiotic complexity. Only one third of the participants were able to reach a conclusion regarding the research question.

3.2   Teachers' Pedagogical Content Knowledge

There has been very little research about teachers' conceptions of statistical graphs and the role these conceptions play in instruction. Rouan (2002), in a study about teachers' conceptions, gave a questionnaire to 221 Moroccan secondary mathematics teachers and conducted 15 interviews in order to establish their conceptions concerning the objectives of teaching statistical graphs and the way they used them in class. Conceptions about the roles of statistical graphs held by the teachers were grouped in the following ways: a *formal conception* related to the calculation of statistical parameters but with little attention to the context, a view that did not stimulate statistical reasoning; a *synthetic conception*, that considered the graph as a summary of the data but did not take account of any loss of information; a *predictive conception* that emphasised the inferential role of a graph at the expense of seeing it as a way to simplify and present data; and a *visual static conception* that included both descriptive and inferential aspects but did not take account of loss of information, or of statistical reasoning.

The research also explored the participants' conceptions related to the reading of statistical graphs, which included the verbal translation of data represented in a diagram, extracting information from the graph, responses that showed confusion between reading and interpretation and conceptions about the interpretation of statistical graphs. These were classified into three categories: a *predictive conception* that was related to inferential operations, a *descriptive conception* that only described aspects of the data and a *stochastic conception* that included descriptive and inferential operations and were largely demonstrated by participants who had taken more advanced courses in statistics. Rouan concludes that the conceptions held by the teachers indicate weaknesses in their understanding of both statistical content and of the importance of statistical reasoning.

González and Pinto (2008) explored the pedagogical content knowledge and conceptions of statistical graphs held by four pre-service secondary mathematics teachers. In this study, the subjects had to classify 20 problems related to statistical graphs taken from different secondary textbooks. Problems were selected according to two criteria: the type of graphs (histograms, pie charts, bar charts, stem and leaf diagrams and frequency polygons) and the different levels of statistical thinking involved (reading the data, reading between the data, reading beyond the data and reading behind the data) (Friel et al., 2001). With regard to content knowledge, the authors found that the pre-service teachers did not recognise some of the graphs, for example, the stem and leaf graphs. In relation to the teaching of the topic, participants felt that the construction and interpretation of graphs was a very simple task. One of them did not consider statistical graphs as proper statistical knowledge. Indeed no participant knew anything about the process of teaching about statistical graphs or the difficulties faced by pupils in relation to statistics and graphs, and they were unable to recognise different levels of graphical understanding.

The common picture emerging from these studies is that many teachers, and pre-service teachers, even at secondary level, may themselves lack graphical competence. A possible explanation is that this is a consequence of inadequate learning opportunities in the teachers' own schooling. Unfortunately these studies suggest that teachers are not well prepared to teach their own students in more appropriate ways. Below we include some reflections about how teachers' current levels of competence and confidence in teaching about statistical graphs may be improved.

4   Implications for Teacher Education

There is a wide range of research describing models of pedagogical content knowledge for mathematical knowledge (see, for example, models described in chapters by Godino, Ortiz, Roa, & Wilhelmi and by Watson, in this book). Graeber and Tirosh (2008) suggested that teacher education must take into account at least three basic components:

- Content knowledge of the subject to be taught (both ordinary and specialised knowledge);
- Knowledge of the students' learning processes, attitudes, strategies, etc.;
- Knowledge of curricula, instructional materials, and approaches.

The following recommendations are made mindful of the fact that the time available to focus on graphical competence in both pre-service and in-service teacher education will be limited, and that in many contexts statistics will be covered within mathematics, rather than as a separate topic.

The *content knowledge of the subject* obviously includes a sound technical knowledge of how different kinds of statistical graphs are constructed and a sense of how this knowledge links to other aspects of the mathematics curriculum, for example, graphs of functions in algebra or the use of scales in measuring instruments. In addition, some epistemological perspective on how graphs have developed historically and how they are used in a range of professional contexts may also be very valuable for teachers.

A common epistemological assumption in mathematics education is that mathematical objects emerge as part of the solution to problems. Looking at situations in which statistical graphs are used (in areas such as Biology, Medicine, Administration, Economics, Geography, or Psychology) could provide relevant teaching resources to learn about the nature of these graphs and to realise that technical knowledge alone is not sufficient to enable a real interpretation of the data represented.

Graphs taken from media sources may also prove a valuable resource for focusing teachers' awareness on the role and uses of graphs in society and the ways in which technical knowledge must be used in combination with contextual knowledge and experience in reading graphs (Monteiro & Ainley, 2007). Examples of media graphs that are poorly presented, or technically inaccurate or incomplete, may also be a useful resource for drawing attention both to aspects of statistical knowledge and to the ways in which graphs can be used to manipulate the presentation of data (Watson, 1997).

The knowledge of students' learning processes refers to both students' understanding of graphs and the difficulties, errors and obstacles that prevent them from using graphs effectively. There is now a considerable body of research that draws attention to a variety of common difficulties learners may have in either reading graphs (e.g., Shaughnessy, 2007; Bruno & Espinel, 2009) or when constructing their own graphs (de Corte, 1996; Shaughnessy, 2007). Teachers need to be aware of the key findings of this research and also to understand that there are various ideas that relate to graphical representation, such as scale, origin, axes, variable, independence, dependence, coordinates, discrete and continuous quantities, which often involve their own difficulties. For example, in relation to scales, many children are capable of reading a scale, but the process of constructing scales is difficult; children may have difficulty choosing an appropriate scale for graphing a particular data set (Friel et al., 2001, p. 141).

Knowledge of instructional tools focuses on the identification of good teaching examples, appropriate instructional approaches, including the use of technology, and the ability to analyse textbooks and curriculum documents that would be relevant to teaching graphs. It is also important to offer teachers frameworks, developed through research, that suggest ways in which learning about different types of graphs and the kinds of questions that might be asked about them might be sequenced (Friel et al., 2001). For example, Kramarski (2004) recommended the design of learning environments in which metacognitive instruction is embedded with a diagnostic teaching approach centred on students' alternative conceptions, so these might be addressed to facilitate graphical competence.

For many pre-service teachers, their own school experiences may have consisted mainly of textbook exercises related to the construction and reading of decontextualized graphs. Therefore, understanding the use of graphs as part of a statistical problem-solving cycle and project work that involves the five dimensions of Wild and Pfannkuch's (1999) model of statistical thinking may be new to them. Ainley and Pratt's work in developing a pedagogic approach known as *Active Graphing* offers a model for developing problem-solving projects in which graphs are used as analytic tools and demonstrates their effectiveness in supporting the development of graphical competence in young children (Ainley, Nardi, & Pratt, 2000; Ainley, 2001; Ainley, Pratt, & Nardi, 2001).

Active Graphing draws on the power of software to store and present data, and in doing so, challenges some previously accepted views about the relative difficulty of some types of graphs: graphs that may be complex to draw by hand turn out to be much less difficult for children to interpret in meaningful contexts. Teacher education in this field needs to include an element of "technology pedagogical content knowledge" (TPCK) (Lee & Hollebrands, 2008, this book) to enable teachers to incorporate the use of technology in their classrooms (Pfannkuch, 2008). It is important that teachers are introduced to the possibilities of a range of resources, including generic software such as spreadsheets and environments developed specifically for the exploration of data such as *TinkerPlots* or *Fathom,* and have opportunities to use and evaluate them (see Lee & Hollebrands, this book). Data sets available on the Internet are powerful resources from which to develop meaningful work in the classroom that allows the development of argumentation, understanding of different representations and the transformation from one representation to another (Carrión & Espinel, 2006; Hall, this book).

5   Implications for Research

Research discussed in this chapter highlights the limitations of the graphical competence of many teachers. There is now a need of further research in order to have a better understanding of the support teachers may need to teach graphical competence effectively in their classrooms in ways that are appropriate for the curricula of the twenty-first century. Such research might address how teachers transform their knowledge about graphs into knowledge to be taught, the strategies they use, how they select appropriate examples and different classroom situations. The rapid development of educational technology is changing what is possible; the traditional sequencing of graphs is being questioned, and attention is being focused on interpretation and developing the critical use of graphs rather than their construction. Research is also needed that explores the implications of these changes for teacher education in relation to content knowledge of the subject to be taught, students' learning processes and instructional approaches.

References

Ainley, J. (2001). Transparency in graphs and graphing tasks: An iterative design process. *Journal of Mathematical Behavior, 19*, 365–384.

Ainley, J., & Monteiro, C. (2008). Comparing curricular approaches for statistics in primary school in England and Brasil: A focus on graphing. In C. Batanero, G. Burrill, C. Reading, & A. Rossman (2008).

Ainley, J., Nardi, E., & Pratt, D. (2000). The construction of meanings for trend in active graphing. *International Journal of Computers for Mathematical Learning, 5*(2), 85–114.

Ainley, J., Pratt, D., & Nardi, E. (2001). Normalising: Children's activity to construct meanings for trend. *Educational Studies in Mathematics, 45*, 131–146.

Aoyama, K. (2007). Investigating a hierarchy of students' interpretations of graphs. *International Electronic Journal of Mathematics Education, 2*(3). Online: www.iejme.com/

Batanero, C., Arteaga, P., & Ruiz, B. (2010). Statistical graphs produced by prospective teachers in comparing two distributions. In V. Durand-Guerrier, S. Soury-Lavergne, & F. Arzarello (Eds.), *Proceedings of the Sixth Congress of the European Society for Research in Mathematics Education*. Lyon: ERME. Online: www.inrp.fr/editions/editions-electroniques/cerme6/

Batanero, C., Burrill, G., Reading, C., & Rossman, A. (Eds.). (2008). *Joint ICMI/IASE Study: Teaching Statistics in School Mathematics. Challenges for Teaching and Teacher Education. Proceedings of the ICMI Study 18 and 2008 IASE Round Table Conference*. Monterrey, Mexico: International Commission on Mathematical Instruction and International Association for Statistical Education. Online: www.stat.auckland.ac.nz/~iase/publications

Batanero, C., Garfield, J., Ottaviani, M., & Truran, J. (2001). Building a research agenda for statistics education. *Statistical Education Research Newsletter, 1*(2), 2–6. Online: www.stat.auckland.ac.nz/serj/

Bruno, A., & Espinel, M. C. (2009). Construction and evaluation of histograms in teacher training. *International Journal of Mathematical Education in Science and Technology, 40*(4), 473–493.

Burgess, T. (2002). Investigating the "data sense" of preservice teachers. In B. Phillips (Ed.), *Proceedings of the Sixth International Conference on Teaching Statistics*. Cape Town, South Africa: International Statistical Institute and International Association for Statistics Education. Online: www.stat.auckland.ac.nz/~iase/publications

Carrión, J. C., & Espinel, M. C. (2006). An investigation about translation and interpretation of statistical graphs and tables by students of primary education. In A. Rossman & B. Chance (Eds.), *Proceedings of the Seventh International Conference on Teaching Statistics*. Salvador, Bahia, Brazil: International Statistical Institute and International Association for Statistical Education. Online: www.stat.auckland.ac.nz/~iase/publications

Curcio, F. R. (1989). *Developing graph comprehension*. Reston, VA: National Council of Teachers of Mathematics.

De Corte, E. (1996). Instructional psychology: Overview. In E. De Corte & F. E. Weinert (Eds.), *International encyclopedia of development and instructional psychology* (pp. 33–43). Oxford, UK: Elsevier.

Espinel, M. C., Bruno, A., & Plasencia, I. (2008). Statistical graphs in the training of teachers. In C. Batanero, G. Burrill, C. Reading, & A. Rossman (2008).

Friel, S., Curcio, F., & Bright, G. (2001). Making sense of graphs: Critical factors influencing comprehension and instructional implications. *Journal for Research in Mathematics Education, 32*(2), 124–158.

Gal, I. (2002). Adults' statistical literacy: Meanings, components, responsibilities. *International Statistical Review, 70*, 1–25.

González, M. T., & Pinto, J. (2008). Conceptions of four preservice teachers on graphical representation. In C. Batanero, G. Burrill, C. Reading, & A. Rossman (2008).

Graeber, A., & Tirosh, B. (2008). Pedagogical content knowledge: Useful concept or elusive notion. In P. Sullivan & T. Wood (Eds.), *The international handbook of mathematics teacher education* (Vol. 1, pp. 117–132). Rotterdam: Sense Publishers.

Kramarski, B. (2004). Making sense of graphs: Does metacognitive instruction make a difference on students' mathematical conceptions and alternative conceptions? *Learning and Instruction, 15*(6), 593–619.

Lee, H. S., & Hollebrands, K. F. (2008). Preparing to teach data analysis and probability with technology. In C. Batanero, G. Burrill, C. Reading, & A. Rossman (2008).

Monteiro, C., & Ainley, J. (2006). Student teachers interpreting media graphs. In A. Rossman & B. Chance (Eds.), *Proceedings of the Seventh International Conference on Teaching Statistics.* Salvador, Bahia, Brazil: International Statistical Institute and International Association for Statistical Education. Online: www.stat.auckland.ac.nz/~iase/publications

Monteiro, C., & Ainley, J. (2007). Investigating the interpretation of media graphs among student teachers. *International Electronic Journal of Mathematics Education, 2*(3), 188–207. Online: www.iejme/

Pfannkuch, M. (2008). Training teachers to develop statistical thinking. In C. Batanero, G. Burrill, C. Reading, & A. Rossman (2008).

Rouan, O. (2002). Secondary school math teachers' conceptions of the statistical graphics' functions, reading and interpretation. In B. Phillips (Ed.), *Proceedings of the Sixth International Conference on Teaching Statistics.* Cape Town, South Africa: International Association for Statistics Education. On line: www.stat.auckland.ac.nz/~iase/publications

Shaughnessy, J. M. (2007). Research on statistics learning and reasoning. In F. Lester (Ed.), *Second handbook of research on mathematics teaching and learning* (pp. 957–1049). Greenwich, CT: National Council of Teachers of Mathematics.

Shulman, L. S. (1986). Those who understand: Knowledge growth in teaching. *Educational Research, 15*(2), 4–14.

Tufte, E. R. (2001). *The visual display of quantitative information.* Cheshire, CT: Graphics Press.

Wainer, H. (1992). Understanding graphs and tables. *Educational Researcher, 21*(1), 14–23.

Wainer, H., & Spence, I. (2005). Graphical presentation of longitudinal data. In B. Everitt & D. C. D. Howell (Eds.), *Encyclopedia of statistics in behavioral science* (pp. 762–772). New York: Wiley.

Watson, J. (1997). Assessing statistical literacy through the use of media surveys. In I. Gal & J. Garfield (Eds.), *The assessment challenge in statistics* (pp. 107–121). Amsterdam: IOS Press.

Watson, J., & Callingham, R. (2003). Statistical literacy: A complex hierarchical construct. *Statistics Educational Research Journal, 2*(2), 3–46. Online: www.stat.auckland.ac.nz/serj/

Wild, C., & Pfannkuch, M. (1999). Statistical thinking in empirical enquiry. *International Statistical Review, 67*(3), 223–265.

Wu, Y. (2004, July). *Singapore secondary school students' understanding of statistical graphs.* Paper presented at the Tenth International Congress on Mathematics Education (ICME-10), Copenhagen, Denmark. Online: www.stat.auckland.ac.nz/~iase/publications

Chapter 21
Teachers' Understanding of Averages

Tim Jacobbe and Carolina Carvalho

Abstract The concept of average has been a part of the curriculum for well over 100 years. Consequently, research on students' understanding of average has been one of widest areas explored in mathematics and statistics education research; however, research focused on teachers is still very scarce. In this chapter, research related to school students' understanding is first described and then research dealing with teachers' understanding or professional knowledge is analysed. Some final implications for research on the training of teachers are provided.

1 Introduction

Even before the recent international movements to increase the emphasis on statistics in the curriculum, the concepts of mean, median, and mode have been a part of the curriculum for many years. One of the reasons for this may be that measures of centre are part of everyday life, as they appear in the media and assessment data being reported. In contrast to many other statistical concepts, averages are generally reported as a singular value and thus the topic fits the traditional view in mathematics that there is only one right answer to a particular problem.

Since it has been a part of the curriculum for a long time, research on understanding averages has been one of the most abundant areas in the statistics education research community. Most of these studies have centred on students' understanding of averages and only a limited number of studies have focused on teachers' understanding. Since these concepts are just beginning to appear at a

T. Jacobbe (✉)
University of Florida, 2403 Norman Hall, P.O. Box 117048, Gainesville, FL 32611, USA
e-mail: jacobbe@coe.ufl.edu

C. Carvalho
Universidade de Lisboa, Instituto de Educação, Alameda da Universidade,
1649-013, Lisbon, Portugal
e-mail: cfcarvalho@ie.ul.pt

C. Batanero, G. Burrill, and C. Reading (eds.), *Teaching Statistics in School Mathematics-Challenges for Teaching and Teacher Education: A Joint ICMI/IASE Study*, DOI 10.1007/978-94-007-1131-0_21, © Springer Science+Business Media B.V. 2011

more sophisticated level in the school curriculum and do not appear to be addressed during teacher preparation programmes, an argument can be made that teachers' understanding is not very different than that of students. In fact, several of the studies concerning teachers' understanding of averages have revealed that teachers lack conceptual knowledge of the topic.

In this chapter the difficulties revealed regarding students' understanding of average will be discussed first, followed by a discussion of research related to teachers' understanding. Implications for future research in the context of teacher training will then be presented at the conclusion of the chapter.

2   Students' Understanding

The computational simplicity of averages, coupled with a systematic devaluation of the context of many teaching situations may give students and teachers the illusion that a set of skills is all that is necessary to understand the topic. However, the analysis of the performance in different research described below reveals that the nature of the mental processes underlying the construction of these concepts is wider than just the learning of algorithms. Since research on students is abundant, we only discuss studies that later relate to research focused on teachers that deal with three different themes: (a) students' procedural and conceptual understanding (Russell & Mokros, 1991; Mokros & Russell, 1995; Cai & Moyer, 1995; Gattuso & Mary, 1996; Carvalho, 2001; Garcia & Garret, 2006); and (b) defining levels of cognitive development for the concept of average (Strauss & Bichler, 1988; Watson & Moritz, 1999, 2000). A wider survey of students' understanding of averages is found in Shaughnessy (2007).

2.1   *Procedural and Conceptual Understanding*

Russell and Mokros (1991) (see also Mokros & Russell, 1995) analysed understanding of average in 25 students (10–12 years old). All participants were asked seven open-ended questions that involved both construction and interpretation problems. Construction problems involved participants constructing a set of data that may have a particular measure of centre. Interpretation problems involved participants describing what information a particular measure of centre gives, or what can be thought of as "typical". In their analysis, Russell and Mokros determined that students exhibited four approaches to understanding average: (a) average as modal, (b) average as what's reasonable, (c) average as the midpoint, and (d) average as an algorithmic relationship. Garcia and Garret (2006) confirmed these findings in 94 students (17 years old).

In a study involving 250 students (11–12 years old) in the United States of America, Cai and Moyer (1995) asked participants to examine the conceptual and computational aspects of the algorithm for arithmetic mean. While the majority of the subjects recognised the right algorithm to compute an arithmetic mean, only half of them could apply the concept in order to solve an open-ended problem. For a significant number, the strategy chosen to solve the problem was the direct application of the algorithm.

Gattuso and Mary (1996) discussed the possibility that students' difficulties may be attributed to the mathematical meaning they attach to the data distribution, because to solve tasks related to comparing distributions students are generally asked to apply an algorithm. In order to check this conjecture, Carvalho (2001) conducted a study where 533 students (11–15 years old) were asked to solve tasks related to average where data were provided in tabular and graphical form. The difficulty was higher in tasks given in graphical format, because students were not able to effectively read and interpret the graph. The most likely reason was the fact that the values necessary to solve the task were not readily apparent.

2.2   Cognitive Development

Strauss and Bichler (1988) conducted a study involving 20 students between the ages of 8 and 14, where participants were asked to solve tasks involving some properties of the arithmetic mean. Students understood that the mean was located between extremes, and that particular data values can influence the mean. However, more complex properties like minimising deviations, or that a zero data value must also be included and accounted for when computing the mean, were very difficult for the students. The authors concluded that the understanding of students increases with age, which provides evidence to support the presence of levels of development for different properties of the concept of average.

Watson and Moritz (1999, 2000) further described the longitudinal development of students' understanding of average across various levels. The authors suggested that the mean, median, and mode are complex and these concepts should be developed in classroom activities, during several school years. By introducing students to contextual situations in a variety of grade levels, students could consider the differences between these measures, depending on the nature and context of the data. These ideas are in agreement with that included in the GAISE framework (Franklin et al., 2007) where development levels of understanding statistical concepts are not associated only with age, but rather with experience.

In their attempt to characterise students' development across various levels, Watson and Moritz (2000) interviewed 137 students, who ranged in age from 8 (grade 3) to 15 (grade 9). Responses were categorised using a neo-Piagetian model of cognitive development that classified the responses in a hierarchical manner. This classification scheme resulted in six levels of development:

1. At the *pre-structural* level the students did not use a term for average, even in a colloquial sense.
2. At the *uni-structural* level the students often used colloquial terms for average, such as normal.

3. The *multi-structural* level included students using two or more ideas, such as most, middle, and add-and-divide algorithm to describe average in straightforward situations.
4. At the *relational level*, students refer to add-and-divide algorithm for the mean to describe average in straightforward situations; however, they also relate these to ideas of the notion of most or middle. This level of development involves the students realising the representative nature of average (e.g., prediction, estimation, or representing whole data set).
5. *Application of average in one complex task.*
6. *Application of average in two complex tasks* (level 6). Levels 5 and 6 involve students' dealing with applications of average in multiple contexts.

The evidence that supports students' development of understanding of average highlights the importance of teachers' possessing conceptual understanding of average. It is expected that teachers are able to help students progress in their understanding of average from a pre-structural level in the early grades to a relational level that allows the application of the concept in complex tasks at the higher grades. The literature concerning students' understanding of average reveals their understanding is mainly procedural or algorithmic. This may indirectly shed light on the understanding of teachers, which is the focus of the next section.

3   Teachers' Understanding

Research specifically focused on teachers' understanding of average is very limited. Of the studies that focus on teachers' statistical content knowledge, the majority focus solely on the arithmetic mean (Batanero, Godino, & Navas, 1997; Gfeller, Niess, & Lederman, 1999). The remaining studies include understanding of the mean, median, and mode (Groth & Bergner, 2006; Jacobbe, 2007, 2008) or focus on the general concept of average (Russell & Mokros, 1991; Callingham, 1997; Begg & Edwards, 1999; Estrada, Batanero, & Fortuny, 2004; Leavy & O'Loughlin, 2006), that is, involve the realisation that measures of centre attempt to find a "typical" amount to summarise a particular data set. Only one study has analysed teachers' professional or didactic knowledge (Cai & Gorowara, 2002). Below we summarise each of these types of research.

3.1   Research Related to the Mean

Batanero et al. (1997) analysed the responses of 273 pre-service teachers in Spain to a questionnaire about the arithmetic mean. The results exhibited a lack of understanding of the algorithm for calculating the average (25% of the sample), participants' ignorance of the relationship between mean, median and mode and little or no understanding of the effect of outliers on the mean (34%).

As recommended in the Spanish curriculum for primary and secondary schools, a comparison of two distributions is one of the basic tasks in the exploratory data analysis that must be presented in teaching. However, participants in this study often had incorrect strategies to make this comparison, such as not taking into account spread or using only part of the data. This suggests that there is a need for pre-service teachers to be exposed to such experiences during preparation programmes.

Gfeller et al. (1999) conducted a study in the United States of America involving 13 mathematics and 6 science pre-service secondary teachers' views of the arithmetic mean and the representations they use when solving problems involving the mean. The instrument they used consisted of ten questions – seven involved concrete scenarios and three were set in an abstract setting. The participants' responses were classified according to whether they utilised a computational algorithm, a procedure based on a "fair share" understanding, or knowledge of variation to solve problems related to the mean. Results revealed no significant differences between science and mathematics pre-service teachers' use of the computational algorithm. However, there was a significant difference regarding the use of variation to create a balance point with mathematics pre-service teachers more likely to use such an approach. All participants in the study utilised a balance point approach on at least one of the problems and 17 of the 19 participants held multiple views of the mean. These results suggest that these pre-service teachers may hold a deeper conceptual knowledge of the mean than reported in other studies.

3.2 Research Related to the Mean, Median, and Mode

Groth and Bergner (2006) investigated 46 pre-service elementary and middle school teachers' understanding of the mean, median, and mode. Their report addressed responses to one question that asked teachers to explain how the statistical concepts of mean, median, and mode were different or similar. In order to obtain some more information regarding how teachers were defining "measures of central tendency", participants were asked the following question: Suppose that you used the phrase "measure of central tendency" while you were teaching, and then a student asks you what the phrase means. How would you respond?

Using the SOLO taxonomy from Biggs and Collis (1982), Groth and Bergner distinguished between pre-service teachers' understanding of mean, median, and mode in four categories: (a) unistructural/concrete symbolic, (b) multistructural/ concrete symbolic, (c) relational/concrete symbolic, and (d) extended abstract. There were eight pre-service teachers who fell in the unistructural/concrete symbolic level as their responses only involved the recitation of definitions in describing how the mean, median, and mode are different and similar. Twenty-one pre-service teachers were classified as exhibiting the multistructural/concrete symbolic level of thinking as their responses suggested that these measures of centre represented a mathematical object rather than simply a procedure. In other words, their responses presented the view that these calculations/findings provided some type of value; however, their responses were not extended enough to describe what value they actually provided. Thirteen teachers exhibited the relational/concrete symbolic level of thinking since their responses went beyond the procedure of finding the measurements to include that they represent what is "typical" about a particular set of data. Finally, three pre-service teachers were classified as possessing extended abstract levels of thinking, because their responses went beyond procedural knowledge and included a discussion of when one measure of centre would be more valuable or useful for a certain set of data. In other words, these responses included the fact that the shape of the distribution would impact whether or not one measure of centre would be more representative of the data set than another. As statistics educators we would like to see more teachers at the extended abstract level of thinking; however there may be some limitations to the design of these questions which prevented pre-service teachers from truly revealing their highest level of thinking. In particular, pre-service teachers were not specifically asked to identify situations where one measure of centre would be more useful than another.

Jacobbe (2007, 2008) conducted a case study of three elementary school teachers' understanding of average. Participants in this study were specifically asked to provide an example of when one measure of centre would be more representative for a given set of data than another. For this particular task, teachers were presented with three different distributions (one skewed to the left, one skewed to the right, and one normal) and asked to arrange the distributions according to which would have the smallest to the greatest mean, median, and mode. The findings of this study revealed that although some of the teachers had difficulty in applying the algorithm to various contexts, they were able to use the shape of a distribution (balancing) to determine when one data set would have a greater mean, median, and mode than another. These results continue to suggest that teachers bring a variety of skills to the table, but that they would benefit from more formal training in statistics prior to teaching it in the classroom.

3.3   Research Related to the General Concept of Average

In their study with children, Russell and Mokros (1991) also included 8 teachers that ranged from grade 4 to grade 8 and 2 mathematics coordinators (i.e., in charge of mathematics curriculum for their school or district). When comparing teachers' conceptions with the children's described in Sect. 2, teachers only exhibited the following ideas: average as modal (one teacher), average as midpoint (number not specified), and average as an algorithmic relationship (six teachers). It can be assumed that the majority of teachers fell in the "average as algorithmic relationship" category. This group was categorised by the idea that one simply adds up all the numbers in a data set and divides by the number in order to determine the average. Teachers exhibited varying levels of success with their application of the algorithm to find the mean. Two teachers in particular had tremendous difficulty in answering the questions and became very confused by the numbers. Russell and Mokros concluded "that the introduction of the algorithm as a procedure disconnected from students' informal understanding of mode, middle, and representativeness causes a short-circuit in the reasoning of many children [and adults]" (1991, p. 313).

Callingham (1997) investigated 100 pre-service and 36 in-service teachers' multimodal functioning in relation to the concept of average, using a four-question survey. Pre-service teachers ranged from early childhood to high school teachers whereas in-service teachers were either primary or high school teachers. The first question involved identifying the average given a set of data. The second and third questions involved participants using a graphical display (bar graph) of data to determine whether one group performed better than another. Finally, the fourth question involved the concept of a weighted mean. Teachers performed relatively well on the first three questions (96.3%, 79.4%, and 73.3% correct, respectively). The most difficult question was the weighted mean question where only 58% of respondents answered correctly. In reference to questions that were presented graphically in order to have participants make a comparison, most of the teachers "did not generally accept a judgment based solely on the appearance of the data in the context of deciding on the better spelling group. Instead they preferred to use some sort of numerical basis to justify their judgment" (1997, p. 215). This finding seems to suggest that teachers may be more prone to applying an algorithm when solving problems concerning average.

Begg and Edwards (1999) conducted a study regarding 22 in-service and 12 pre-service elementary school teachers in New Zealand. The researchers investigated the teachers' ideas about teaching statistics, including some ideas related to average. The data that was presented was based on "unstructured, semi-structured, and clinical interviews; survey (Likert) scales that provided a guide with respect to the efficacy of the research" (Begg & Edwards, 1999, p. 2). Many of the in-service teachers had substantial teaching experience (mean number of years not reported). In considering the teachers' ideas of average, or measures of centre, most teachers were not familiar with the mathematical definitions of the terms mean, median, and mode. When asked about the word average, the most common response given was that it "was in the middle". However, when pressed about their understanding regarding specific measures, the teachers possessed better understanding of the mean than the median or mode (Begg & Edwards, 1999).

Estrada et al. (2004) conducted a study aimed to assess the statistical knowledge of 367 pre-service primary teachers. The instrument used was adapted from a similar instrument developed by Konold and Garfield (1992). From the instrument, items were selected that related to what these pre-service teachers would be required to teach. Of the pre-service teachers, 75% had previous training in statistics; however many still exhibited errors concerning the average, such as not being conscious of the effect of outliers on the mean; routine application of the algorithm without taking into account the context; not being able to invert the algorithm and confusing mean, median and mode. Their study found that there was no significant difference in pre-service teachers' understanding based on the number of years they had been in the programme. This may have been a result of training in the preparation programme that was focused on calculations rather than the application of the concepts. The authors recommend that future teachers specifically take statistical courses during their preparation programmes. These courses should emphasise the study of the properties of distribution as a statistical concept, measures of central tendency, differences of dispersion when comparing two distributions, as well as highlight the importance of sample size during a statistical study.

A study conducted by Leavy and O'Loughlin (2006) in Ireland explored 263 pre-service elementary school teachers' understanding of the mean. All participants were asked to respond to 5 tasks and clinical interviews were held with 25 participants in an attempt to capture further information regarding the participants' conceptual knowledge. In the first task, which involved choosing the mean as a functional tool to compare data sets, 57% of the participants were able to accurately choose and find the mean in order to compare the data sets. An additional 5% selected the mean; however, they made some type of computational error. No participants selected the median, which suggests that this type of question may lead itself to a common misconception among pre-service elementary school teachers that the only suitable measure of centre or method of comparison is the mean. On the second task, only 21% of the participants were able to find the weighted mean. Other conceptual difficulties (such as confusing mean and mode) were found on tasks 3 and 4.

On the final task, participants were given a line plot and asked to indicate if a mean could be found based solely on the information displayed. If they indicated that a mean could be found, they were asked to find that value. Participants were also asked to explain what the mean represents in a distribution. In the first part of the task, only 3% of respondents utilised an approach that showed some level of conceptual understanding of the mean. Both parts of this task revealed that many participants confused the concept of the mean with the other measures of centre (median and mode). It also revealed that 52% participants view the term "mean" as synonymous with "average". Leavy and O'Loughlin (2006) replicate those by Callingham (1997), but both seem to contradict the findings of Gfeller et al. (1999).

3.4   Research Related to Teachers' Professional Knowledge

Cai and Gorowara (2002) investigated 12 inexperienced versus 11 experienced teachers' conceptions and construction of representations for teaching the concept of mean. The inexperienced teachers were pursuing teaching as a second career and/or possessed a degree in mathematics or psychology. The experienced teachers were either grade 6 or 7 teachers that had taken on leadership roles within their schools. Data collection involved teachers: (a) turning in a lesson plan focused on mean, (b) responding to possible ways grade 6 or 7 students might respond to a series of 5 questions involving the mean, and (c) evaluating students' responses to two questions posed in part (b) described above.

The results revealed that all teachers – both experienced and inexperienced – were able to solve the five questions in part (b) of data collection above. Eight of the experienced teachers and only two inexperienced teachers provided multiple solution strategies for solving the problems. With the exception of one inexperienced teacher, all of the inexperienced participants solved the problems using an algorithmic approach. Although the tasks asked teachers to indicate ways students might respond, the inexperienced teachers did not discuss possible student misconceptions whereas the experienced teachers did so, on almost all tasks. The other tasks involving a lesson plan and evaluation of students' work revealed similar results. Overall, the results show that as teachers gain experiences and take on leadership roles they are able to reflect upon students' misconceptions to enhance their own understanding as well as their ability to address the misconceptions students reveal.

4   Implications for Training Teachers and Research

Research analysed in this chapter suggests that teachers' understanding of average seems to be similar to that of students. Consequently the place to begin making an impact on students' understanding seems to be by addressing teachers' understanding. The research shows an exaggerated reliance upon procedural algorithms and a general lack of conceptual understanding by both students and teachers.

This situation is neither the fault of the students nor of the teachers. Rather, teacher preparation programmes should do a better job addressing statistical concepts. A teacher that feels uncomfortable with statistics may have a tendency to reduce or omit statistical topics from their enacted curriculum. Teachers are being asked to teach statistical content at a level that exceeds their level of preparation. As a result, opportunities must be provided during teacher preparation and professional development programmes for pre-service and in-service teachers to be successful. Since it is well known that students and teachers struggle with concepts related to average, we recommend that the field move towards a solution rather than the continued identification of misconceptions.

One possible way to enhance teachers' preparation to teach statistics is through the use of collaborative work (Carvalho, 2008). Teachers can benefit through the exchange of ideas and thus will be more motivated and confident to teach statistics. Another way to address the concept of average during preparation or professional development programmes is to use the media (see Watson, 1997) or to use assessment data as the context for discussing statistical concepts (see Makar & Confrey, 2004; see also Chap. 16).

Another issue for the research community to consider once these concepts become a part of preparation and professional development programmes is that research concerning teachers' understanding of average must involve more than understanding the arithmetic mean. Clearly, understanding of average is far more involved than the concept of the mean. Research studies concerning teachers' understanding of average should at least include the median and mode; furthermore, research similar to the work of Russell and Mokros (1991) would further capture teachers' conceptions concerning the general concept of average. We, therefore, propose the following research questions:

1. How could we use research on children's understanding to develop courses directed to increase the professional knowledge of teachers?
2. What is the influence of in-service professional development programme focussed on teaching statistics to children on teachers' understanding of average?
3. What is the relationship between teachers' understanding of average and students' understanding of average?

Clearly there is a gap regarding teachers training to teach the concept of average and in research related to teachers' knowledge of the topic. Hopefully the statistics education community can work towards increasing knowledge in regard to addressing the questions above. The answers to these questions will help lead to finding solutions that will move the understanding of all teachers forward.

References

Batanero, C., Burrill, G., Reading, C., & Rossman, A. (Eds.). (2008). *Joint ICMI/IASE Study: Teaching Statistics in School Mathematics. Challenges for Teaching and Teacher Education. Proceedings of the ICMI Study 18 and 2008 IASE Round Table Conference.* Monterrey, Mexico: International Commission on Mathematical Instruction and International Association for Statistical Education. Online: www.stat.auckland.ac.nz/~iase/publications

Batanero, C., Godino, J., & Navas, F. (1997). Concepciones de maestros de primaria en formción sobre promedios (Primary school teachers' conceptions on averages). In H. Salmerón (Ed.), *Actas de las VII Jornadas LOGSE: Evaluación Educativa* (pp. 310–340). Granada, Spain: University of Granada.

Begg, A., & Edwards, R. (1999). Teachers' ideas about teaching statistics. *Proceedings of the 1999 Combined Conference of the Australian Association for Research in Education and the New Zealand Association for Research in Education.* Melbourne: Australian Association for Research in Education. Online: www.aare.edu.au/99pap/

Biggs, J. B., & Collis, K. F. (1982). *Evaluating the quality of learning: The SOLO taxonomy.* New York: Academic.

Cai, J., & Gorowara, C. C. (2002) Teachers' conceptions and constructions of pedagogical representations in teaching arithmetic average. In B. Phillips (Ed.), *Proceedings of the Sixth International Conference on Teaching Statistics.* Cape Town, South Africa: International Statistical Institute and International Association for Statistical Education. Online: /www.stat.auckland.ac.nz/~iase/publications

Cai, J., & Moyer, J. (1995). Beyond the computational algorithm: Students' understanding of the arithmetic average concept. In L. Meira (Ed.), *Proceedings of the 19th Psychology of Mathematics Education Conference* (Vol. 3, pp. 144–151). Recife, Brasil: Universidade Federal de Pernambuco.

Callingham, R. (1997). Teachers' multimodal functioning in relation to the concept of average. *Mathematics Education Research Journal, 9*(2), 205–224.

Carvalho, C. (2001). *Interacções entre pares: Contributos para a promoção do desenvolvimento lógico e do desempenho estatístico no 7° ano de escolaridade (Students interactions: A contribution to promotion of logical development and statistical knowledge in 7th grade).* Lisboa: Associação de Professores de Matemática.

Carvalho, C. (2008). Collaborative work in statistics classes: Why do it? In C. Batanero, G. Burrill, C. Reading & A. Rossman (2008).

Estrada, A., Batanero, C., & Fortuny, J. M. (2004). Un estudio sobre conocimientos de estadística elemental de profesores en formación (Prospective teachers' knowledge on elementary statistics). *Educación matemática, 16*(1), 89–111.

Franklin, C., Kader, G., Mewborn, D., Moreno, J., Peck, R., Perry, M., & Scheaffer, R. (2007). *Guidelines for assessment and instruction in statistics education (GAISE) report*. Alexandria, VA: American Statistical Association.

Garcia, C., & Garret, A. (2006). On average and open-end questions. In A. Rossman & B. Chance (Eds.), *Proceedings of the Seventh International Conference on Teaching Statistics*. Salvador (Bahia), Brasil: International Statistical Institute and International Association for Statistical Education. Online: www.stat.auckland.ac.nz/~iase/publications

Gattuso, L., & Mary, C. (1996). Development of concepts of the arithmetic average from high school to university. *Proceedings of the Twentieth International Conference for the Psychology of Mathematics Education* (Vol. 2, pp. 401–408). Valencia, Spain: University of Valencia.

Gfeller, M. K., Niess, M. L., & Lederman, N. G. (1999). Preservice teachers' use of multiple representations in solving arithmetic mean problems. *School Science and Mathematics, 99*(5), 250–257.

Groth, R. E., & Bergner, J. A. (2006). Preservice elementary teachers' conceptual and procedural knowledge of mean, median, and mode. *Mathematical Thinking and Learning, 8*, 37–63.

Jacobbe, T. (2007). Elementary school teachers' understanding of essential topics in statistics and the influence of assessment instruments and a reform curriculum upon their understanding. Online: www.stat.auckland.ac.nz/~iase/publications

Jacobbe, T. (2008). Elementary school teachers' understanding of the mean and median. In C. Batanero, G. Burrill, C. Reading & A. Rossman (2008).

Konold, C., & Garfield, J. (1992). *Statistical reasoning assessment: Part 1. Intuitive thinking*. Amherst, MA: Scientific Reasoning Research Institute.

Leavy, A., & O'Loughlin, N. (2006). Preservice teachers understanding of the mean: Moving beyond the arithmetic average. *Journal of Mathematics Teacher Education, 9*, 53–90.

Makar, K., & Confrey, J. (2004). Secondary teachers' statistical reasoning in comparing two groups. In D. Ben-Zvi & J. Garfield (Eds.), *The challenges of developing statistical literacy, reasoning, and thinking* (pp. 353–374). Dordrecht, The Netherlands: Kluwer.

Mokros, J., & Russell, S. J. (1995). Children's concepts of average and representativeness. *Journal for Research in Mathematics Education, 26*, 20–39.

Russell, S. J., & Mokros, J.R. (1991). What's typical? Children's ideas about average. In D. Vere-Jones (Eds.), *Proceedings of the Third International Conference on Teaching Statistics* (pp. 307–313). Voorburg, The Netherlands: International Statistical Institute.

Shaughnessy, J. M. (2007). Research on statistics learning and reasoning. In F. Lester (Ed.), *Second handbook of research on mathematics teaching and learning* (pp. 957–1010). Greenwich, CT: Information Age Publishing and National Council of Teachers of Mathematics.

Strauss, S., & Bichler, E. (1988). The development of children's concepts of the arithmetic average. *Journal for Research in Mathematics Education, 19*(1), 64–80.

Watson, J. M. (1997). Assessing statistical thinking using the media. In I. Gal & J. Garfield (Eds.), *The assessment challenge in statistics education* (pp. 107–121). Amsterdam: IOS Press and International Statistical Institute.

Watson, J. M., & Moritz, J. B. (1999). The developments of concepts of average. *Focus on Learning Problems in Mathematics, 21*, 15–39.

Watson, J. M., & Moritz, J. B. (2000). The longitudinal development of understanding of average. *Mathematical Thinking and Learning, 2*, 11–50.

Chapter 22
Teachers' Understanding of Variation

Ernesto Sánchez, Cláudia Borim da Silva, and Cileda Coutinho

Abstract The aim of this chapter is to summarise the major studies related to teachers' understanding of variability in both data analysis and chance contexts. Since there is a relation between this research and previous studies dealing with students, some results on students' reasoning on variation are also described. At the end some recommendations for teaching and research are presented.

1 Introduction

There is a growing recognition among statisticians and statistics educators that variation is at the heart of statistics. The analysis made by Moore (1990) emphasised the omnipresence of variability and the importance of modelling and measure variation in statistics. Wild and Pfannkuch (1999) proposed the perception of variation as one fundamental kind of statistical thinking, and Watson, Kelly, Callingham, and Shaughnessy (2003, p. 1) pointed out that "statistics requires variation for its existence".

Variation is a very complex concept and it is difficult to find a simple definition in the literature. The terms *variability* and *variation* are used as synonyms by

E. Sánchez (✉)
Departamento de Matemática Educativa, Centro de Investigación y de Estudios Avanzados del Instituto Politécnico Nacional, Av. Instituto Politécnico Nacional 2508, Col. San Pedro Zacatenco, Cp. 07360, México D.F., Mexico
e-mail: esanchez@cinvestav.mx

C.B. da Silva
University São Judas Tadeu, Sao Paulo, Brazil
e-mail: dasilvm@uol.com.br

C. Coutinho
Pontifícia Universidade Católica de São Paulo,
Programa de Estudos Pósgraduados em Educação Matemática Campus
Consolação Rua Marquês de Paranaguá, 111 CEP 01303-050 São Paulo - SP - Brasil
e-mail: cileda@pucsp.br

C. Batanero, G. Burrill, and C. Reading (eds.), *Teaching Statistics in School Mathematics-Challenges for Teaching and Teacher Education: A Joint ICMI/IASE Study*, DOI 10.1007/978-94-007-1131-0_22, © Springer Science+Business Media B.V. 2011

some authors, while others distinguish between them. According to Reading and Shaughnessy (2004) variability is a property of an entity to change or vary (variable), and refers to data that can vary, while *variation* would be the description or measure of change in the variable. In this chapter, we will not distinguish between these two terms, rather variability and variation will be used interchangeably.

Variation is linked to many other fundamental statistical ideas. In school mathematics, Watson, Callingham, and Kelly (2007) suggest that understanding variation involves perceiving uncertainty, anticipated change, unanticipated change, and outliers. Konold and Pollatsek (2002) emphasised the importance of jointly considering variability (noise) and centre (signal) because both ideas are needed to find meaning when analysing data. Garfield and Ben-Zvi (2008) remarked that "understanding the ideas of spread or variability of data is a key component of understanding the concept of distribution, and is essential for making statistical inferences" (p. 203). Wild (2006) suggested that the notion of distribution "underlies virtually all statistical ways of reasoning about variation" (p. 11). Moreover, a *big picture* of variation can be found when it is linked to other important statistical concepts such as data, chance, sample, centre, and distribution, and when making inferences. Several of these concepts and the teachers' understanding of them are discussed in other chapters of this book.

In spite of this relevance, the curricula of many countries do not include the study of variation until high school, although research results indicate the possibility of developing intuitive notions of variation in earlier grades. It is also important to note that statistics courses in high school and undergraduate levels usually include variation measures (such as range, interquartile range, and standard deviation). The contrast between the role of variation in statistics and the lack of research on students' understanding of this concept until very recently was pointed out in Shaughnessy (1997). Since then, an increasing number of studies on students' understanding of variation have been published. Although research on students' understanding of variability is more extensive today, in this chapter we do not intend to review all this literature. The focus of this chapter is research on teachers' understanding of variation, that has analysed language, comparing groups, technology, standard deviation, and variation in random situations. Some studies on students' understanding of variation that focussed on these same topics are reviewed in Sects. 2 and 3. In the conclusion, some recommendations for teaching and research are made.

2   Students' Understanding of Variation in Data Contexts

Although research on students' understanding of variability is relatively recent, the related literature is extensive. Therefore, only a few studies related to themes that will emerge in research on teachers' understanding of variability have been reviewed in this chapter. Papers analysed in this section focus on the language of variability (Watson & Kelly, 2003; Bakker, 2004), perception of variability in comparing groups (Watson & Moritz, 1999; Watson et al., 2007), role of technological tools (Bakker, 2004; Ben-Zvi, 2004), and understanding standard deviation (delMas & Liu, 2005, 2007; Lehrer, Kim, & Schauble, 2007). In the next section, studies on variation in chance contexts are summarised.

When teaching statistics in school, teachers tend to focus on performing operations and give little attention to the description of terminology and to explanations. However, encouraging students to express statistical ideas in their own language has been considered a relevant way to help them make sense of statistical ideas. Watson and Kelly (2003) explored the understanding of the word "variation" among 379 students in grades 7 and 9 with the following questions: (a) What does variation mean?; (b) Use the word variation in a sentence; and (c) Give an example of something that varies. Later, Watson (2006) constructed a hierarchy that resulted from an analysis of students' responses to these questions. She observed that less than 10% of the middle school students in her sample were likely to respond at the highest level in a scale of four. In this level students should relate the idea of change with appropriate responses to the above questions, such as "Something that differs from its previous state", "There is a big variation in the results", and "The weather varies" (Watson, 2006, p. 220).

In a study oriented to promote reasoning about variability, sampling, data, and distributions, with 30 students in grades 7 and 8, Bakker (2004) discussed the use of informal language to articulate ideas of variation. Expressions such as "average", "range", and "spread" were used in a rather unconventional way by the students during the activities, whereas other non-statistical terms such as "majority", "semicircle", and "pyramid" were also used to express some statistical features of distributions.

The task of comparing groups, with data often presented in graphical format, was used to investigate children's thinking about centres (Gal, Rothschild, & Wagner, 1989, 1990), although in the last decade, this kind of task became more paradigmatic for studying several intertwined statistical concepts such as centre, variability, and distribution. Watson (2006) remarked that this task can help observe how much students are aware of variation: "there are differences between the two sets that should be noted but these contrasts need to be made with respect to the shape and spread within the two sets themselves" (p. 239). However, students encounter serious difficulties in comparing groups. For example, Watson and Moritz (1999) found that the majority of 88 students (in grades 3, 6, and 9) could compare data sets of equal size, but were unable to compare sets of unequal sizes, because the latter requires proportional reasoning skills.

Using the same task with 73 students, Watson et al. (2007) identified only one student who reached the highest level of variation understanding. This student could observe different features of distribution such as differences on sample sizes, mean and variation around the mean. It seems that the difficulties in comparing data sets stem from the students' inability to understand a data set as a whole entity with features such as centre, spread, and shape (Konold & Higgins, 2003).

Shaughnessy (1997, p. 13) remarked that technological tools have encouraged the emergence of research on students' understanding of variation: "nowadays it is very easy to actually 'draw samples' […] by introducing a simulation […] of the experiment using some sort of sample-resample software". The graphing possibilities of some software have been also explored in research on variation. For example,

Ben-Zvi (2004) studied two middle school students' reasoning about variability by asking them to compare the lengths of Hebrew and American surnames using a spreadsheet (Excel). As a result from the activity both students evolved from using local information towards a global point of view of describing and explaining the variability between the groups. Bakker (2004) also used software minitools to represent and analyse data sets. The responses to the task "growing a sample" was strongly influenced by the minitools, however, the author was unsure of whether it limited or fostered the students' capabilities. Pfannkuch (2005) analysed the studies on variation published in a special issue of *Statistics Education Research Journal* and suggested that all of them used tabular, graph, or diagrammatic representations; in three of the five papers the activities were also immersed in a computerised environment.

Standard deviation is the most popular measure of variation or spread, but at the same time, is a very difficult concept and being a rather formal concept, its study has almost been confined to high school and university level. Some investigations about how university students learn standard deviation have been reported by delMas and Liu (2005, 2007), who suggest that the concepts of distribution, mean, and deviation from the mean are fundamental to construct the notion of standard deviation.

The ideas of error in measurement and deviation from the mean seem adequate for pre-college students and have been explored by Lehrer and Schauble (2002) who conducted a teaching study with 22 grade 4 students. The study developed during 2 years, and involved measurement tasks aimed at helping students perceive measurement error as distributed around an unknown true value. Similarly, Lehrer et al. (2007) investigated children's development of the concept of variation in measures using TinkerPlots. All the students participating in this study – grades 5 and 6 – measured the same object and realised that the true length of the object was obtained when there was no error. Students were then asked to establish a degree of precision for their measurements (closeness or agreement). Although their precision criteria were unconventional, students began to see variation as a relation between data and its distance to a centre, which is often the sample mean or median. The differences in the students' solutions supported the central idea that many measurements (data) are necessary to make an accurate estimation of the unknown quantity. The authors suggested that *TinkerPlots* afforded that students invent different statistics, some of them being not conventional, activity that makes the subsequent introduction of conventional measures of variation as range, inter-quartile range and standard deviation more meaningful.

3   Students' Understanding of Variation in Chance Contexts

Other authors developed tasks that allow students to display their perception of variation in chance settings. The paradigmatic task asks participants for predictions in repeated trials of simple experiments such as tossing dice or throwing a coin. Two features of variation affect the distribution of the outcomes that appear over repeated trials: the first is the expected variation indicated by the theoretical distribution (for example, uniform or binomial distributions), and the second is the unexpected variation, or departure from the theoretical expectations.

The "Lollies task" was used by Shaughnessy, Watson, Moritz, and Reading (1999) in order to investigate the conceptions of variability in a sample of 324 students in grades 4, 5, 6, and 12. A bowl has 50 red, 30 blue, and 20 yellow lollies. Someone pulls out ten lollies, and the teacher counts the number of red lollies; this action is repeated five times. The students were asked to predict the number of red lollies if the experiment of selecting ten lollies at random was repeated a given number of times using three different formats: (a) In the *range* version students were asked for the maximum and the minimum number of red lollies in the five experiments, (b) in the *choice* version they were asked to choose the most likely list of results among several possible lists, and (c) in the *list* version they were asked to write a possible list of outcomes in the five experiments. Students' responses were categorised both on the basis of their use of centres and variation. Students were classified in *high, "five", and low* according to the magnitude of the centre in their responses and as *narrow, reasonable, and wide*, according to the range proposed in their responses. The authors observed a "steady growth throughout grades 4–12 on the centring scale, but there was considerable oscillation across grades on the spread scale, and even a dip down in performance among our grades 12 students" (p. 21).

Another task in a chance setting is predicting the number of occurrences of each number when a six-sided die is rolled 60 times (Watson et al., 2003; Sanchez & Trujillo, 2009). Watson et al. proposed this task to 746 students in grades 3, 5, 7, and 9, and the responses were classified in five levels (0–4). A response was classified at level 0 (20.5%) when the sum of frequencies was greater than 60 or included a frequency greater than 21. Level 1 (30.2%) was assigned to responses in which students provided frequencies with a total of 60 but included subjective patterns. Level 2 (31.8%) involved frequencies reflecting strict probability or unusual variation with reasoning that reflected some understanding of the context. At Level 3 (8.4%) responses showed either too narrow or too wide variation with appropriate reasoning. At Level 4 (9.2%) the students responses showed appropriate frequencies and reasoning. Using the same task Sanchez and Trujillo (2009) found that 18% of 327 middle school students, 20% of 214 high school, and 27% of 74 college students displayed a *realistic appearance* variation in their predictions, since the frequencies totalled 60 and each particular frequency fell in a range from 4 to 16.

Watson et al. (2003) explored the perception of variation of 189 students in grade 7 and 197 in grade 9 with the following task: A 50–50 (black and white) spinner was spun 50 times and the results for the number of times it landed on the black part were recorded. Students were given three hypothetical results of these experiments and were asked to select which of the three plots they considered a reasonable result of the experiment. One of the plots represented results generated by simulating the uniform distribution of integer numbers from 0 to 50; another plot represented the theoretical binomial distribution that models the situation; finally, the third plot represented a distribution generated from a simulation of the experiment, which was close to a theoretical binomial distribution but with some deviations from the expected frequencies. The authors reported that 52.6% students in grades 7 and 9 judged correctly that the first two distributions were made up and the last one was real. In a similar experiment, Sanchez and Trujillo (2009) found that 16% of 327 middle school students, 16% of 214 high school students and 30% of 74 college students judged correctly that the first two distributions were hypothetical while the last one was real.

Shaughnessy and Ciancetta (2002) used a task with two spinners, each having equal area zone of black and white. Students were told that a participant won a price if both arrows landed in the black zone, and were asked to decide if there was 50–50 chance of winning or not. Only 20% of the 273 students in grades 6–8 and 43% of 90 students in grade 9 said that the chance of winning was less than 50%. These authors concluded that "there is a connection between the concept of sample space in probability and the expected variation for the value of a random variable in statistics" (p. 5).

4   Teachers' Understanding of Variation in Data Contexts

Research on teachers' understanding of variation is scarcer than that dealing with students and most of it has been carried out with pre-service teachers. All these studies have training or intervention component where the learners were teachers, since an observation context had to be created that would allow the teachers' ideas about variation to emerge. Two of these studies involving variation in data context are described in this section and other two involving variation in a chance context will be presented in Sect. 5.

Makar and Confrey (2005) examined how 17 prospective secondary mathematics and science pre-service teachers articulated notions of variation as they compared two empirical distributions. The teachers were enrolled in a one-semester undergraduate course on assessment and data analysis. The authors interviewed the teachers in the first and the last weeks of the course while they analysed a pair of dot plots of real data, taken from students in an enrichment class, and a regular eighth grade class. Teachers were asked to compare the relative improvement of students in the two groups, and then were given related questions if their responses needed clarification.

Makar and Confrey were interested to see how the teachers would use the measures of centre, variation, and spread to compare groups. The study showed how prospective teachers expressed important ideas connecting variation with distributions using standard and nonstandard language. Some examples of standard statistical expressions used by respondents were: "proportion", "mean", "maximum/minimum", "sample size", "outliers", "range", "shape", and "standard deviation". The inclusion of standard statistical terms in the teachers' responses increased from the first week to the end of the course. It is remarkable that the term "standard deviation" was used by none of participants in the first interview and by only two of them in the second. In the teachers' attempt to express ideas of variation, two categories of nonstandard terms emerged in the interviews with underlying intuitive ideas of spread ("clustered", "spread out") and distribution ("triads", "modal clumps"); the use of these terms also increased in the second interview.

Silva and Coutinho (2008) studied nine in-service mathematics teachers' reasoning about variation with a univariate distribution. Teachers were asked to conduct a survey and use the data obtained about age of participants in the survey to create a distribution. The authors analysed the teachers' reasoning on variation when they analysed the distribution of ages. After organising data in a frequency table and drawing a histogram, teachers were asked to think of a different way to represent the set of ages. Some excerpts from the teachers' discussions revealed different levels of variation reasoning, which were classified along the levels adapted from Garfield (2002) model of statistical reasoning, which is consistent with SOLO taxonomy of learning outcomes (Biggs & Collis, 1991).

At the first level (idiosyncratic), teachers calculated the mean and mentioned the standard deviation without understanding its meaning. In the second level (verbal reasoning), four stages were observed: (a) perception of variation, (b) understanding standard deviation as a measure of differences between values, (c) understanding that small standard deviation is always better, and (d) recognising that some quantities of data fall inside the interval with extremes at one standard deviation from the mean. In the third level (transitional reasoning), teachers used more than one summary statistics to describe the ages, for example, the maximum, minimum, and modal group. At the fourth level (procedural reasoning), teachers grasped the meaning of mean and deviations from the mean, and started to consider the meaning of intervals measured in standard deviations around the mean. Finally, in the fifth level (process reasoning), which no teacher in this research attained, participants would relate the mean, deviations from the mean, the interval of k standard deviations around the mean, and the density estimation of frequency in that interval. Most of the nine teachers' reasoning fell in the second level (verbal reasoning), where standard deviation is understood as a measure of sample homogeneity.

5 Teachers' Understanding of Variation in Chance Contexts

In the context of a pre-service course in probability and statistics in which 30 pre-service teachers participated, Canada (2006) studied the responses of 11 participants who volunteered to be interviewed before the course (pre-survey). During the course, series of activities which were designed to offer opportunities to investigate and discuss variation were conducted. After the course, the teachers were interviewed again (post-survey). The activities during the course were centred on data and graphs, sampling and probability situations. The surveys contained three tasks. The first was a "one set" context where teachers were asked to predict the number of heads in 50 flips of a fair coin. In the "compare sets" task, teachers were asked to predict the results of a second set of experiments in comparison with the results of the first set of trials. Finally, in the "six sets" task, teachers were asked to describe what would happen in each of the 6 sets of 50 flips each. Canada classified the teachers' responses to the one set question according to whether teachers believed the result would be "a number near 25", "just 25", or "an interval of values around 25" (levels 0, 1, and 2 respectively). The teachers' reasonings were classified according to whether they gave no explanation or a vague reasoning for their response, used additive reasoning, used proportional reasoning, or used proportional reasoning with suggestions about what else might happen (levels 0, 1, 2, and 3). There was clear progress from the responses on the pre-survey to the responses on the post-survey, indicating a better understanding of variation. An interesting contribution of this work is an "evolving framework" to help characterise elementary pre-service teachers' thinking about variation.

Sanchez and Garcia (2008) were interested in the problems teachers face to construct the elementary notions needed to use variation in order to make predictions. They gave a questionnaire to six in-service middle school teachers in Mexico before and after performing some simulation activities with Fathom as a part of a professional training course. One task asked the teachers to predict the number of times each number would occur if a dice was rolled 60 times. After predicting results from ten experiments the teachers compared their responses with the simulation results. Two types of reasoning were observed in the teachers who responded "10, 10, 10, 10, 10, and 10" in the pre-test; while some of them perceived the variability in the results but were unable to express their ideas, other teachers really expected the theoretical expected outcomes. In the post-test, all the teachers predicted sequences with reasonable variation.

Another question in the post-test asked to predict what would happen if the die was rolled 1,000 times. Despite having predicted good variation for 60 rolls, 3 teachers predicted a very small variation for 1,000 rolls. Even when teachers recognised that there was variation, they presumed it would decrease as the number of repetitions increased. The authors identified an intuitive misunderstanding of the law of large numbers, because these teachers expected that as the number of repetitions increased, the absolute frequencies for each face will tend to be equal instead of understanding that relative frequencies of each face would converge to the theoretical probability.

6 Implications for Teaching and Research

The studies reviewed in this chapter describe some teachers' responses when faced with variation tasks in data and in chance settings. In the four studies reviewed where some activities were carried out through classroom instructions, technology had a role in three of them, and real- or classgenerated data were analysed by participants in all the studies.

These features agree with some principles of the Statistical Reasoning Learning Environment described by Garfield and Ben-Zvi (2008), and based on Cobb and McClain (2004): (a) Using real and motivating data sets, (b) using classroom activities to support the development of students' reasoning, (c) integrating the use of appropriate technological tools, (d) focusing on development of central ideas, and (e) promoting classroom discourse and using of assessment to uncover what students know. Therefore, the Statistical Reasoning Learning Environments is a suitable framework for training teachers as well as students. The recommendations for teaching variation should also include (Shaughnessy, 2007):

- Emphasise variability as one of the primary issues in statistical thinking and statistical analysis,
- Build on students' intuitive notions of centre and variability, and
- Introduce comparison of data sets much earlier to students, prior to formal statistics (p. 1002).

Another conclusion is that the same type of research was carried out in most studies of teachers' and students' reasoning about variation even though the problems of training teachers and students are different. While the studies reviewed contribute to our understanding of teachers' statistical reasoning on variation, the study of teachers' professional knowledge and teachers' practices while teaching variation is an urgent need.

References

Bakker, A. (2004). Reasoning about shape as pattern in variability. *Statistics Education Research Journal, 3*(2), 64–83. Online: www.stat.auckland.ac.nz/serj/

Batanero, C., Burrill, G., Reading, C., & Rossman, A. (Eds.). (2008). *Joint ICMI/IASE Study: Teaching Statistics in School Mathematics. Challenges for Teaching and Teacher Education. Proceedings of the ICMI Study 18 and 2008 IASE Round Table Conference.* Monterrey, Mexico: International Commission on Mathematical Instruction and International Association for Statistical Education. Online: www.stat.auckland.ac.nz/~iase/publications

Ben-Zvi, D. (2004). Reasoning about variability in comparing distributions. *Statistics Education Research Journal, 3*(2), 42–63. Online: www.stat.auckland.ac.nz/~iase/serj/

Biggs, J. B., & Collis, K. F. (1991). Multimodal learning and the quality of intelligent behaviour. In H. A. H. Rowe (Ed.), *Intelligence, reconceptualization and measurement* (pp. 57–76). Hillsdale: Erlbaum.

Canada, D. (2006). Elementary pre-service teachers' conceptions of variation in a probability context. *Statistics Educations Research Journal, 5*(1), 36–63. Online: www.stat.auckland.ac.nz/serj/

Cobb, P., & McClain, K. (2004). Principles of instructional design for supporting the development od students' statistical reasoning. In D. Ben-Zvi & J. Garfield (Eds.), *The challenge of developing statistical literacy, reasoning and thinking* (pp. 375–395). Dordrecht, The Netherlands: Kluwer Academic.

delMas, R., & Liu, Y. (2005). Exploring student's conceptions of the standard deviation. *Statistics Education Research Journal, 4*(1), 55–82. Online: www.stat.auckland.ac.nz/serj/

delMas, R., & Liu, Y. (2007). Students' conceptual understanding of the standard deviation. In M. C. Lovett & P. Shah (Eds.), *Thinking with data* (pp. 87–116). New York: Erlbaum.

Gal, I., Rothschild, K., & Wagner, D. A. (1989, April). *Which group is better? The development of statistical reasoning in elementary school children.* Paper presented at the Meeting of the Society for Research in Child Development, Kansas City, MO.

Gal, I., Rothschild, K., & Wagner, D. A. (1990, April). *Statistical concepts and statistical reasoning in school children.* Paper presented at the Meeting of the American Educational Research Association, Boston.

Garfield, J. (2002). The challenge of developing statistical reasoning. *Journal of Statistics Education, 10*(3), 1–11. Online: www//amstat.org/publications/jse/

Garfield, J., & Ben-Zvi, D. (2008). *Developing students' statistical reasoning. Connecting research and teaching practices.* New York: Springer.

Konold, C., & Higgins, T. L. (2003). Reasoning about data. In J. Kilpatrick, W. G. Martin, & D. Schifter (Eds.), *A research companion to principles and standards for school mathematics* (pp. 193–215). Reston, VA: National Council of Teachers of Mathematics.

Konold, C., & Pollatsek, A. (2002). Data analysis as the search for signals in noisy processes. *Journal for Research in Mathematics Education, 33*(4), 259–289.

Lehrer, R., Kim, M., & Schauble, L. (2007). Supporting the development of conceptions of statistics by engaging students in measuring and modeling variability. *International Journal of Computers for Mathematical Learning, 12*, 195–216.

Lehrer, R., & Schauble L. (2002). Distribution: A resource for understanding error and natural variation. In B. Phillips (Ed.), *Proceedings of the Sixth International Conference on Teaching Statistics*. Cape Town, South Africa: International Statistical Institute and International Association for Statistics Education. Online: www.stat.auckland.ac.nz/~iase/publications

Makar, K., & Confrey, J. (2005). "Variation-talk": Articulating meaning in statistics. *Statistics Education Research Journal, 4*(1), 27–54. Online: www.stat.auckland.ac.nz/serj/

Moore, D. S. (1990). Uncertainty. In L. A. Steen (Ed.), *On the shoulders of giants: New approaches to numeracy* (pp. 95–137). Washington, DC: National Academy Press.

Pfannkuch, M. (2005). Thinking tools and variation. *Statistics Education Research Journal, 4*(1), 83–91. Online: www.stat.auckland.ac.nz/serj/

Reading, C., & Shaughnessy, J. M. (2004). Reasoning about variation. In D. Ben-Zvi & J. Garfield (Eds.), *The challenge of developing statistical literacy, reasoning and thinking* (pp. 201–226). Dordrecht, The Netherlands: Kluwer.

Sanchez, E., & Garcia, J. (2008). Acquisition of notions of statistical variation by in-service teachers. In C. Batanero, G. Burril, C. Reading, & A. Rossman (2008).

Sanchez, E., & Trujillo, K. (2009). Exploring students' notions of variability in chance settings. In S. L. Swars, D. W. Stinson, & S. Lemons-Smith (Eds.), *Proceeding of the 31st Annual Meeting of the North American Chapter of the International Group for the Psychology of Mathematics Education* (pp. 671–680). Atlanta, GA: Georgia State University.

Shaughnessy, J. M. (1997). Missed opportunities in research on the teaching and learning of data and chance. In F. Biddulph & K. Carr (Eds.), *People in mathematics education. Proceedings of the 20th Annual Meeting of the Mathematics Education Research Group of Australasia* (Vol. 1, pp. 6–22). Waikato, New Zealand: Mathematics Education Research Group of Australasia.

Shaughnessy, J. M. (2007). Research on statistical learning and reasoning. In F. K. Lester (Ed.), *Second handbook of research mathematics teaching and learning* (pp. 957–1010). Charlotte, NC: Information Age Publishing and National Council of Teachers of Mathematics.

Shaughnessy, J. M., & Ciancetta, M. (2002). Students' understanding of variability in a probability environment. In B. Phillips (Ed.), *Proceedings of the Sixth International Conference on Teaching Statistics*. Cape Town, South Africa: International Statistical Institute and International Association for Statistics Education. Online: www.stat.auckland.ac.nz/~iase/publications

Shaughnessy, J. M., Watson, J., Moritz, J., & Reading, C. (1999, April). School mathematics students' acknowledgment of statistical variation. In C. Maher (Chair), *There's more to life than centers*. Presession Research Symposium, 77th Annual National Council of Teachers of Mathematics Conference, San Francisco, CA.

Silva, C., & Coutinho, C. (2008). Reasoning about variation of a univariate distribution: A study with secondary mathematics teachers. In C. Batanero, G. Burril, C. Reading, & A. Rossman (2008).

Watson, J. M. (2006). *Statistical literacy at school. Growth and goals.* Mahwah: Erlbaum.

Watson, J. M., Callingham, R. A., & Kelly, B. A. (2007). Students' appreciation of expectation and variation as a foundation for statistical understanding. *Mathematical Thinking and Learning, 9*(2), 83–130.

Watson, J. M., & Kelly, B. (2003). The vocabulary of statistical literacy. *Educational Research, Risks & Dilemmas: Proceedings of the Joint Conferences of New Zealand Association for Research in Education and the Australian Association for Research in Education*. Auckland: NZARE & AARE. Online: www.aare.edu.au/03pop/alpha.htm

Watson, J. M., Kelly, B., Callingham, R., & Shaughnessy, M. (2003). The measurement of school students' understanding of statistical variation. *International Journal of Mathematical Education in Science and Technology, 34*(1), 1–29.

Watson, J. M., & Moritz, J. B. (1999). The beginning of statistical inference: Comparing two data sets. *Educational Studies in Mathematics, 37*, 145–168.

Wild, C. (2006). The concept of distribution. *Statistics Education Research Journal, 5*(2), 10–26. Online: www.stat.auckland.ac.nz/serj/

Wild, C., & Pfannkuch, M. (1999). Statistical thinking in empirical enquiry. *International Statistical Review, 67*(3), 223–265.

Chapter 23
Teachers' Knowledge of Distribution

Chris Reading and Dan Canada

Abstract To teach statistics effectively teachers need to have a well-developed knowledge of distribution. As a key concept in an intricate web of statistical knowledge, distribution depends on, and is depended on by many other statistical concepts. Various frameworks have been developed as researchers strive to describe the cognitive development of knowledge of distribution. Considering the professional learning continuum that a teacher needs to traverse, research studies are reported that have focused, from the perspective of teachers as learners, on the development of teacher knowledge of distribution both before teaching and while teaching. Recommendations are made for teacher learning and future research into teachers' knowledge of distribution.

1 Introduction

Knowledge of distribution relies on a coordination of key concepts such as centre, spread, and shape (Bakker & Gravemeijer, 2004; Reading & Reid, 2006), many of which have been the focus in other chapters. Through a synthesis of these concepts, distribution of data emerges as a fundamental concept in its own right. Reasoning about distribution, the central focus of a special issue of the *Statistics Education Research Journal*, was described as a "complex and challenging research topic" (Pfannkuch & Reading, 2006, p. 5).

C. Reading (✉)
SiMERR National Centre, University of New England, Education Building,
Armidale, NSW 2351, Australia
e-mail: creading@une.edu.au

D. Canada
Department of Mathematics, Eastern Washington University, Cheney, WA 99004, USA
e-mail: mathaction@gmail.com

C. Batanero, G. Burrill, and C. Reading (eds.), *Teaching Statistics in School Mathematics-Challenges for Teaching and Teacher Education: A Joint ICMI/IASE Study*, DOI 10.1007/978-94-007-1131-0_23, © Springer Science+Business Media B.V. 2011

The purpose of this chapter is to highlight research on knowledge of distribution, with a particular focus on teacher (both pre-service and in-service) knowledge. After defining distribution and explaining its place in a web of statistical knowledge, emerging frameworks to describe the knowledge are shared. Taking the perspective of teachers as learners, research studies that have investigated the knowledge development of teachers both while training and while teaching are presented. Finally, recommendations are made for future teacher learning and for future research.

2 Distribution in a Web of Statistical Knowledge

A vast collection of interconnected concepts forms a web of statistical knowledge that is critical to functioning statistically. Such connectedness has previously been identified by other researchers (e.g., Shaughnessy, 2007). The concept of distribution is an important node in this web, depending on some statistical concepts and being depended on by others. To facilitate learning about statistics, teachers need to develop a strong knowledge of distribution, as well as a clear understanding of the role of distribution in this web. Distribution is usefully defined as the arrangement of values of a variable along a scale of measurement resulting in a representation of the observed frequencies or the theoretical probability of a range of values of the variable (modified from Leavy, 2006, p. 90). This concept is fundamental to statistical reasoning (Wild, 2006) and is complex (Shaughnessy, 2007), despite its relatively straightforward definition.

Knowing about distribution is critical to teachers preparing to teach about distribution, and an awareness of the major distinction between empirical and theoretical distributions is essential to knowing the concept. Empirical distributions are what are seen in the data by way of the frequencies of the variables. Theoretical distributions are what are imagined to exist by way of random variables. The first step to knowing about distribution is being able to describe how data are distributed (empirical). The second, more difficult, step from empirical to theoretical distributions can be facilitated by providing experiences in distinguishing between sample and population distributions, such as linking real data to the theoretical distribution (Batanero, Tauber, & Sanchez, 2004) and using multivariate data (Wild, 2006).

To establish the critical role of distribution in the web of statistical knowledge, both the concepts on which distribution depends, and the concepts that depend on distribution, need to be identified. A critical role for distribution was identified by Friel, O'Connor, and Mamer (2006) who included among the big ideas for developing statistical skills being able to deal with variability and centre when analysing data distributions (i.e., concepts on which distribution depends), and being able to view new data as a distribution for inferential work (i.e., concepts which depend on distribution).

2.1 Concepts on Which Distribution Depends

Nine key concepts on which the concept of distribution depends have been clearly identified: centre, variability, shape, density, skewness, relative frequency, probability, proportionality and causality. The first seven of these concepts are identifiable features of distribution, while the last two are more abstract. Centre, variability (spread) and shape are commonly agreed (see, e.g., Bakker, 2004; Leavy, 2006; Pfannkuch & Reading, 2006; Shaughnessy, 2007; Garfield & Ben-Zvi, 2008) to be core concepts. The notion of a web connecting concepts is reinforced by Shaughnessy's (2007) explanation of the interrelated connections between centre, variability and shape. Connecting natural variation to distribution structures is a major conceptual hurdle when learning statistics (Reading & Reid, 2006; Shaughnessy, 2007). Implicit in the concept of variability is the concept of variable, which Garfield and Ben-Zvi (2008) pointed out, is itself essential to an understanding of distribution. Density, skewness and relative frequency (for data distribution) or probability (for theoretical distributions) have also been shown (see, e.g., Bakker & Gravemeijer, 2004; Batanero et al., 2004; Leavy, 2006) to be critical concepts when dealing with distributions. The potential of distribution to connect probability to statistics has been under-utilised in teaching and curricula (Pfannkuch & Reading, 2006).

Proportionality and causality, the more abstract concepts, are less commonly cited as key concepts on which distribution depends but warrant consideration. Proportionality is important when working with specific distributions. Ciancetta (2007) found that comparing distributions of different sized groups proved to be easier for those who could reason proportionally. This is consistent with Shaughnessy's (2007) research that showed a progression of thinking from additive to proportional and finally to distributional. Focusing on causality is important in co-ordinating the frequency perspective of distribution (i.e., data-centric, paying attention to variation and shape of collected data) and the probability perspective (i.e., modelling, paying attention to randomness and probabilities that mould the outcomes). A deep knowledge of distribution cannot develop until these two perspectives have been co-ordinated. Thinking-in-change when working in a technology-supported environment (Prodromou & Pratt, 2006) has proved useful in assisting such co-ordination.

The notion of a web conveys the intricate interconnectedness of concepts important to statistical knowledge. The necessary interconnectedness of these nine key concepts, to distribution and to other concepts, illustrates the complexity of the knowledge of distribution that teachers need to develop.

2.2 Concepts that Depend on Distribution

Distribution's role in the web of statistical knowledge does not end with its dependence on concepts. Three more sophisticated concepts, sampling distribution, statistical confidence and statistical significance, have been identified as depending on the concept of distribution. Sampling distributions are difficult to understand (Shaughnessy, 2007) and depend heavily on variation (Madden, 2008; Reading & Reid, 2006) but distribution is another key concept that needs to be integrated to develop that understanding (Chance, delMas, & Garfield, 2004; Kadijevich, 2008). This necessary integration further illustrates the complexity of the web, for example, sampling distribution is connected to variability by broadening its scope to include study-to-study variation, as well as within-study variation (Wild, 2006; see also the chapter by Harradine, Batanero, & Rossman in this book). Distribution is also a key concept for both statistical confidence and statistical significance (Chance et al., 2004), which highlights its importance when drawing inferences.

The development of teacher knowledge must be supported from two perspectives. First, teachers need to have a well-developed knowledge of the concepts on which distribution depends to allow them to develop their own knowledge of distribution. Second, they need to expand their knowledge to include concepts that depend on distribution to allow them to effectively facilitate student learning about distribution.

2.3   Relating Reasoning and Understanding to Knowledge

There is a complex relationship between knowledge, a term commonly used in curriculum documents, and reasoning and understanding which are more commonly used in statistics education research. Reasoning is an important cognitive process that is necessary when acquiring knowledge but one cannot reason until a certain level of understanding has been achieved. Further, understanding requires a relational (interconnected) set of links between relevant elements before increased cognitive activity can occur. For example, in their two-cycle hierarchy of reasoning about distribution, Reading and Reid (2006) found that the first cycle of levels involved "understanding" of key elements and then the second, more cognitively sophisticated, cycle of levels involved "using" those elements. When reporting on teacher knowledge of distribution in the following sections, research into teacher understanding of, and reasoning about, distribution are also considered relevant. For information about teachers' knowledge of other concepts in the web of statistical knowledge see other chapters in this book.

3   Frameworks to Describe Knowledge of Distribution

To better describe teachers' knowledge of distribution and how it develops, it is necessary to consider frameworks proposed by researchers to explain such knowledge, including understanding the concept of distribution and using the concept of distribution.

3.1   *Frameworks Dealing with Understanding the Concept of Distribution*

Three levels of understanding of distribution were described by Bakker and Gravemeijer (2004). At the first least sophisticated level, a distribution is simply viewed as a set of data values. At the second level, a distribution is viewed in terms of its underlying characteristics, expressed as summary statistics such as centre, spread and skewness. At the final level, the conceptual entity of distribution is recognised with data viewed as an aggregate. These levels were supported by Prodromou and Pratt's (2006) in-depth study of "thinking-in-change" about distribution and by others (e.g., Friel et al., 2006).

Four levels of increasing cognitive sophistication described by Ciancetta (2007) expanded on Bakker and Gravemeijer's (2004) levels. The first and second Ciancetta levels, similar to Bakker and Gravemeijer's first two levels, involved a local view of data that only allowed additive reasoning. The third Ciancetta level involved proportional reasoning and initial recognition of the global aspects of data, and the fourth (final) Ciancetta level involved integration of multiple aspects, indicating a global view of the distribution. These last two levels provide an expanded view of Bakker and Gravemeijer's third level.

Researchers generally agree on the levels required in the development of an understanding of distribution. However, for teachers the full knowledge of distribution must also include levels of development of reasoning about distribution, that is, how distribution can be used.

3.2   *Frameworks Dealing with Using the Concept of Distribution*

Understanding of a concept is more apparent when problems can be solved using that concept. A more cognitively sophisticated level of knowledge of distribution becomes apparent when the distribution concept appears in frameworks describing the development of knowledge of other statistical concepts. One such instance is based on an underlying framework for solving mathematical problems where Batanero et al. (2004) found that emphasis on distribution helped the transition from data analysis to statistical inference (a process which depends on distribution). A second instance is the Franklin and Garfield (2006) four-level developmental model for the statistical problem-solving process. Distribution figures prominently in the first three levels which deal with using characteristics of distribution in the context of a specific example, as tools for analysis, and in analysis as a global concept, respectively. The fourth (final) level does not refer specifically to distribution. These levels show increasing cognitive sophistication and are more about "using" than "understanding" the concept of distribution.

3.3   Framework for the Cognitive Development of Knowledge of Distribution

The various frameworks considered thus far have been proposed for developing an understanding, or for steps in reasoning or for applying in specific tasks involving distribution. While these frameworks have some commonality, each only covers part of the continuum of knowledge of distribution, from dealing with concepts that distribution depends on, that is, understanding distribution, to dealing with concepts that depend on distribution, that is, using distribution. A framework that covers a fuller spectrum of this knowledge (Reading & Reid, 2006) has levels of increasing cognition arranged into two separate cycles. The framework, based on the Structure of Observed Learning Outcome (SOLO) Taxonomy, was developed from assessment-task responses given by tertiary students, some of whom were training to be teachers. The first cycle of levels described increasing cognition in understanding distribution and the second cycle in using distribution for statistical inference. This second cycle is important for teachers as they need to have a well-developed knowledge of how to use distribution if they are to teach effectively about distribution. Reading and Reid (2006) found that supporting activities were needed to assist those with poor knowledge of variation to move through the first cognitive cycle of understanding distribution to the second cycle of using distribution to engage in inference.

Distribution's complex role in the web of statistical knowledge means that the process of developing a framework to describe the knowledge of distribution is equally complicated. Of the frameworks proposed to date, the Reading and Reid (2006) framework is the only one that provides detailed levels of cognitive development to describe both understanding and using distribution. These necessary levels provide a good focus for teacher learning about distribution, both while still being trained to teach and while teaching.

4   Teachers as Learners

Teachers need to build on what they know and hence grow as learners. On the one hand, teachers need to learn about the concept of distribution. On the other hand, they need to learn about the pedagogy associated with how students come to know that content. Given the concerns over gaps in teachers' statistical knowledge (Watson, 2001; Shaughnessy, 2007), it is not unexpected that research studies are now addressing teachers' learning about the concept of distribution and investigating the need for teachers to develop a better awareness of the difficulties students have in developing the concept of distribution (Garfield & Ben-Zvi, 2008). Although teachers are capable of building upon their students' own intuitive ideas, for example, using informal language such as referring to "clumps" or "bumps" of data when describing a distribution, Bakker and Gravemeijer (2004) noted that some classroom teachers had difficulty in maintaining class discussions adequate to elicit student understanding. Teachers need to learn by listening and observing how their students reason about distribution.

Two research studies about the statistical reasoning of an elementary school teacher (Mickelson & Heaton, 2004) and secondary school teacher (Pfannkuch, 2006) centred on different aspects of distributional reasoning but shared the commonality that they firmly cast the teacher in the role of learner. Mickelson and Heaton (2004) focused on one third grade teacher's reasoning about distribution as that teacher applied her evolving knowledge to conduct statistical investigation lessons. As context varied, the teacher inconsistently applied both exemplary and naïve statistical reasoning about distribution and seemed less able to demonstrate a deep understanding of distribution when the classroom investigation involved a truly open-ended statistical inquiry.

Pfannkuch (2006) focused on the reasoning articulated by a secondary teacher when making informal inferences based on comparing box plots and interpreting box plot distributions. The research developed a model comprising ten distinct elements to categorise the teacher's thinking and to describe the nature and type of informal inferential reasoning when students reason while comparing box plot distributions. The teacher was actively learning while teaching.

Both studies reinforced the notion that teachers need to be considered as learners and to consider themselves as learners, learning the knowledge of distribution itself as well as learning how students learn about distribution. Now the lens of teachers as learners is used to examine more deeply knowledge development *before* teaching, that is, as a part of pre-service teacher training and knowledge development *while* teaching, that is, as a part of professional development activities or through in-class research.

5   Knowledge Development Before Teaching

Recognising that there are different points along the professional learning cycle at which knowledge development occurs, attention is turned to research situated in pre-service teacher training. Studies involving elementary (i.e., primary and middle school) level pre-service teachers are described, followed by studies involving secondary level pre-service teachers. One common feature of these studies is that the pre-service teachers had at least one course in their programme that afforded statistical education opportunities.

5.1   *Elementary Pre-service Teacher Knowledge Development*

Three important research studies (Leavy, 2006; Canada & Ciancetta, 2007; Canada, 2008) have investigated the knowledge development of elementary pre-service teachers. Leavy's (2006) research took place in a teaching method class and investigated the approaches 23 pre-service teachers used to compare distributions of data. A shift from exclusive focus on calculating statistics (especially the mean) to inclusion of graphical representations occurred as the pre-service teachers considered variability as well as centres. Leavy (2006) cautioned that pre-service teachers need to examine the structure of their own knowledge to know what it means to understand a concept such as distribution and that this has not been encouraged by traditional teaching of pedagogy.

Canada and Ciancetta (2007) asked 58 pre-service teachers to explain their reasoning in deciding whether there was a real difference between trip times for two trains based on data with equal sample means. The research analysis framework considered how centre (average) and spread (variation) were incorporated in the responses. More limited comparisons of distributions were made when the pre-service teachers focused solely on centres, with richer comparisons being made by the distributional reasoners who also attended to variation.

Extending these findings, Canada (2008) compared 58 pre-service teacher responses with 50 middle school student responses, using the Canada and Ciancetta (2007) framework. Overall, more than three times as many pre-service teachers as students had distributional responses, that is, attending to both centre and spread. Despite this, there were qualitative similarities among the responses of both groups. Specifically, the informal language used by the pre-service teachers and students was very similar to that reported by Bakker and Gravemeijer (2004). For example, a pre-service teacher emphasised that distributions differed because one was "very spread out" and the other was "clustered together".

These three research studies involved elementary pre-service teachers and emphasised the concepts on which distribution depends, such as the integration of centre and variation while reasoning about graphical representations of data. Although pre-service teachers have been shown to be more distributional in their thinking than their students, they must still be encouraged to examine the structure of their own knowledge of distribution to better inform their teaching.

5.2   Secondary Pre-service Teacher Knowledge Development

Two important research studies (Makar & Canada, 2005; Ciancetta, 2007) have investigated the knowledge of secondary school teachers. Makar and Canada (2005) interviewed, before and after their course about assessment, 23 pre-service teachers (17 of them training for secondary teaching) after showing them 2 stacked dot plots of real data. The pre-service teachers were asked to determine the effectiveness of a course by comparing the exam mark improvement for students in, and not in the, course. In the evidence used by the pre-service teachers to support claims made, the words used to describe distribution were often very much like those used by elementary pre-service teachers, and both resembled the language students used. Teacher training needs to utilise familiar language to assist teachers to become more comfortable with mathematical language, to learn to reason about distribution and to learn how to delve into what their students understand (Makar & Canada, 2005).

Ciancetta's (2007) research involved data comparison tasks that focused on proportionality and included both equal and unequal sample sizes. Overall, the 275 pre-service teachers had difficulties with proportional reasoning. They reached different conclusions depending on which feature of distribution was the focus and they relied on additive reasoning when numerical summaries or graphs were not used. Ciancetta was careful to acknowledge the important role of context as the different inferences made varied according to the data context.

As with the elementary pre-service teachers, the research involving secondary pre-service teachers incorporates knowledge of multiple concepts connected to distribution in the web of statistical knowledge. However, these latter studies are distinctive in recognising a greater emphasis on proportionality. Importantly, pre-service teacher conclusions varied depending on what features of the distribution were the focus of their analyses.

6   Knowledge Development While Teaching

Research on the development, while teaching, of teacher knowledge of distribution falls across three categories focused on: specific teachers' learning while teaching in the classroom; teachers learning during professional development activities occurring outside the classroom and teachers learning while working with their students during in-class activities. Having previously described examples of studies (Mickelson & Heaton, 2004; Pfannkuch, 2006) from the first category, examples of studies in the second and third categories are now offered.

6.1   Knowledge Development During Professional Development Involving Teachers

Two studies (Makar & Confrey, 2004; McClain, 2008) have investigated knowledge development while teachers are engaged in professional development. Makar and Confrey (2004) described how four secondary teachers used statistical learning software to compare distributions to help decide whether two groups were different. A tremendous improvement was seen in teacher understanding but teachers were using informal language to convey an intuitive recognition of variation within a group as well as between groups. Makar and Confrey (2004) noted that teachers struggled to distinguish between within- and between-group variation, which is of critical importance when comparing distributions. It was also underscored how difficult it was to learn about teacher reasoning as compared to student reasoning, in part because teachers often see themselves as experts. Yet, because their knowledge of statistics is often weak, teachers may be more open to learning about distributional reasoning as opposed to other mathematical content domains.

McClain (2008) described professional development activities with 17 middle school teachers that involved comparing distributions using computer technology. Despite initially focusing on specific data points, the teachers were able to reason about the data as an aggregate, after engaging in professional development activities. In particular, for one task McClain (2008) suggested that the teachers were "attempting to coordinate the differences in the relative densities of the data sets as they clarified their arguments" (p. 5).

These studies focused on work with teachers through their participation in professional development activities. Taking place outside the teachers' classroom environment, this research involved, to some extent, a comparison of distributions to further develop teachers' knowledge. Teachers were found to have increased understanding after the professional development activities.

6.2   Knowledge Development During Professional Development Involving Teachers and Their Students

Hammerman and Rubin's (2004) research encompassed 11 middle- and high-school teachers as they participated in professional development seminars over 2 years. The researchers observed the teachers working in their classrooms and participating in a six-week sixth-grade teaching experiment. Both teachers and students analysed distributions of data via "cut points" (values which separate the data into groups) and "slices" (groups produced by multiple cut points) using the dynamic statistical program *Tinkerplots* (Hammerman & Rubin, 2004). The researchers found that different representations of a distribution gave rise to different questions and justifications about the data. Dynamic software facilitated these representations because "*seeing*" the distribution helped realisation that a measure of centre may not be representative of the entire distribution. This methodology allowed teachers and their students to be cast in the role of learners as they negotiated a deeper understanding of distribution.

Including students as part of the teacher professional development process allowed researchers to show that teachers have representations of distributions, and related questions and conclusions, that differed from those of their students. The variety of professional learning, within or outside the classroom, and involving only the teacher or the teacher and students, and associated research gives a broad perspective on how teachers can develop a deeper knowledge of distribution.

7   Conclusion

An important step to improving the way that teachers teach about statistical concepts, such as distribution, is recognition that the teacher needs a deep knowledge of the concept, including a functional understanding of the web of statistical knowledge. A better knowledge of distribution is critically connected to knowledge of concepts on which distribution depends and also to knowledge of concepts which depend on distribution. Teachers need to be moved beyond just understanding the concept of distribution to being able to use the concept in statistical activities. More professional learning experiences that allow examination of personal knowledge development of concepts, and of the connections between the concepts must be offered to teachers. It is recommended that representations such as concept maps (see, e.g., Afamasaga-Fuata'I & Reading, 2007) be considered as a visual aid to assist teachers to organise the interconnections of distribution and other concepts.

Further research is recommended to support teacher professional learning by refining a framework to explain the cognitive development of knowledge of distribution and by evaluating approaches to supporting the improvement of teacher knowledge of distribution. Consideration should be given to providing more opportunities for teachers to be involved in research and to encouraging teachers to co-learn with their students. In particular, teachers should be encouraged to research their practice and to share their findings with colleagues. Until researchers are able to better inform teachers' professional learning about distribution, it is unreasonable to expect that teachers' pedagogical approaches to teaching about distribution in particular and statistics, more generally, will improve.

References

Afamasaga-Fuata'i, K., & Reading, C. (2007). Using concept maps to assess pre-service teachers' understanding of connections between statistical concepts. In B. Phillips & L. Weldon (Eds.), *Assessing student learning in statistics: Proceedings of the IASE/ISI Satellite Conference on Statistical Education*. Voorburg, The Netherlands: International Association for Statistical Education. Online: www.stat.auckland.ac.nz/~iase/publications

Bakker, A. (2004). Reasoning about shape as a pattern in variability. *Statistics Education Research Journal, 3*(2), 64–83. Online: www.stat.auckland.ac.nz/serj/

Bakker, A., & Gravemeijer, K. P. E. (2004). Learning to reason about distribution. In D. Ben-Zvi & J. Garfield (Eds.), *The challenges of developing statistical literacy, reasoning, and thinking* (pp. 327–352). Dordrecht, The Netherlands: Kluwer.

Batanero, C., Burrill, G., Reading, C., & Rossman, A. (Eds.). (2008). *Joint ICMI/IASE Study: Teaching Statistics in School Mathematics. Challenges for teaching and teacher education. Proceedings of the ICMI Study 18 and 2008 IASE Round Table Conference*. Monterrey: International Commission on Mathematical Instruction and International Association for Statistical Education. Online: www.stat.auckland.ac.nz/~iase/publications

Batanero, C., Tauber, L. M., & Sanchez, V. (2004). Students' reasoning about the normal distribution. In D. Ben-Zvi & J. Garfield (Eds.), *The challenges of developing statistical literacy, reasoning, and thinking* (pp. 257–276). Dordrecht, The Netherlands: Kluwer.

Canada, D. (2008). Conceptions of distribution held by middle school students and preservice teachers. In C. Batanero, G. Burrill, C. Reading, & A. Rossman (2008).

Canada, D., & Ciancetta, M. (2007). Elementary preservice teachers' informal conceptions of distribution. In T. Lamberg & L. Wiest (Eds.), *Proceedings of the 29th Annual Meeting of the North American Chapter of the International Group for the Psychology of Mathematics Education* (pp. 960–967). Lake Tahoe, NV: University of Nevada, Reno.

Chance, B., delMas, R., & Garfield, J. (2004). Reasoning about sampling distributions. In D. Ben-Zvi & J. Garfield (Eds.), *The challenges of developing statistical literacy, reasoning, and thinking* (pp. 295–323). Dordrecht, The Netherlands: Kluwer.

Ciancetta, M. (2007). *Students' reasoning when comparing distributions of data.* Unpublished Ph.D. dissertation, Portland State University, Portland.

Franklin, C. A., & Garfield, J. (2006). The GAISE Project: Developing statistical education guidelines for grades pre-K-12 and college courses. In G. F. Burrill (Ed.), *Thinking and reasoning with chance and data: 68th NCTM Yearbook (2006)* (pp. 345–375). Reston, VA: National Council of Teachers of Mathematics.

Friel, S. N., O'Connor, W., & Mamer, J. D. (2006). More than "meanmodemedian" and a bar graph: What's needed to have a statistical conversation? In G. F. Burrill (Ed.), *Thinking and reasoning with chance and data: 68th NCTM Yearbook (2006)* (pp. 117–137). Reston, VA: National Council of Teachers of Mathematics.

Garfield, J. B., & Ben-Zvi, D. (2008). *Developing students' statistical reasoning: Connecting research and teaching practice.* New York: Springer.

Hammerman, J., & Rubin, A. (2004). Strategies for managing statistical complexity with new software tools. *Statistics Education Research Journal, 3*(2), 17–41. Online: www.stat.auckland.ac.nz/serj/

Kadijevich, D. (2008). Towards a suitably designed instruction on statistical reasoning: Understanding sampling distribution through technology. In C. Batanero, G. Burrill, C. Reading, & A. Rossman (2008).

Leavy, A. (2006). Using data comparisons to support a focus on distribution: Examining preservice teachers' understandings of distribution when engaged in statistical inquiry. *Statistics Education Research Journal, 5*(2), 27–45. Online: www.stat.auckland.ac.nz/serj/

Madden, S. (2008). *High school mathematics teachers' evolving understanding of comparing distributions.* Unpublished Ph.D. dissertation, Western Michigan University, Kalamazoo.

Makar, K., & Canada, D. (2005). Pre-service teachers' conceptions of variation. In H. Chick & J. Vincent (Eds.), *Proceedings of the 29th Conference of the International Group for the Psychology of Mathematics Education* (pp. 273–280). Melbourne, Australia: University of Melbourne.

Makar, K. M., & Confrey, J. (2004). Secondary teachers' reasoning about comparing two groups. In D. Ben-Zvi & J. Garfield (Eds.), *The challenges of developing statistical literacy, reasoning, and thinking* (pp. 327–352). Dordrecht, The Netherlands: Kluwer.

McClain, K. (2008). The evolution of teachers' understandings of distribution. In C. Batanero, G. Burrill, C. Reading, & A. Rossman (2008).

Mickelson, W., & Heaton, R. (2004). Primary teachers' statistical reasoning about data. In D. Ben-Zvi & J. Garfield (Eds.), *The challenges of developing statistical literacy, reasoning, and thinking* (pp. 327–352). Dordrecht, The Netherlands: Kluwer.

Pfannkuch, M. (2006). Comparing box plot distributions: A teacher's reasoning. *Statistics Education Research Journal, 5*(2), 27–45. Online: www.stat.auckland.ac.nz/serj/

Pfannkuch, M., & Reading, C. (2006). Reasoning about distribution: A complex process. *Statistics Education Research Journal, 5*(2), 4–9. Online: www.stat.auckland.ac.nz/serj/

Prodromou, T., & Pratt, D. (2006). The role of causality in the co-ordination of two perspectives on distribution within a virtual simulation. *Statistics Education Research Journal, 5*(2), 69–88. Online: www.stat.auckland.ac.nz/serj/

Reading, C., & Reid, J. (2006). An emerging hierarchy of reasoning about distribution: From a variation perspective. *Statistics Education Research Journal, 5*(2), 46–68. Online: www.stat.auckland.ac.nz/serj/

Shaughnessy, J. M. (2007). Research on statistics learning and reasoning. In F. Lester (Ed.), *Second handbook of research on mathematics teaching and learning* (pp. 957–1010). Greenwich, CT: Information Age Publishing and National Council of Teachers of Mathematics.

Watson, J. (2001). Profiling teachers' competence and confidence to teach particular mathematics topics: The case of chance and data. *Journal of Mathematics Teacher Education, 1*(2), 1–33.

Wild, C. (2006). The concept of distribution. *Statistics Education Research Journal, 5*(2), 10–26. Online: www.stat.auckland.ac.nz/serj/

Chapter 24
Students and Teachers' Knowledge of Sampling and Inference

Anthony Harradine, Carmen Batanero, and Allan Rossman

Abstract Ideas of statistical inference are being increasingly included at various levels of complexity in the high school curriculum in many countries and are typically taught by mathematics teachers. Most of these teachers have not received a specific preparation in statistics and therefore, could share some of the common reasoning biases and misconceptions about statistical inference that are widespread among both students and researchers. In this chapter, the basic components of statistical inference, appropriate to school level, are analysed, and research related to these concepts is summarised. Finally, recommendations are made for teaching and research in this area.

1 Introduction

Statistical inference, in the simplest possible terms, is the process of assessing strength of evidence concerning whether or not a set of observations is consistent with a particular hypothesised mechanism that could have produced those observations. It is an essential tool in management, politics and research; however, people's understanding of statistical inference is generally flawed. The application and interpretation of standard inference procedures is often incorrect

A. Harradine (✉)
Potts-Baker Institute, Prince Alfred College,
P.O. Box 571, Kent Town, SA 5071, Australia
e-mail: aharradine@pac.edu.au

C. Batanero
Departamento de Didáctica de la Matemática, Universidad de Granada,
Facultad de Ciencias de la Educación, Campus de Cartuja, 18071 Granada, Spain
e-mail: batanero@ugr.es

A. Rossman
California Polytechnic State University, San Luis Obispo, USA
e-mail: arossman@calpoly.edu

C. Batanero, G. Burrill, and C. Reading (eds.), *Teaching Statistics in School Mathematics-Challenges for Teaching and Teacher Education: A Joint ICMI/IASE Study*, DOI 10.1007/978-94-007-1131-0_24, © Springer Science+Business Media B.V. 2011

(see, e.g., Harlow, Mulaik, & Steiger, 1997; Batanero, 2000; Cumming, Williams, & Fidler, 2004).

Because of the relevance and importance of statistical inference, education authorities in some countries include a basic study of statistical inference in the curriculum of the last year of high school (17–18 year olds). For example, South Australian and Spanish students learn about statistical tests and confidence intervals for both means and proportions (Senior Secondary Board of South Australia [SSABSA], 2002; Ministry of Education and Sciences, 2007). New Zealand students learn about confidence intervals, resampling and randomisation (Ministry of Education, 2007).

Some of the fundamental elements of basic inference are implicitly or explicitly included in various middle school curricula, as well. For example, the National Council of Teachers of Mathematics (NCTM) Standards (2000) suggest that grades 6–8 students should use observations about differences between two or more samples to make conjectures about the populations. NCTM further recommends that grades 9–12 should use simulations to explore the variability of sample statistics from a known population and to construct sampling distributions; they also should understand how a sample statistic reflects the value of a population parameter and use sampling distributions as the basis for informal inference. More recently, the American Statistical Association's Guidelines for Assessment and Instruction in Statistics Education (GAISE; Franklin et al., 2005) highlights the need for students to look beyond the data when making statistical interpretations in the presence of variability and urges that students in middle grades recognise the feasibility of conducting inference and that high school students learn to make inferences both with random sampling from a population and with random assignment to experimental groups.

This chapter analyses the basic elements of statistical inference and then summarises part of the wider research that is relevant to teaching this topic (see Vallecillos, 1999; Batanero, 2000; Castro-Sotos, Vanhoof, Noortgate, & Onghena, 2007 for an expanded survey). The chapter finishes with some implications for teaching and research.

2   Statistical Inference – A Rich Melting Pot

Classical statistical inference consists primarily of two types of procedures, hypothesis testing and confidence intervals. These techniques build on a scheme of interrelated concepts including probability, random sampling, parameter, distribution of values of a sample statistic, confidence, null and alternative hypothesis, p-value, significance level and the logic of inference (Liu & Thompson, 2009).

Consequently, statistical inference consists of three distinct, but interacting, fundamental elements: (a) the reasoning process, (b) the concepts and (c) the associated computations. Because the computations are often easily learned by students, and can be facilitated by user-friendly software, teachers of statistics must teach the three components and not just the mechanics of inference, because the main difficulties in understanding statistical inference lie within the other two elements.

2.1   The Reasoning Process

Garfield and Gal (1999) suggest that, across the primary, middle and high school years, teachers must develop students' statistical reasoning – the processes people use to reason with statistical ideas and make sense of statistical information. This process is supported by concepts such as distribution, centre, spread, association, uncertainty, randomness and sampling, some of which have been analysed in other chapters in this book. While most students may be able to perform the calculations associated with an inferential process, many students hold deep misconceptions that prevent them from making an appropriate interpretation of the result of an inferential process (Vallecillos, 1994; Batanero, 2000; Castro-Sotos et al., 2007). In addition, Garfield (2002) remarks that some teachers do not specifically teach students how to use and apply types of reasoning but rather teach concepts and procedures and hope that the ability to reason will develop as a result. As a consequence, students reach their first inferential reasoning experience with a reasoning-free statistical background, giving rise to a mind-set that statistics is solely about the computation of numerical values. One possible reason for this unfortunate circumstance is that teachers responsible for teaching statistics at a high school level may have serious deficiencies in their knowledge that lead to inadequate understandings of inference (Liu & Thompson, 2009).

2.2   The Concepts

Central to learning statistical inference is understanding that the variation of a given statistic (e.g., the mean) calculated from single random samples is described by a probability distribution – known as the sampling distribution of the statistic. When thinking about statistical inference it is necessary to be able to clearly differentiate between three distributions:

- The *probability distribution* that models the values of a *variable* from the population/process. This distribution usually depends on some (typically unknown) parameter values. For example, a normally distributed population is specified by two parameters – its mean and standard deviation, often denoted by μ and σ .
- The *data distribution* of the values of a *variable* for a *single* random sample taken from the population/process. From this sample, statistics such as the mean and standard deviation, often denoted by $\bar{x}$ and s can be used in the process of estimating the unknown values of the population parameters.
- The *probability distribution* that models the variability in values of a statistic from 'all' potential random samples taken from the population/process, called the sampling distribution. One example is the sampling distribution of a sample mean, which in many circumstances has an approximately normal distribution with mean μ and standard deviation $\frac{\sigma}{\sqrt{n}}$, where n represents the sample size. This result provides the basis for much of classical statistical inference.

Sampling distributions are more abstract than the distribution of a population or a sample and so are typically very challenging for students to understand (see Sect. 3.2). One reason for this difficulty is that when thinking about both the population distribution and the single random sample's distribution, the unit of analysis (case) is an individual object. This is in stark contrast to the sampling distribution where the case is a single random sample (Batanero, Godino, Vallecillos, Green, & Holmes, 1994). The object of interest for each distribution might be the mean, for example, but in each case the distribution's mean has a different interpretation and a different behaviour. One strategy for helping students to understand these distinctions is to engage in activities that involve repeatedly taking random samples from a population. When working with such activities, high school students often struggle with moving between the various levels of imagery (Saldanha & Thompson, 2002). Proper application and interpretation of statistical inference requires mastery of the knowledge and techniques specific to each distribution and understanding of the rich links among these distributions.

3   Difficulties in Understanding Statistical Inference

Research reviewed in this section deals with understanding sampling and the sampling distribution, hypothesis tests and confidence intervals.

3.1   Understanding Sampling

Research on inferential reasoning started with the *heuristics and biases* programme of research in psychology (Kahneman, Slovic, & Tversky, 1982), which established that most people do not follow the normative mathematical rules that guide formal scientific inference when they make a decision under uncertainty. Instead, people tend to use simple judgmental heuristics that sometimes cause serious and systematic errors, and such errors are resistant to change. For example in the *representativeness* heuristics, people tend to estimate the likelihood for an event based on how well it represents some aspects of the parent population. An associated fallacy that has been termed *belief in the law of small numbers* is the belief that even small samples should exactly reflect all the characteristics in the population distribution.

Most curricula at a high school level include some instruction on random sampling, which is mostly theoretical and includes descriptions of different methods of random sampling. The core message of such instruction is that if a sample is chosen in a *suitable random manner* and is *sufficiently big*, it will be representative of the population from which it has been drawn. Students therefore learn to think about a random sample as a *mini-me* of the population and that the purpose of drawing a random sample is to ensure *representativeness* in order to gain knowledge about the population from the sample. This conception constrains students' thinking to a single random sample only and provides no avenue to appreciate the range of possible samples that *might have been drawn* and the variability across that range.

Understanding the purpose of drawing a single random sample in the context of hypothesis tests and confidence intervals, requires the assimilation of "two apparently antagonistic ideas: sample representativeness and (sampling) variability" (Batanero et al., 1994). In these situations the purpose of drawing a single sample is to quantify that sample's level of unusualness relative to the many other samples that could have been drawn. Saldanha and Thompson (2002) observed that, without a suitable sense of the variation across many possible samples, which extends to the notion of the distribution of a statistic, grade 11 and 12 students tended to judge a sample's representativeness only in relation to the population parameter. Hence, when required to decide how rare a sample was, these students did so based on how different they thought it was to the underlying population parameter and not "on how it might compare to a clustering of the statistic's values" (Saldanha & Thompson, 2002).

3.2   Understanding Sampling Distributions

Reasoning about sampling distributions requires students to integrate several statistical concepts and to be able to reason about the hypothetical behaviour of many samples – an intangible thought process for many students (Chance, Delmas, & Garfield, 2004). According to these authors, many students fail to develop a deep understanding of the sampling distribution concept and as a result can only manage a mechanical knowledge of statistical inference, leaving such tasks as interpreting a *p*-value well beyond those students.

Saldanha and Thompson (2002) studied the understandings of high school students when engaged in activities that used computer applets to simulate repeated random sampling from a population. The activity required students to randomly draw a sample from a population, compute a sample proportion and then repeat this process over and over. They found that most students had extreme difficulty in conceiving of repeated sampling in terms of three distinct levels: population, sample and collection of sample statistics. These difficulties led many students to misinterpret a simulation's result as a percentage of people rather than a percentage of sample proportions.

Chance et al. (2004) found that while students were able to observe behaviours and notice patterns in the behaviour (e.g., larger the sample size smaller the variation) shown by random sampling applets, they did not understand why the behaviour occurred. The authors noted that, after exposure to applets, students were unable to suggest plausible distributions of samples for a given sample size and agreed with Saldanha and Thompson that students did not have a clear distinction between the distribution of one sample of data and the distribution of means of samples.

Simply being exposed to the applets was not sufficient to render a learning gain. The authors concluded that: (a) students need to become more familiar with the process of sampling, (b) activities associated with applets need to be both structured and unstructured and (c) students need to discuss their observations after an activity so they could become focussed on what observations are most important, what important observations they did not make and how the important observations are connected.

3.3   Understanding the Null and Alternative Hypotheses

Errors and misinterpretations in hypothesis tests can lead to a paradoxical situation, where, on one hand, a significant result is often required to get a paper published in many journals and, on the other hand, significant results are misinterpreted in these publications (Falk & Greenbaum, 1995). There is confusion between the roles of the null and alternative hypotheses as well as between the statistical alternative hypothesis and the research hypothesis (Chow, 1996). Vallecillos (1994) reported that many students in her research, including 6 out of 31 pre-service mathematics teachers, believed that correctly carrying out a test proved the truth of the null hypothesis, as in the case of a deductive procedure. Vallecillos (1999) described four different conceptions regarding the type of proof that hypotheses tests provide: (a) as a decision-making rule, (b) as a procedure for obtaining empirical support for the hypothesis being researched, (c) as a probabilistic proof of the hypotheses and (d) as a mathematical proof of the truth of the hypothesis. While the two first conceptions are correct, many students in her research, including some pre-service teachers, held either conception (c) or (d).

Belief that rejecting a null hypothesis means that one has proven it to be wrong was also found in the research by Liu and Thompson (2009) when interviewing 8 high school statistics teachers, who seemed not to understand the purpose of statistical tests as mechanisms to carry out statistical inferences.

3.4   Understanding Statistical Significance and p-values

Two particularly misunderstood concepts are the significance level and the p-value. The significance level is defined as the probability of falsely rejecting a null hypothesis. The p-value is defined as the probability of observing the empirical value of the statistic or a more extreme value, given that the null hypothesis is true. The most common misinterpretation of these concepts consists of switching the two terms in the conditional probability: interpreting the level of significance as the probability that the null hypothesis is true once the decision has been made to reject it or interpreting the p-value as the probability that the null hypothesis is true, given the observed data. For example, Birnbaum (1982) reported that his students found the following definition reasonable: "A level of significance of 5% means that, on average, 5 out of every 100 times we reject the null hypothesis, we will be wrong". Falk (1986) found that most of her students believed that α was the probability of being wrong when rejecting the null hypothesis at a significance level α. Similar results were found by Krauss and Wassner (2002) in university lecturers involved in the teaching of research methods. More specifically they found that four out of every five *methodology instructors* have misconceptions about the concept of significance, just like their students. Vallecillos (1994) carried out extensive research on students' misconceptions related to statistical tests ($n = 436$ students from different backgrounds) that included 31 pre-service mathematics teachers (students graduating in mathematics), 13 of whom interpreted the level of significance as the probability that the null hypothesis is true, once the decision to reject it has been made.

Liu and Thompson (2009) remark that the ideas of probability and unusualness are central to the logic of hypothesis testing, where one rejects a null hypothesis when a sample from a population is judged to be sufficiently unusual in light of the null hypothesis. However, they found that teachers "conceptions of probability (or unusualness) were not grounded in a conception of distribution and thus did not support thinking about distributions of sample statistics and the fraction of the time that a statistic's value is in a particular range" (p. 16). While a single random sample is a critical part of statistical inference, probably more important is an appreciation of the "could-have-been" – all the other random samples that could have been drawn but were not. "Sampling has not been characterised in the literature as a scheme of interrelated ideas entailing repeated random selection, variability and distribution". (Saldanha & Thompson, 2002, p. 258).

3.5   Understanding Confidence Intervals

Fidler and Cumming (2005) asked a sample of 55 undergraduates and postgraduate science students to interpret statistically non-significant results and gave the results in two different ways (first as p-values and then as confidence intervals or vice versa). Students were asked to indicate whether the results provided support for the null hypothesis (considered as a misconception), provided support against the null hypothesis or neither. The authors found that students misinterpreted p-values twice as often as they misinterpreted confidence intervals. There was also evidence that students who were given the confidence interval results first gave the correct answer on the p-value presentation more often than students who were given the p-value results first. The author concluded there are benefits of teaching inference via confidence intervals rather than hypothesis tests.

Cumming et al. (2004) reported an internet study in which researchers were given results from an experiment (simulated in an applet) and were asked to show where they thought the ten means from ten 'new' samples could plausibly fall. The results suggested that a majority of the researchers held a misconception that an $r\%$ confidence interval will, on average, capture $r\%$ of the means of the 'new' samples.

4 Implications for Teaching and Research

Castro-Sotos (2009) reported slightly lower percentages of students with certain misconceptions related to hypothesis testing when compared to similar studies from earlier years. The author suggests that innovation in statistics education in the last decade may be resulting in some level of improved understanding of statistical inference. While this is merely conjecture, it highlights the idea that students must develop an understanding of many challenging probabilistic and statistical concepts and the relationships between them before meeting statistical inference. Given the difficulty learners have integrating the concepts involved in statistical inference, it makes sense that the underpinning ideas need to be developed over years, not weeks.

4.1 Inference-Friendly Views of a Sample

Statistical inference is applied to a wide variety of situations. However, understanding why it can be validly applied to one situation does not mean learners will understand why it can (or cannot) be validly applied to another, for example, a situation involving the mean of a finite population compared to a situation involving measurement error (where a population does not exist, but a true value of the measurement does). Students need to hold multiple views of a sample, appreciating the source(s) of the variability that give rise to the sample characteristics, to deeply understand statistical inference and its many applications. Context is clearly critical in supporting a student to develop different views of a sample. Konold and Lehrer (2008) discuss three contexts from which samples are produced: measurement error, manufacturing processes and natural variation.

A critical view of a sample is as the result of a *target-error* process, which aims to consistently produce a single value but fails due to the unavoidable variation in the process (e.g., the machine process that aims to cut fruit bars to be exactly 7 cm long). This can be referred to as the *target-error-view* of sample. Opportunities to develop this view are rarely, if ever, provided at a school level. Natural variation contexts (e.g., the weight of all female quokkas on Rottnest Island) are the most common contexts students meet at school but do not help in developing this critical view of a sample.

Students also need opportunities, over a period of years, to develop a *view* of a sample as a single instantiation of the random sampling process from a population and to develop the appreciation that each possible random sample carries with it an associated level of unusualness (the probability of being drawn). This is referred to as the *population-view* of a sample. While this is the most common view, and current school curricula attempt to develop this using contexts associated with natural variation, it is possible that the target-error-view of a sample should be developed prior to the population-sample view. Konold, Harradine, and Kazak (2007) describe activities in which middle school students build *data factories* with the aim of assisting in the development of the target-error-view. Their approach also develops the notion that data result from chance based processes and as such makes explicit the relationship between data and chance; a relationship critical to understanding statistical inference and that has been lost (or was never present) in many current school curricula (Konold & Kazak, 2008). Without such views of sample, it is difficult to develop a deep understanding of, and validly apply, statistical inference.

4.2 Developing an Understanding of the Population-View of a Sample

Many interactive applets are now available that provide dynamic, visual environments within which students can engage in the construction of sampling distributions. Chance et al. (2004) reported on a series of studies that investigated the impact that interacting with such applets had on students' understanding when learning about sampling distributions. In the first studies, students tended to look for rules when answering test items and did not understand the underlying relationships that caused the visible patterns they noticed as a result of using the applets. In later studies, the authors asked the students to make predictions about sampling distributions of means before using the applets to validate their predictions. This strategy proved to be useful in improving the students' reasoning about sampling distributions.

4.3 Alternative Ways to Introduce Statistical Inference

Most students' first introduction to statistical inference is via a first course in classical statistical inference. In recent years the literature has included thinking about what is termed *informal inference*. While informal inference, as a concept, is not yet universally agreed upon, a consistent feature of informal inference is that suggested activities engage students in the reasoning process of statistical inference without relying on probability distributions and formulas.

Some see informal inference as the collection of the fundamental ideas that underpin the understanding of classical statistical inference. These fundamentals include discriminating between signal and noise in aggregates, understanding sources of variability, recognising the effect of sample size and being able to identify tendencies and sources of bias (Rubin, Hammerman, & Konold, 2006). Other views of informal inference include (Zieffler, Garfield, Delmas, & Reading, 2008): (a) reasoning about possible characteristics of a population from a sample of data, (b) reasoning about possible differences between two populations from observed differences between two samples of data and (c) reasoning about whether or not a particular sample statistic is likely or unlikely given a particular expectation about the population.

Cobb (2007) proposes teaching the logic of inference with randomisation tests rather than using normal distributions as approximate models for sampling distributions, noting that such an approach is what Ronald Aylmer Fisher advocated, but which was not realistic in his day due to the absence of computers. Rossman (2008) claims that teachers could use randomisation tests to connect the randomness that students perceive in the process of collecting data to the inference to be drawn. He provides examples of how such a randomisation-based approach might be implemented, while Scheaffer and Tabor (2008) propose such an approach for the secondary curriculum and provide relevant examples.

4.4 Teacher Knowledge

Research results summarised in this chapter primarily concern students' misconceptions and difficulties in learning about statistical inference. The little research available about teachers' understanding of statistical inference (Vallecillos, 1994, 1999; Krauss & Wassner, 2002; Liu & Thompson, 2009) indicates it is possible that some teachers share the same misconceptions as the students. In addition, teachers who have not studied statistical inference prior to having to teach it are likely to have the same difficulties in learning the concepts as students do. If this is the case, and the situation is not addressed, then it is unlikely that widespread improvement in student understanding will be seen any time soon.

4.5 Some Research Priorities

The valid application of statistical inference is of critical importance in a broad range of human endeavours. Areas in which research attention is needed include:

- The creation and critical evaluation of a curriculum that systematically develops the key ideas that underpin statistical inference across a number of years in the middle and high school years, so a proper foundation is laid for the formal instruction of statistical inference.
- The study of the current level of understanding and professional knowledge, both at a school and university level, of those teachers charged with teaching statistical inference.
- The critical evaluation of the use of alternative methods (e.g., randomisation tests) when first introducing statistical inference. Great care should be taken in this area given the widespread and long-term use of classical statistical inference.

References

Batanero, C. (2000). Controversies around significance tests. *Mathematical Thinking and Learning,* 2(1–2), 75–98.

Batanero, C., Godino, J. D., Vallecillos, A., Green, D. R., & Holmes, P. (1994). Errors and difficulties in understanding elementary statistical concepts. *International Journal of Mathematics Education in Science and Technology,* 25(4), 527–547.

Birnbaum, I. (1982). Interpreting statistical significance. *Teaching Statistics, 4*, 24–27.

Castro-Sotos, A. E. (2009). How confident are students in their misconceptions about hypothesis tests? *Journal of Statistics Education, 17*(2), 1–19. Online: www.amstat.org/publications/jse/

Castro-Sotos, A. E., Vanhoof, S., Noortgate, W., & Onghena, P. (2007). Students' misconceptions of statistical inference: A review of the empirical evidence from research on statistics education. *Educational Research Review, 2*, 98–113.

Chance, B., delMas, R. C., & Garfield, J. (2004). Reasoning about sampling distributions. In D. Ben-Zvi & J. Garfield (Eds.), *The challenge of developing statistical literacy, reasoning and thinking* (pp. 295–323). Amsterdam: Kluwer.

Chow, L. S. (1996). *Statistical significance: Rationale, validity and utility*. London: Sage.

Cobb, G. (2007). The introductory statistics course: A Ptolemaic curriculum? *Technology Innovations in Statistics Education, 1*(1). Online: www.repositories.cdlib.org/uclastat/cts/tise/

Cumming, G., Williams, J., & Fidler, F. (2004). Replication, and researchers' understanding of confidence intervals and standard error bars. *Understanding Statistics, 3*, 299–311.

Falk, R. (1986). Misconceptions of statistical significance. *Journal of Structural Learning, 9*, 83–96.

Falk, R., & Greenbaum, C. W. (1995). Significance tests die hard: The amazing persistence of a probabilistic misconception. *Theory and Psychology, 5*(1), 75–98.

Fidler, F., & Cumming, G. (2005). Teaching confidence intervals: Problems and potential solutions. *Proceedings of the International Statistical Institute 55th Session*. Sydney, Australia: International Statistical Institute. Online: www.stat.auckland.ac.nz/~iase/publications

Franklin, C., Kader, G., Mewborn, D., Moreno, J., Peck, R., Perry, M., & Scheaffer, R. (2005). *Guidelines for assessment and instruction in statistics education (GAISE) report: A pre K-12 curriculum framework*. Alexandria, VA: American Statistical Association. Online: www.amstat.org/Education/gaise/

Garfield, J. B. (2002). The challenge of developing statistical reasoning. *Journal of Statistics Education, 10* (3). Online: http://www.amstat.org/publications/jse/

Garfield, J., & Gal, I. (1999). Teaching and assessing statistical reasoning. In L. Stiff (Ed.), *Developing mathematical reasoning in grades K-12* (pp. 207–219). Reston, VA: National Council Teachers of Mathematics.

Harlow, L. L., Mulaik, S. A., & Steiger, J. H. (1997). *What if there were no significance tests?* Mahwah, NJ: Lawrence Erlbaum Associates.

Kahneman, D., Slovic, P., & Tversky, A. (1982). *Judgment under uncertainty: Heuristics and biases*. New York: Cambridge University Press.

Konold, C., Harradine, A., & Kazak, S. (2007). Understanding distributions by modeling them. *International Journal of Computers for Mathematical Learning, 12*(3), 217–230.

Konold, C., & Kazak, S. (2008). Reconnecting data and chance. *Technology Innovations in Statistics Education, 2*(1). Online: www.repositories.cdlib.org/uclastat/cts/tise/

Konold, C., & Lehrer, R. (2008). Technology and mathematics education: An essay in honor of Jim Kaput. In L. D. English (Ed.), *Handbook of international research in mathematics education* (2nd ed., pp. 49–71). New York: Routledge.

Krauss, S., & Wassner, C. (2002). How significance tests should be presented to avoid the typical misinterpretations. In B. Phillips (Ed.), *Proceedings of the Sixth International Conference on Teaching Statistics*. Cape Town, South Africa: International Statistical Institute and International Association for Statistical Education. Online: www.stat.auckland.ac.nz/~iase/publications

Liu, Y., & Thompson, P. W. (2009). Mathematics teachers' understandings of proto-hypothesis testing. *Pedagogies, 4*(2), 126–138.

Ministry of Education. (2007). *The New Zealand curriculum*. Wellington, New Zealand: Learning Media Limited.

Ministry of Education and Sciences. (2007). *Real Decreto 1467/2007, de 2 de noviembre, por el que se establece la estructura del bachillerato y se fijan sus enseñanzas mínimas* (Royal Decree establishing the structure of high school curriculum).

National Council of Teachers of Mathematics. (2000). *Principles and standards for school mathematics*. Reston, VA: Author.

Rossman, A. (2008). Reasoning about informal statistical inference: One statistician's view. *Statistics Education Research Journal, 7*(2), 5–19. Online: www.stat.auckland.ac.nz/serj/

Rubin, A., Hammerman, J. K. L., & Konold, C. (2006). Exploring informal inference with interactive visualization software. In B. Phillips (Ed.), *Proceedings of the Sixth International Conference on Teaching Statistics.* Cape Town, South Africa: International Association for Statistics Education. Online: www.stat.auckland.ac.nz/~iase/publications

Saldanha, L., & Thompson, P. (2002). Conceptions of sample and their relationship to statistical inference. *Educational Studies in Mathematics, 51*, 257–270.

Scheaffer, R., & Tabor, J. (2008). Statistics in the high school mathematics curriculum: Building sound reasoning under uncertainty. *Mathematics Teacher, 102*(1), 56–61.

Senior Secondary Board of South Australia. (2002). *Mathematical studies curriculum statement.* Adelaide, Australia: Author.

Vallecillos, A. (1994). *Estudio teórico-experimental de errores y concepciones sobre el contraste estadístico de hipótesis en estudiantes universitarios* (*Theoretical and experimental study on errors and conceptions about hypothesis testing in university students*). Unpublished Ph.D., University of Granada, Spain.

Vallecillos, A. (1999). Some empirical evidence on learning difficulties about testing hypotheses. *Proceedings of the International Statistical Institute 52nd Session.* Helsinki: International Statistical Institute. Online: www.stat.auckland.ac.nz/~iase/publications

Zieffler, A., Garfield, J. B., delMas, R., & Reading, C. (2008). A framework to support research on informal inferential reasoning. *Statistics Education Research Journal, 7*(2), 5–19. Online: www.stat.auckland.ac.nz/serj/

Chapter 25
Correlation and Regression in the Training of Teachers

Joachim Engel and Peter Sedlmeier

Abstract Although the notion of functional dependence of two variables is fundamental to school mathematics, teachers often are not trained to analyse statistical dependencies. Many teachers' thinking about bivariate data is shaped by the deterministic concept of a mathematical function. Statistical data, however, usually do not perfectly fit a deterministic model but are characterised by variation around a possible trend. Therefore, understanding regression and correlation requires, apart from basic knowledge about functions, an appreciation of the role of variation. In this chapter, common errors and fallacies related to the concepts of correlation and regression are revisited and suggestions on how teachers may overcome some of these difficulties are provided.

1 Introduction

Since Felix Klein declared in the year 1905 that the notion of "functional reasoning" was one of the overarching mathematical ideas (Inhetveen, 1976), the concept of function has pervaded the school curriculum in mathematics. Even though the term 'function' is not explicitly introduced before middle grades of secondary school, students encounter situations where elements from one set are put into correspondence with elements from another set since grade 1. The concept of association or statistical dependence extends the notion of functional dependence and is fundamental for many statistical methods. Although a function is defined by uniquely assigning each element of a domain to an element of the range, the situation is more complicated

J. Engel (✉)
University of Education, Reuteallee 46, 71634 Ludwigsburg, Germany
e-mail: engel@ph-ludwigsburg.de

P. Sedlmeier
University of Technology, 09107 Chemnitz, Germany
e-mail: peter.sedlmeier@psychologie.tu-chemnitz.de

C. Batanero, G. Burrill, and C. Reading (eds.), *Teaching Statistics in School Mathematics-Challenges for Teaching and Teacher Education: A Joint ICMI/IASE Study*, DOI 10.1007/978-94-007-1131-0_25, © Springer Science+Business Media B.V. 2011

when dealing with bivariate statistical data: Often it is difficult to differentiate between the dependent and independent variable with only an association that does not allow an attribution of cause and effect in contrast to a deterministic situation.

In this chapter, some common misconceptions and errors with respect to bivariate relationships are first described. Then some mathematical problems associated with the understanding of correlation and regression are discussed, and the problematic issue of variation is treated in more detail by interpreting the analysis of functional data as search for signals in noisy processes. Finally, research on pre-service teachers' learning of association is summarised, and its implications for training teachers as well as for future research are discussed.

2 Research on Understanding Correlation and Regression

2.1 Psychological Biases

Psychological research has shown that making judgments about associations, in both bidirectional (correlations) and unidirectional (regression) analyses, is not always easy and that even people with statistical backgrounds sometimes experience difficulties in assessing and interpreting associations (Nisbett & Ross, 1980). This is due to a variety of factors:

- *Influence of previous beliefs*: Adults tend to base their judgments on their previous beliefs about the type of association that ought to exist between the variables that are to be studied rather than on the empirical contingencies presented in the data. A well known experimental demonstration of this phenomenon was given by Chapman and Chapman (1969), who presented to participants drawings of faces combined with some psychiatric diagnoses. Even if objectively there was no association between the drawing and the diagnosis, participants "noticed" some correlation if the drawing fit common expectations (or prejudices) about patients with a given diagnosis. For instance, a face with huge eyes was often paired with the diagnosis "paranoia".
- *Illusory correlation*: Sometimes people perceive a correlation where there is none if the base rate is high. In studies with artificial diseases and potential symptoms, where prior expectations did not exist, small correlations between a disease and a symptom tended to be seen as substantial if the base rate of the disease was high (Smedslund, 1963; Vallee-Tourangeau, Hollingsworth, & Murphy, 1998). A related common finding in social psychology research is an illusory correlation between group size (majority and minority groups) and the social adequacy of behaviour (positive and negative behaviour): The majority group is judged more positively even when the two groups do not differ in their behaviour. This effect is also due to differences in base rates, that can also be "created" by participants' selective attention (Fiedler, Brinkmann, Betsch, & Wild, 2000; Fiedler, Walther, Freytag, & Plessner, 2002) and can be well simulated with models of associative learning (e.g., Sedlmeier, 2006).

- *Misjudgement of strength of covariation*: Implicit theories about covariation seem to have a strong impact on the correlation people perceive, and therefore, correlation is likely to be overestimated when previous theories about associations exist. However, a strong correlation between data is necessary to detect the association when previous theories do not exist, and in this case correlations tend to be underestimated (Jennings, Amabile, & Ross, 1982). Also the format of presentation seems to have an impact on the perceived amount of covariation since a graphical format of the data tends to induce judgments of stronger correlation than a tabular format (Lane, Anderson, & Kellam, 1985). Moreover, positively correlated variables are more likely to be perceived correctly than negatively correlated data (Erlick & Mills, 1967).

- *Confounding variables*: The Simpson's paradox occurs when neglecting an explanatory third variable or confounder which causes a reversal of an association (e.g., Freedman, Pisani, & Purves, 1998). For example, a German newspaper reported that students who progress slowly through their academic programme make more money in their first year on a job than those students who graduate in shorter time. In the example (see Fig. 25.1), the confounding or lurking variable is the field in which the degree was obtained. Although it usually takes the longest time to get a diploma in chemistry, within the field, the ones who finish faster earn more. When regressing salary on time enrolled for the whole data, a positive slope is obtained, although the slope is negative when differentiating according to the field of study.

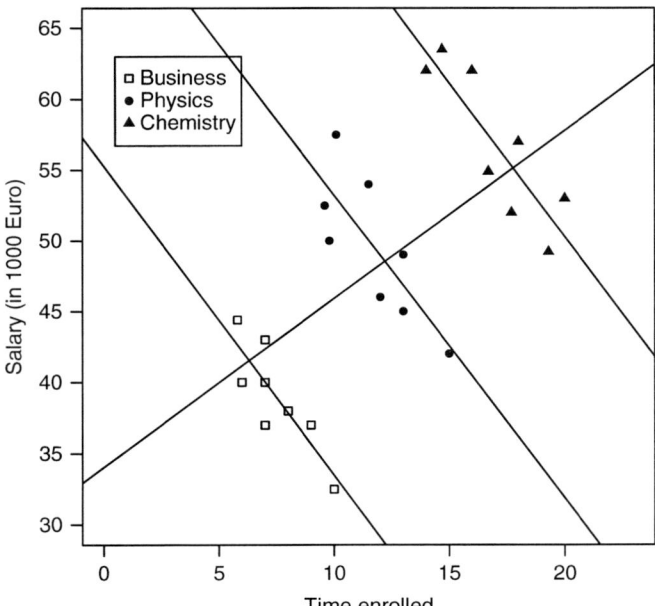

Fig. 25.1 Time enrolled until graduation (in semesters) and salary in first year of employment (in thousand €)

- *Regression effect regarded as real effect*: In any test-retest situation, a variable that is extreme on its first measurement will tend to be closer to the centre of the distribution on a later measurement *even if there is no effect of intervention*. To prevent wrong inferences due to the *regression fallacy* this effect has to be taken into account when designing experiments and interpreting empirical data. A regression effect cannot be avoided unless two variables correlate perfectly: the smaller the correlation, the larger the regression effect. Some laypeople and researchers have a poor understanding of the regression effect and tend to interpret it as a "real" effect of some treatment related to the variable (Jennings et al., 1982; Stelzl, 1982).
- *Transitivity misconception*: Some people believe that when two variables X and Z are positively correlated with a third variable Y, then X and Z have positive correlation (Castro-Sotos, Van Hoof, Van den Noortgate, & Onghena, 2009). Mathematical analysis, however, shows this to be true only under very restrictive conditions. Specifically, denoting the correlations respectively by r_{xy}, r_{yz} and r_{xz}, then it holds that $r_{xz} > 0$ if and only if $r_{xy}^2 + r_{yz}^2 > 1$.

2.2   Mathematical Difficulties

Although mathematically trained people such as mathematics teachers can usually derive the formulae for the regression line and for the correlation coefficient, the meaning of the formulae and their relationship is not always readily understood, as shown in the following common difficulties:

- *Association instead of dependence*: Although statistical variables may be in a causal relationship, there might be many other reasons for their covariation, and in many situations it is not possible to identify one variable as independent and the other as dependent (e.g., Freedman et al., 1998). Moreover, although in correlation analysis the relationship between the two variables is completely symmetrical, in regression, the role of the two variables is not symmetrical. One variable is termed *predictor* or *explanatory*, expressing the idea of predicting Y for a given X, the other is called *criterion* or *response* variable. The formulae for both regression and correlation are very similar, but their meaning and their use are quite different. Correlation is meaningful, if both X and Y are considered random, whereas a regression line (as a function fitted to the data) makes sense in either a fixed or a random design, that is, regardless if the values of the predictor variable are set to a set of fixed values or if predictor and response variables are both random.
- *High correlation does not imply validity of a linear model*: Many investigators would consider values of r close to $+1$ or -1 as proof that the linear model is valid. However, this is not always the case. The squared correlation coefficient is simply a measure of how much of the variation (variance) of the Y variable is accounted for by the mathematical model and as a consequence, the coefficient alone is an inappropriate mean for evaluating linearity. In Fig. 25.2, four data

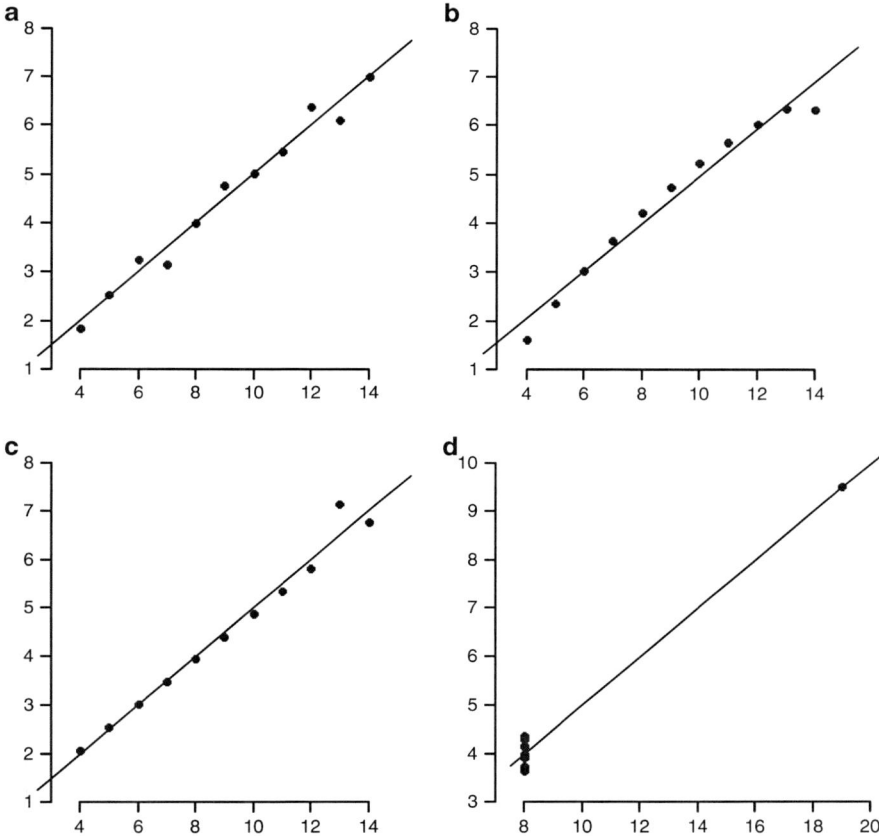

Fig. 25.2   Four data sets fitted to the same linear model, $y=0.5x$ with $r=0.99$

sets taken from a famous example by Anscombe (1973) with small modifications
are presented. All four data sets in Fig. 25.2 have identical correlation coefficients
(to a certain decimal place) of $r=0.99$ and regression lines $y=0.5x$. Whereas
data set (**a**) represents an ideal pattern with randomly scattered residuals around
the regression line, the data in (**b**) are noise-free but curved. The data in (**c**) and
(**d**) illustrate the effect of a single outlier. These four examples illustrate
impressively the importance of looking at graphical representations instead of
relying on computed parameters alone.

• *Interpreting the correlation coefficient*: Many people believe that the correlation
  between two variables X and Y is positive if as X increases Y also increases, or
  equivalently, if large values of X often correspond with large values of Y.
  Kharsikar and Kunte (2002) give a striking example showing that such
  statements are not always correct. A more precise statement characterising a
  positive correlation is that *above-average* values of X correspond to *above-average*
  values of Y.

2.3 Difficulties with the Functional Understanding of Associations

Many misconceptions regarding correlation and regression are related to a deterministic world view. The tendency to search for specific causes is very deep-seated and leads people to search for causes if the data are quite within the bounds of what would be expected in random variation. In particular, secondary school students' adherence to a mechanistic-deterministic view of the world is well documented and does not seem to fade with increasing years of schooling (see Engel & Sedlmeier, 2005).

The concept of a function is usually acquired in a deterministic set-up where X is the single cause of Y and to every value x of X there corresponds exactly one value y of Y. However, when dealing with statistical data, due to variation, it is natural to have several y's observed with the same x-value – strictly speaking a violation of the defining property of a function. Therefore, this deterministic framework may add to the difficulty in analysing bivariate statistical data and learning about correlation and regression. Estepa and Batanero (1996) studied pre-university students' strategies when evaluating correlations between numerical variables and identified the following misconceptions of statistical association:

- *Deterministic conception of association*: Some students expected a correspondence that assigned only a single value to the dependent variable for each value of the independent variable. When this was not so, they assumed there was no dependency between the variables. That is, the correspondence between the variables had to be, from the mathematical point of view, a function.
- *Unidirectional conception of association*: Some students perceived the association only when the sign was positive and considered an inverse association as independence.
- *Local conception of association*: Students used only part of the data in their judgment and generalised their conclusion to the whole dataset.
- *Causal conception of association*: Some students recognised an association between variables only if they believed there was a causal explanation between them.

Estepa and Sanchez Cobo (2001) identified another deterministic conception of functional dependence inappropriate for statistical data. They tested undergraduate university students after an introductory statistics course. Most of the students in their study assumed that if a functional relationship between X and Y exists, the correlation coefficient must be ± 1, whereas they considered a deterministic dependence only if the correlation was different from zero, although there can be perfectly deterministic *non-linear* relationships that yield $r = 0$. Both statements indicate a lack of awareness of nonlinear relations and a tendency to not understand functional data as a combination of deterministic trend and random variation (see also Fig. 25.2b).

2.4   A Framework for a Better Understanding of Statistical Associations

As with any statistical concept, to understand regression requires an appreciation of the role of variation in statistical data. Responses or *Y*-values vary because of an explained and an unexplained component. The trend in the data in linear regression is expressed through the slope and intercept of the regression line and more generally through a parameterised function *f*. This trend represents the explained part of variation whereas the difference between the regression curve and the data represent the unexplained variation. On a conceptual level, the relationship between a functional dependency and a bivariate statistical relationship can be expressed by the signal-noise metaphor. According to Konold and Pollatsek (2002) data analysis can be seen as the search for signals in noisy processes. In its generic form, data are thought of as a structural component plus residuals, that is, Data = Signal + Noise. This metaphor helps to deal with an overwhelming amount of relevant and irrelevant information contained in the observed data. Figure 25.3 shows different versions of expressing the signal-noise idea.

In the context of modelling scatter plot data $(x_1, y_1),...,(x_n, y_n)$, the signal-noise idea translates to the formula $y_i = f(x_i) + e_i$, $i = 1,...,n$, where the function *f* is the structure to be recovered, while the e_i's represent randomness. The y_i's are perceived as a signal *f* evaluated at x_i, perturbed by a noise e_i. Whereas approximation theory focuses on retrieving the function *f*, in stochastics we model both model fitting and residuals. When analysing bivariate numerical data, the signal-noise metaphor is very useful to bridge the gap between a deterministic view of a function and a statistical perspective that appreciates variation. The signal or structure *f* captures the explained part of the variation, while the noise comprises the unexplained part of the variation. In the scatter plot of the data it is the unexplained part of the variation that is the reason for several *y*'s associated with a single *x*.

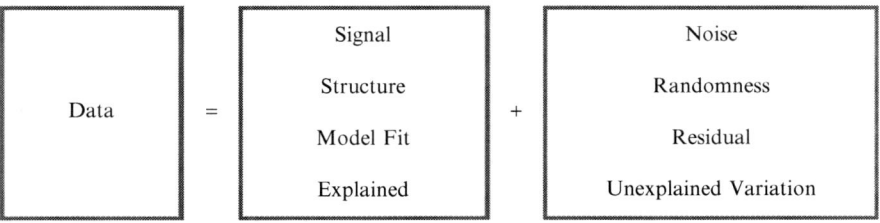

Fig. 25.3 Different versions of the signal-noise representation of data

3   Training Teachers with Respect to Correlation and Regression

There is scarce research on teacher understanding of correlation of regression. This section summarises two specific studies carried out with pre-service teachers that might be used as a starting point for teacher training.

3.1   Enhancing Understanding of Association in a Computer-Based Teaching Environment

Batanero, Estepa, and Godino (1997) investigated pre-service teachers' understanding of association before and after they attended a first year course in exploratory data analysis extending over 21 class sessions. During seven of these meetings the experimental group of 19 students was instructed in a computer-based teaching environment where they analysed different real datasets provided by the teacher or collected by themselves with the support of interactive computer software. The datasets were carefully chosen to cover a wide range of the students' interests and were sufficiently rich to ensure that questions of didactic interest would arise. Progress in acquiring concepts was assessed through a pre-test and post-test. In addition, the tests were also given to a control group of 213 students to compare how typical the experimental students' responses were. In addition, to obtain more in-depth insights into students' thinking processes, the interactions of two students with the computer were recorded and analysed together with their written responses and discussions during the problem solving process. Batanero et al. (1997) concluded in their study that most students overcame a deterministic conception of association, accepting random dependence. The local conception of association was also eradicated as the students noticed the importance of taking into account the complete dataset to evaluate association. Most students used all the different conditional distributions in the contingency tables, and gave up the additive procedure, using multiplicative comparison of the different frequencies in the table instead. The unidirectional conception of association was corrected only by some students, while others continued considering the inverse association independence. Finally, there was no improvement at all concerning the causal conception of association. Most students did not realise that a strong association between two variables is not enough to draw conclusions about cause and effect.

3.2   Learning to Model Scatter Plot Data

Engel, Sedlmeier, and Wörn (2008) investigated whether 78 pre-service teachers' appreciation of random variation can be enhanced in a course on functional relationships with an emphasis on modelling. The goal of the study was to find out how an emphasis on the signal-noise concept helps students to overcome their deterministic view of functional relationships and allow them a broader statistical perspective on analysing bivariate numerical data. The study was devised as a treatment-control group study with second year university students preparing to be secondary teachers. Participants attended two different courses in applied mathematics. While the control group attended a class with a more traditional syllabus (e.g., elementary functions, linear optimisation, no analysis of real data, no residual analysis or considerations of variation in data), the course for the treatment group followed strictly a pathway of technology-supported modelling of functional relationships of real data (Engel, 2009). Students were instructed about standard functions (e.g. polynomial, exponential, trigonometric and logistic) and learned through projects how to fit them to real data sets in a computerised learning environment based on the software Fathom. At first they selected appropriate parameters by adjusting a slider, later they minimised a least-squares criterion in case of a linear structure. In case of nonlinear scatter plots data had to be linearised through a suitable transformation. Throughout the course students were challenged to discuss deviations between model and data and to analyse residual plots, paying increasing attention to the concept "Data = Signal + Noise". Although statistical concepts were not explicitly taught in the course, statistical thinking skills and in particular the handling of variation in scatter plot data improved greatly, as shown in a pre-test–post-test comparison. Unlike in the pre-test and different from the control group very few students of the treatment group interpolated the data – in the given context an indication of a deterministic mindset – when being asked to sketch a free-hand curve over a scatter plot. Also the treatment group improved significantly in change-point detection tasks, requiring a judgement on whether a system has or has not changed over time (for more details, see Engel et al., 2008).

4 Implications for Teacher Training and Research

Research summarised in this chapter suggests that reasoning about correlation and regression is not always easy for laypeople, and even researchers sometimes struggle with a correct understanding. Fallacies and misunderstandings occur even in seemingly simple situations and are challenging. Teachers have to be aware that they cannot solve the problem of teaching by providing mere definitions and computational procedures. Students (and maybe also sometimes teachers) need to be confronted with their potential misconceptions, and they need to understand that correlation and regression are not only found in "textbook problems" but play an important part in daily life. Therefore, there is a need to include activities that help students more deeply reflect on and accommodate their views. Below, we recommend four approaches that should be helpful in fostering students' understanding of correlation and regression (see also chapters by Pratt, Davies, & Connor, and by Ridgway, Nicholson, & McCusker, in this book):

1. *Make use of fallacies and misunderstandings.* Students' conceptions change if their conceptions conflict with situations they do not understand or with previously held concepts. Fallacies and misunderstandings in statistical reasoning are a challenge for students because of the outcomes are often surprising. In the context of correlation, the regression effect and Simpson's paradox are of particular interest. Levin (1993), for instance, provides instructive activities for students to explore and experience the regression effect. Effective instruction may include simulating situations as well as analysing real stories from newspapers or research reports where people unaware of these fallacies drew wrongful conclusions.

2. *Make use of real data.* Genuine data are an authentic source of information. Bearing witness of real situations, real data give legitimacy and meaning to dealing with statistics. To find out how two or more quantities co-vary with each other is an important question for human inquiry. Analysing real data sets that are of genuine interest to teachers and their students has a great potential to motivate and encourage project work and scientific investigations.

3. *Make use of experience in modelling.* Models, by their nature, are not the real thing but an oversimplification of the complexity and disorder that reality throws at us. In modelling functional relationships we fit a curve to scatter plot data. In regression we do alike, but we model at the same time the structure as well as the residuals - the latter as a random process. Since this task is more challenging than curve fitting without stochastic modelling, the most introductory textbooks in statistics restrict regression to the case of linear data. While nonlinear and nonparametric regression are much more demanding when working out the exact mathematics behind the methods, a conceptual approach to nonlinear regression and basic ideas of smoothing opens up many realistic problems because in the world many things are non-linear. Furthermore, within school mathematics, such an approach connects regression with elementary functions such as inverse, quadratic, polynomial, exponential and so on.

4. *Make use of technology.* Plotting bivariate data, selecting appropriate functional models, fitting curves, and drawing and analysing residual plots are important activities that help students and teachers alike to connect their mathematical knowledge about functions with their growing statistical competencies. Technology allows the retrieval and use of real data on almost any subject of interest. The possibilities for multiple linked representations in dynamic interactive software such as Fathom offer a major potential to hands-on activities and to constructivist learning. For instance, residual plots can be drawn by one mouse click and allow the user to investigate the deviation between model and data. In traditional teaching, the calculation of the regression line and the correlation coefficient, by hand or with a calculator, is usually the focus of the work. Due to time needed for training students in computational methods, only a few examples are solved, but for effective and conceptual learning such an approach may be

counterproductive. Quantitative complexity is inherent in statistics, and there are many statistical concepts and methods for analysing the relationship between two variables. With the time saved by computation and graphical representation because of the availability of computers, activities should be designed to allow the student to encounter a variety of different situations, thus supporting the development of conceptual learning (Batanero et al., 1997).

Because of the scarce empirical research on learning about regression and correlation, many of the following recommendations are based on teaching experience but lack a foundation in empirical research. While there is a large community of researchers focusing on mathematical applications and modelling, classroom studies that evaluate the impact of the recommendations given above on students' understanding of correlation and regression are almost nonexistent.

How can fallacies and misunderstandings and real data be used to their best? How do modelling competencies, as described e.g., by Blum, Galbraith, Henn, and Niss (2002), in the context of modelling with functions relate to statistical literacy? How is content knowledge about functions related to mastering tasks involving regression and correlation? Under which circumstances can the use of computer based simulations support learning about association? Does the inclusion of nonlinear relationships impede or help learners to appreciate linear regression and correlation? It is highly desirable to have empirically supported evidence to answer these questions in order to enable mathematics teachers to design learning experiences that will develop a sound understanding of correlation and regression as part of everybody's knowledge.

References

Anscombe, F. J. (1973). Graphs in statistical analysis. *American Statistician, 27,* 17–21.

Batanero, C., Estepa, A., & Godino, J. (1997). Evolution of students' understanding of statistical association in a computer-based teaching environment. In J. Garfield & G. Burrill (Eds.), *Research on teaching statistics and new technologies* (pp. 191–206). Voorburg, The Netherlands: International Statistical Institute.

Blum, W., Galbraith, P., Henn, H., & Niss, M. (2002). *Modelling and applications in mathematics education: The 14th ICMI Study.* New York: Springer.

Castro-Sotos, A. E., Van Hoof, S., Van den Noortgate, W., & Onghena, P. (2009). The transitivity misconception of Pearson's correlation coefficient. *Statistics Education Research Journal, 8*(2), 33–55.

Chapman, L. J., & Chapman, J. P. (1969). Illusory correlation as an obstacle to the use of valid psychodiagnostic signs. *Journal of Abnormal Psychology, 74,* 271–280.

Engel, J. (2009). *Anwendungsorientierte Mathematik: Von Daten zur Funktion (Application oriented mathematics: From data to functions).* Heidelberg: Springer.

Engel, J., & Sedlmeier, P. (2005). On middle-school students' comprehension of randomness and chance variability in data. *Zentralblatt für Didaktik der Mathematik, 37*(3), 168–179.

Engel, J., Sedlmeier, P., & Wörn, C. (2008). Modeling scatterplot data and the signal-noise metaphor: Towards statistical literacy for pre-service teachers. In C. Batanero, G. Burrill, C. Reading, & A. Rossman (Eds.), *Joint ICMI/IASE Study: Teaching Statistics in School*

Mathematics. Challenges for Teaching and Teacher Education. Proceedings of the ICMI Study 18 and 2008 IASE Round Table Conference. Monterrey, Mexico: International Commission on Mathematical Instruction and International Association for Statistical Education. Online: www.stat.auckland.ac.nz/~iase/publications

Erlick, D. E., & Mills, R. G. (1967). Perceptual quantification of conditional dependency. *Journal of Experimental Psychology, 73*(1), 9–14.

Estepa, A., & Batanero, C. (1996). Judgments of correlation in scatter plots: An empirical study of students' intuitive strategies and preconceptions. *Hiroshima Journal of Mathematics Education, 4*, 25–41.

Estepa, A., & Sanchez Cobo, F. (2001). Empirical research on the understanding of association and implications for the training of researchers. In C. Batanero (Ed.), *Training researchers in the use of statistics* (pp. 37–51). Granada, Spain: International Association for Statistical Education.

Fiedler, K., Brinkmann, B., Betsch, R., & Wild, B. (2000). A sampling approach to biases in conditional probability judgments: Beyond base-rate neglect and statistical format. *Journal of Experimental Psychology: General, 129*, 1–20.

Fiedler, K., Walther, E., Freytag, P., & Plessner, H. (2002). Judgment biases in a simulated classroom: A cognitive-environmental approach. *Organizational Behavior and Human Decision Processes, 88*, 527–561.

Freedman, D., Pisani, R., & Purves, R. (1998). *Statistics* (3rd ed.). New York: Norton.

Inhetveen, H. (1976). *Die Reform des gymnasialen Mathematikunterrichts zwischen 1890 und 1914 (The reform of mathematics instruction for upper level schools between 1980 and 1914)*. Bad Heilbrunn: Klinkhardt.

Jennings, D. L., Amabile, T. M., & Ross, L. (1982). Informal covariation assessment: Data-based versus theory-based judgments. In D. Kahneman, P. Slovic, & A. Tversky (Eds.), *Judgment under uncertainty: Heuristics and biases* (pp. 211–230). New York: Cambridge University Press.

Kharsikar, A. V., & Kunte, S. (2002). Understanding correlation. *Teaching Statistics, 24*(2), 66–67.

Konold, C., & Pollatsek, A. (2002). Data analysis as the search for signals in noisy processes. *Journal for Research in Mathematics Education, 33*(4), 259–289.

Lane, D. M., Anderson, C. A., & Kellam, K. L. (1985). Judging the relatedness of variables: The psychophysics of covariation detection. *Journal of Experimental Psychology, 11*(5), 640–649.

Levin, J. (1993). An improved modification of a regression-towards-the-mean demonstration. *American Statistician, 47*, 24–26.

Nisbett, R., & Ross, L. (1980). *Human inference: Strategies and shortcomings of social judgment*. New Jersey: Prentice Hall.

Sedlmeier, P. (2006). Intuitive judgments about sample size. In K. Fiedler & P. Juslin (Eds.), *Information sampling and adaptive cognition* (pp. 53–71). Cambridge: Cambridge University Press.

Smedslund, J. (1963). The concept of correlation in adults. *Scandinavian Journal of Psychology, 4*, 165–173.

Stelzl, I. (1982). *Fehler und Fallen der Statistik (Errors and fallacies in statistics)*. Bern: Huber.

Vallee-Tourangeau, F., Hollingsworth, L., & Murphy, R. (1998). Attentional bias in correlation judgements? Smedslund (1963) revisited. *Scandinavian Journal of Psychology, 39*, 221–233.

Chapter 26
Teacher Knowledge of and for Statistical Investigations

Tim A. Burgess

Abstract Increasingly, statistics investigations are being advocated for teaching school statistics, even from beginning primary school levels. Successful adoption of this approach in the classroom is dependent on the teacher, and specifically on teacher knowledge. In this chapter, a framework for identifying and describing teacher knowledge, that reveals the extent of what teacher knowledge is needed in the classroom for teaching statistics through investigations, is briefly described. Some examples are given of how particular aspects of teacher knowledge, or absence of these, impact on the learning opportunities in the classroom. Implications are considered for teacher education regarding how to develop comprehensive teacher knowledge for teaching statistics through investigations.

1 Introduction

The current strong advocacy for teaching school statistics through investigations stems from Tukey's pioneering ideas of exploratory data analysis (EDA) back in the 1970s. Worldwide, many school curricula have moved from a more skills based curriculum to one that recommends students working with real (or at least, realistic) data for a purpose. However, it is recognised that this is not necessarily easy for teachers, as such an approach requires much more from the teachers in various ways.

In this chapter, the teacher knowledge needed to successfully implement the teaching of statistics through investigations is discussed. A framework is presented that breaks down teacher knowledge into components and links each of these with aspects of statistical thinking. Implications for teacher education are suggested, and recommendations given for ways of addressing these.

T.A. Burgess (✉)
School of Curriculum and Pedagogy, Massey University, Private Bag 11-222,
Palmerston North 4442, New Zealand
e-mail: t.a.burgess@massey.ac.nz

C. Batanero, G. Burrill, and C. Reading (eds.), *Teaching Statistics in School
Mathematics-Challenges for Teaching and Teacher Education: A Joint ICMI/IASE Study*,
DOI 10.1007/978-94-007-1131-0_26, © Springer Science+Business Media B.V. 2011

2   Using Investigations to Teach Statistics

In the 1970s, Tukey's foundational ideas of exploratory data analysis initiated a move away from formal methods for statistics education towards a more real-data based approach. Such an approach to the teaching and learning of statistics is seen as an important and necessary precursor to the more formal statistical ideas and approaches at the more advanced levels (for a more in-depth discussion of teaching through investigations, see MacGillivray & Pereira-Mendoza, in this book). Cobb and Moore (1997) suggested that such a role for EDA is significant, in that it encourages an examination of data, thereby revealing insights from the meaningful patterns (and not just patterns *per se*) that are found in the data. These insights can provide important clues for subsequent analysis of the data, since without exploring the data prior to more formal analysis, a person can easily miss significant information that could help inform which analytical procedures might be appropriate for the data. In this sense, Cobb and Moore (1997, p. 48) warned that:

> Most real data contain surprises, some of which can invalidate or force modification of the inference that was planned. This is one reason why running data through a sophisticated (and therefore automated) inference procedure before exploring them carefully is the mark of a statistical novice.

Another justification for adopting an EDA approach stems from the differences between mathematics and statistics (an extensive discussion is presented in Gattuso & Ottaviani, in this book). One essential difference between mathematics and statistics comes back to numbers and how they are "viewed". In mathematics, the general development through the school levels is towards an abstract, analytical, and deductive approach. According to Cobb and Moore (1997), the "numbers" are important with respect to concepts and relationships between them, which tend to be the focus, and giving too much attention to real-world contexts can obscure the mathematical patterns and structure. Cobb and Moore also suggest that, in comparison, statistics needs the context from which the numbers (or data) were derived as an essential part of "telling the story" of the data or of "getting inside the data". The context provides the meaning through which the data patterns can be examined and maybe explained.

2.1   Statistical Thinking and Its Components

Throughout the world, many school mathematics curricula now include statistics and advocate or require that students be involved in investigations. The New Zealand curriculum (Ministry of Education, 2007) additionally requires students to be engaged in thinking mathematically and statistically, at all levels from junior primary through to senior secondary.

Wild and Pfannkuch's (1999) description of statistical thinking covers some of the "big" ideas of statistics, and provides a useful way of examining and therefore encouraging students' use of statistical thinking. The components of statistical thinking in Wild and Pfannkuch's model include: (a) recognising the need for data, rather than relying on anecdotal evidence, and realising that the more data the better the conclusions that can be drawn; (b) acknowledging and understanding variation in data; (c) being able to transnumerate the data in various ways to help with making more sense of the data; (d) using "models" for reasoning further about the problem; and (e) considering the context of the problem and how this context links with statistical knowledge. Approaching a problem using only mathematical thinking and not statistical thinking avoids the "messiness" of data and the associated subjectivity. However, as Groth (2006) suggests, students must be exposed to the messiness of real data as this encourages them to grapple with statistical claims in everyday life and the media, and to develop the thinking needed for evaluating such claims.

2.2   The Investigative Process

Getting students involved in EDA is therefore seen as a way of helping them develop statistical literacy, statistical reasoning, and statistical thinking (Garfield & Ben-Zvi, 2008). According to these authors, such an approach to statistics fits with the investigative process of: (a) specifying a problem, planning, posing a question or formulating a hypothesis; (b) collecting and producing data from a variety of sources (survey, experiments); (c) processing, analysing, and representing the data; and (d) interpreting the results, discussing, and communicating conclusions. Thinking in relation to this investigative process or cycle along with the interrogative cycle (in which the statistician is in constant "dialogue" with the data and the problem), and having a questioning, skeptical, and open disposition while problem solving are considered by Pfannkuch and Wild (2004) as other important dimensions of their statistical thinking framework.

Engaging students in statistical investigations therefore opens the opportunity to develop students' thinking and problem solving in statistics. Students must have firsthand experience with investigations (Moore, 1998) so that their statistical reasoning will be encouraged. Statistical investigations in the classroom can be approached from two different starting points. One starting point is giving data to the students. With these data, the teacher and/or the students can either pose a question to be answered or problem to be solved, or the students can be given the freedom to conduct a more open investigation, posing questions to be answered once they have some sense of what the data contains and therefore what that data might feasibly reveal. This approach enables the students (particularly those with less experience of statistical investigations) to focus on the analysis and conclusion phases of the investigative process, and to avoid some of the logistical complexity around the data collection phase. Once the students have some experience with this type of "reduced" investigation, another approach can be used. The second way is to start the investigative cycle with a problem, question, or hypothesis, and from there move to the collection of data, followed by the subsequent phases involving analysis, interpretation and conclusions. Irrespective of which approach is taken, and in spite of the pervasive and compelling arguments for teaching with investigations, there are challenges for the teacher that an investigative approach presents.

3 Teacher Knowledge of Statistics Investigations

All teaching requires teachers to have knowledge, of both the content to be covered, and of effective ways to teach it. The work of Shulman (1987) provided researchers with a language and categorisation for teacher knowledge that enabled further investigation into the types of knowledge needed for effective teaching and the links with classroom practice and student outcomes. Because statistics has generally been a more recent addition to the primary school mathematics curriculum, the knowledge required to teach this topic has however only come under the researchers' "spotlights" in more recent years. In this section a summary is given of research looking at teachers' knowledge of statistics investigation, while in Sect. 4, a model for the professional knowledge teachers need to teach statistics investigations is presented.

3.1 Teacher Statistical and Pedagogical Content Knowledge

To teach statistics effectively, teachers require good knowledge of statistics, and this fact led to some research that has looked at teacher knowledge in specific statistics content areas, such as the mean and/or median (e.g., Jacobbe, 2008) or graphing (e.g., Gonzalez & Pinto, 2008; see Chaps. 20–25 that describe teacher knowledge in relation to specific statistical topics). In addition, Greer (2000) suggests that because of the changing emphasis from the development of statistical skills (or literacy) towards statistical reasoning and thinking, teachers will be required to develop ways of encouraging greater conceptual understanding of statistics in their students. For example, when students are confronted with data that exhibit variation, they are more likely to notice the trends or patterns in the data than the variation (Ben-Zvi, 2004), or the individual features in the data over and above the global features (e.g., Konold & Higgins, 2003). Teachers need knowledge of such challenges that students face in developing a good conceptual understanding in statistics, as well as knowledge of how to develop students' conceptual understanding.

3.2 Teacher Knowledge of Statistical Investigations

In addition to the understanding of various statistical concepts, teachers need experience and understanding of the investigative process itself. A summary is given below of work by a number of researchers, all of whom examined teachers' competence in dealing with statistics investigations.

Burgess (2002) engaged 30 primary school pre-service teachers in an open-ended investigation using multivariate data that required a written report with evidence to support the findings. The author found that pre-service teachers who examined the dataset in relation to more than one variable were more likely to include generalisations in their report than those who focused on only one variable at a time. The latter group also tended to include more descriptive statements about the data rather than generalisations.

Heaton and Mickelson (2002) found, similar to Burgess, that pre-service teachers tended to lose sight of the goal of their investigations and instead focused on the production of a graph. The 44 pre-service teachers had insufficient understanding and knowledge of the process of statistical investigations to properly complete their investigations. Heaton and Mickelson claimed that their teaching effectiveness was affected, as in a subsequent teaching episode, the student teachers tended to teach only what they themselves knew. The researchers advocated that teachers must have multiple opportunities to be involved in statistics investigations as well as to develop understanding of children's statistical thinking and understanding.

In a study involving the professional development of 29 middle school teachers, Lee and Mojica (2008) engaged nine teachers in a probability-based investigation within their own classrooms. Lee and Mojica suggest that although some teachers instructed the students to use bar graphs, the graphing became the focus rather than the graphs being used as tools to notice features of the distribution of the empirical data. The authors found that the teachers missed significant opportunities for deepening their students' statistical reasoning in probability, particularly in relation to the latter parts of the investigative cycle, namely the analysing and interpreting phases. For instance, although the effect of sample size on components of distribution could have been a major learning focus for the students, the teachers did not steer the lesson towards useful ways of representing the data in order to notice the features of distribution and to compare empirical results with theoretical probabilities. The use of proportions and fractions in simplest form obscured the effect of sample size on the relative frequencies and the approximation of these to theoretical probabilities. Many teachers interpreted the variation in results as an indication that theoretical probabilities were more "reliable" than empirical results, and not as an effect of sample sizes. The teachers missed the chance to develop students' understanding of the frequentist approach to probability because of their own limited knowledge.

Groth (2006) describes other challenges for teaching through investigations: first, managing classroom discourse, and second, assessing students' understanding while students are engaged in investigations. However, rather than using these challenges as an excuse not to teach in this way, Groth encourages teachers to face them and get the rewards from such teaching. In the next section, the ways in which these types of challenges for a teacher link with teacher knowledge are described.

4 Teacher Knowledge to Teach Statistical Investigations

There are different types of knowledge that teachers need to teach statistics, particularly through investigations (see Chaps. 27–29 that describe different models of professional knowledge to teach statistics). Watson (2001) reported on a profiling tool that was used to explore teacher knowledge across the seven knowledge bases that Shulman (1987) had identified. However, the model was not used directly to do research on the practice of the classroom.

To make a more direct link with what happens in the classroom, Burgess (2006) developed a framework for a classroom-based investigation of teacher knowledge needed and/or actually used for teaching through statistics investigations. This framework (see Fig. 26.1) was based on two different models: (a) the teacher knowledge work of Ball and colleagues (Ball, Thames, & Phelps, 2005; Hill, Schilling, & Ball, 2004) in mathematics education, who differentiated the following types of knowledge: common knowledge of content, specialised knowledge of content, knowledge of content and students, and knowledge of content and teaching; and (b) the statistical thinking model of Wild and Pfannkuch (1999). The framework was used to map teachers' knowledge as they taught statistics across the four categories in the Ball et al. model and in relation to six components of statistical thinking from Wild and Pfannkuch (namely, the four fundamental thinking types of transnumeration, variation, reasoning with models, and integration of statistical and

| | | Statistical knowledge for teaching | | | |
| | | Content knowledge | | Pedagogical content knowledge | |
		Common knowledge of content (CKC)	Specialized knowledge of content (SKC)	Knowledge of content and students (KCS)	Knowledge of content and teaching (KCT)
Thinking	Transnumeration				
	Variation				
	Reasoning with models				
	Integration of statistical and contextual				
Investigative cycle					
Interrogative cycle					

Fig. 26.1 Framework for teacher knowledge to teach statistics through investigations

contextual knowledge, along with general thinking linked to the investigative cycle and the interrogative cycle). The aim was to describe what knowledge was needed in the classroom, and whether there were aspects of knowledge on the framework that were unnecessary for the teacher's work in the classroom.

In later work Burgess (2007) used this model to analyse the knowledge of four teachers while actively teaching statistics investigations. The analysis of teacher knowledge obtained by this framework revealed descriptions of different types of knowledge that were either needed and used, or needed but not used, during teaching in four upper primary level classrooms (Years 5–8). An extensive description of the teacher knowledge in relation to each cell of the framework, and specific examples from the classroom episodes have been given in Burgess (2009). The next section has an analysis of some examples that illustrate the types of knowledge needed, based on two cells of the framework.

4.1   Use of Different Types of Knowledge to Foster Students' Learning

Specialised knowledge of content is the type of knowledge of statistics that a teacher needs over and above what an educated person might know (which would correspond to *common knowledge of content*). Then, *specialised knowledge of the content: variation* would link to a teacher having to judge, from a statistical point of view, whether a student's response in relation to variation in data was reasonable.

An example of this type of knowledge appeared in Burgess's (2007) research, where one class had been given data about the TV programme preferences for boys (based on two possible choices) which showed only a small proportion of boys choosing one of the options. The teacher asked the children if they thought that many boys in the school would choose that programme and one student responded: "Don't know; she hasn't asked all the classes yet". The teacher had to evaluate whether this response was statistically appropriate in relation to the student's knowledge of variation. The teacher's subsequent explanation indicated that she had considered the important factors from a statistical point of view and that the student's explanation was statistically naïve and in need of further development. This was therefore evidence of the teacher *specialised knowledge of the content: variation*.

Another type of teacher knowledge is *knowledge of content and students*, which includes knowledge of typical challenges for students and their common misconceptions. Again in Burgess's (2007) research, when considering potential issues prior to moving into data collection, one teacher anticipated that students may face a challenge when answering the data collection question of what position they were in the family: youngest, middle, or eldest. The teacher pre-empted questions such as how a person in a family of four or more children, or someone who was a twin, may answer the question. This was an example of *knowledge of content (integration of contextual and statistical knowledge) and students*. The teacher had thought about the statistical implications in relation to the data and in relation to various family types that the children may come from. Consequently, based on this knowledge, the teacher had considered how to deal with this in the classroom (by engaging the students in a class discussion around the issues with various family types and the link with the statistical data collection question). This showed that he had *knowledge of content (integration of contextual and statistical knowledge) and teaching.*

4.2 Using Professional Knowledge to Provide Learning Opportunities

Burgess (2007) found that across the group of four teachers in this study, all aspects of knowledge included in the model described in Fig. 26.1 were needed in the classroom. Burgess (2008) compared the teacher knowledge profile of two teachers taking part in the above study, and found significant differences in spite of both teachers basing their lessons on the same unit plan, and having the same amount of teaching experience (both in their second year of teaching). One teacher's knowledge profile was reasonably comprehensive across all "cells" of the framework, while the other teacher's profile revealed some interesting patterns in "missing" aspects of knowledge. Situations were identified for this teacher for which teaching and learning opportunities were missed.

Some of these missed opportunities were around transnumerating the data in order to make more sense of the data or to reveal the information within the data. Other missed opportunities were linked to the teacher's *knowledge of content and students*, and *knowledge of content and teaching* (which includes knowing how to sequence the learning and finding good representations or models to use for particular concepts).

For example, a missed opportunity in relation to *specialised knowledge of the content: transnumeration* happened when the teacher missed opportunities to assist the students with how to sort the data or how to represent it in another way. The teacher did not ask for clarification from a student who suggested that they could "add them together and do averages" in relation to a question they were investigating where an average would have been meaningless. A missed opportunity linked to *specialised knowledge of the content: reasoning with models* was identified in another situation, when the students suggested how they could sort their data cards. The teacher did not recognise or follow up with the students that such a sorting would not assist with answering their investigative question. One more example was when the teacher recognised the inappropriateness of some students' comparison of two unequal groups by finding the sums of heights and their invalid conclusion. The teacher was unable to address that with the students and did not follow this up. This indicated a problem with his *knowledge of the content (transnumeration) and teaching* as well as *knowledge of the content (reasoning with models) and teaching.*

In comparison with the missed opportunities from that teacher, the second teacher had more comprehensive teacher knowledge. As an example, the teacher predicted that the students might have trouble with sorting their data cards appropriately (unlike the first teacher), so she had considered a way to support the students with their sorting. This showed that first, the teacher had *knowledge of the content (transnumeration) and students*, and second, because of having anticipated their problems and considered how to handle this, she had *knowledge of content (transnumeration) and teaching*.

The examples outlined above give some indication of the range of challenges that teachers face in using investigations in the classroom, in relation to categories of teacher knowledge. It was clear from Burgess's (2008) study that teachers needed knowledge across all categories and in relation to all components of statistical thinking. The missed opportunities in the classroom could be linked to particular cells of the framework. This gave some insight into why, for example, the younger students, who had the teacher with the more comprehensive teacher knowledge, progressed further with their investigations, and made more sophisticated and appropriate statements based on the data than the older students, who had the teacher with significant knowledge gaps.

Analysis of the missed opportunities also revealed one interesting theme linked to teacher "listening" and responding to students. Part of the core work of teachers involves responding to students' questions, and evaluating students' responses to questions or tasks. Without appropriate teacher knowledge, such work can be compromised. For example, when a student is making a statement about what the data shows, unless a teacher is able to *reason with a statistical model* that teacher may not be able to evaluate whether the student's statement is valid for the data, or knowing that the statement is not valid, the teacher may not know how to deal with that in the classroom.

The example given earlier of the teacher not knowing how to respond when the students used sums of heights to compare unequal groups indicated a problem with *knowledge of the content: reasoning with models and teaching*. In contrast, when the second teacher struggled to make sense of what a student was saying about the data and therefore whether the statement was valid, she developed an alternative representation of the data from which she was then able to evaluate the student's statement. This indicated *specialised knowledge of the content: reasoning with models* (to evaluate the somewhat incomplete statistical statement from the student) and *knowledge of the content: reasoning with models and teaching* (when she illustrated on the board how to use the representation to check the statement's validity). These examples illustrate how different teacher knowledge components can impact the teacher's ability to respond to students and to evaluate their statistical claims.

5 Implications for Teacher Education

Adoption of investigations for teaching statistics based on widespread advocacy for such a teaching approach is, on its own, unlikely to be effective. As has been discussed, the success of this approach is dependent on comprehensive teacher knowledge across four different categories of knowledge and across the components of statistical thinking. In this section, recommendations are given for educating teachers to teach statistics investigations, which complement other recommendations in the chapter by Makar and Fielding-Wells in this book. Teacher educators, whether working with pre-service students or with practising teachers through professional development, need to consider how comprehensive teacher knowledge can be developed. Some of the examples given above show that different aspects of teacher knowledge are closely linked, and do not act in isolation from one another.

Engaging teachers in their own investigations, with explicit attention to the phases of the investigative cycle, and with appropriate support, is one way in helping them develop their *common knowledge of content*. This approach has been recommended or used by various researchers (e.g., Heaton & Mickelson, 2002). The other three components of teacher knowledge are linked specifically to students and the classroom and are not likely to be found in people outside of teaching. Therefore they need to be addressed in other ways as they will not develop from teachers undertaking their own investigations.

To develop the two components of teacher knowledge that are intimately connected with engaging with students' responses, namely *specialised knowledge of content* and *knowledge of content and students*, teachers are required to observe and interact with students, through watching and listening, as the students engage in investigations. Although there is a growing research literature about students' statistical conceptions from which teachers can learn about students' development of statistical ideas and common misconceptions, observing real students in real classroom settings is an important source of learning, as this is where teachers are required to use their knowledge (see the chapter by Ponte in this book). It is known that situations can arise in classrooms where teachers do not use their knowledge. The reality of a teacher's decision making and responding to students in the classroom is affected by a large number of factors, including the need to respond within a "conversationally appropriate" period of time (O'Connor, 2001). This may explain some situations where teachers do not use their knowledge. Even so, having teachers interact in real-time in the classroom will help develop some aspects of knowledge.

For pre-service teachers, a classroom video will provide them with the opportunity to explore the knowledge that is needed within classroom interactions. The *knowledge of content and teaching* category has close connections with and is activated by at least one other category, such as *knowledge of content and students*. Research literature also has a role in helping the development of both types of knowledge.

The discussion above argues that focusing on only one aspect of knowledge would be impractical. The four knowledge categories are closely linked and distinguishing between them at times is difficult. So for teachers to develop the knowledge necessary for teaching through investigations, opportunities are needed to link the various types of knowledge through engaging teachers in investigations, both for themselves and with students, either in real time in the classroom, or through the use of a classroom video. A teacher's broad base of connected knowledge categories in relation to the statistical thinking components will enable the teacher to provide the best opportunities for students' learning through statistical investigations.

References

Ball, D. L., Thames, M. H., & Phelps, G. (2005, April). *Articulating domains of mathematical knowledge for teaching*. Paper presented at the Annual Meeting of the American Education Research Association, Montreal, Canada.

Batanero, C., Burrill, G., Reading, C., & Rossman, A. (Eds.). (2008). *Joint ICMI/IASE Study: Teaching Statistics in School Mathematics. Challenges for Teaching and Teacher Education. Proceedings of the ICMI Study 18 and 2008 IASE Round Table Conference.* Monterrey, Mexico: International Commission on Mathematical Instruction and International Association for Statistical Education. Online: www.stat.auckland.ac.nz/~iase/publications

Ben-Zvi, D. (2004). Reasoning about variability in comparing distributions. *Statistics Education Research Journal, 3*(2), 42–63.

Burgess, T. A. (2002). Investigating the 'data sense' of pre-service teachers. In B. Phillips (Ed.), *Proceedings of the Sixth International Conference on Teaching Statistics*. Cape Town, South Africa: International Statistical Institute and International Association for Statistics Education. Online: www.stat.auckland.ac.nz/~iase/publications

Burgess, T. A. (2006). A framework for examining teacher knowledge as used in action while teaching statistics. In A. Rossman & B. Chance (Eds.), *Proceedings of the Seventh International Conference on Teaching Statistics*. Salvador, Brazil: International Statistical Institute and International Association for Statistical Education. Online: www.stat.auckland.ac.nz/~iase/publications

Burgess, T. A. (2007). *Investigating the nature of teacher knowledge needed and used in teaching statistics*. Unpublished Ed.D. dissertation, Massey University, Palmerston North, New Zealand. Online: www.stat.auckland.ac.nz/~iase/publications

Burgess, T. A. (2008). Teacher knowledge for teaching statistics through investigations. In C. Batanero, G. Burrill, C. Reading, & A. Rossman (2008).

Burgess, T. A. (2009). Teacher knowledge and statistics: What types of knowledge are used in the primary classroom? *The Montana Mathematics Enthusiast, 6*(1&2), 3–24.

Cobb, G. W., & Moore, D. S. (1997). Mathematics, statistics, and teaching. *American Mathematical Monthly, 104*(9), 801–823.

Garfield, J. B., & Ben-Zvi, D. (2008). *Developing students' statistical reasoning: Connecting research and teaching practice*. Berlin: Springer.

Gonzalez, T., & Pinto, J. (2008). Conceptions of four pre-service teachers on graphical representation. In C. Batanero, G. Burrill, C. Reading, & A. Rossman (2008).

Greer, B. (2000). Statistical thinking and learning. *Mathematical Thinking and Learning, 2*(1), 1–9.

Groth, R. (2006). Engaging students in authentic data analysis. In G. Burrill (Ed.), *Thinking and reasoning with data and chance: 68th NCTM Yearbook (2006)* (pp. 41–48). Reston, VA: National Council of Teachers of Mathematics.

Heaton, R. M., & Mickelson, W. T. (2002). The learning and teaching of statistical investigation in teaching and teacher education. *Journal of Mathematics Teacher Education, 5*(1), 35–59.

Hill, H. C., Schilling, S., & Ball, D. L. (2004). Developing measures of teachers' mathematics knowledge for teaching. *Elementary School Journal, 105*(1), 11–30.

Jacobbe, T. (2008). Elementary school teachers' understanding of the mean and median. In C. Batanero, G. Burrill, C. Reading, & A. Rossman (2008).

Konold, C., & Higgins, T. L. (2003). Reasoning about data. In J. Kilpatrick, W. G. Martin, & D. Schifter (Eds.), *A research companion to principles and standards for school mathematics* (pp. 193–215). Reston, VA: National Council of Teachers of Mathematics.

Lee, H. S., & Mojica, G. F. (2008). Examining how teachers' practices support statistical investigations. In C. Batanero, G. Burrill, C. Reading, & A. Rossman (2008).

Ministry of Education. (2007). *The New Zealand Curriculum*. Wellington, NZ: Learning Media.

Moore, D. S. (1998). Statistics among the liberal arts. *Journal of the American Statistical Association, 93*(444), 1253–1257.

O'Connor, M. C. (2001). "Can any fraction be turned into a decimal?" A case study of a mathematical group discussion. *Educational Studies in Mathematics, 46*, 143–185.

Pfannkuch, M., & Wild, C. J. (2004). Towards an understanding of statistical thinking. In D. Ben-Zvi & J. B. Garfield (Eds.), *The challenge of developing statistical literacy, reasoning, and thinking* (pp. 17–46). Dordrecht, The Netherlands: Kluwer.

Shulman, L. S. (1987). Knowledge and teaching: Foundations of the new reform. *Harvard Educational Review, 57*(1), 1–22.

Watson, J. M. (2001). Profiling teachers' competence and confidence to teach particular mathematics topics: The case of chance and data. *Journal of Mathematics Teacher Education, 4*, 305–337.

Wild, C. J., & Pfannkuch, M. (1999). Statistical thinking in empirical enquiry. *International Statistical Review, 67*(3), 223–265.

Chapter 27
Models for Statistical Pedagogical Knowledge

Juan D. Godino, Juan J. Ortiz, Rafael Roa, and Miguel R. Wilhelmi

Abstract The education of statistics teachers should be based on adequate models for pedagogical knowledge that guide the teachers' educators in implementing and assessing the training of teachers. In this chapter, some models that are relevant for mathematics and statistics are analysed, and a new framework that complements the previously described models is proposed. The different facets and levels that should be taken into account when educating mathematics and statistics teachers are highlighted. Some implications for the training of teachers are presented and a formative cycle directed to increase the teachers' statistical and pedagogical knowledge simultaneously is briefly described.

1 Introduction

One main conclusion in the Joint ICMI/IASE Study Conference was the need to elaborate models for statistical pedagogical knowledge that provide a foundation in training teachers to teach statistics. Research related to teacher education, development, and thinking (Philipp, 2007; Sowder, 2007; Wood, 2008) includes diverse theoretical frameworks describing the knowledge that teachers need in order to enhance the

J.D. Godino (✉)
Facultad de Educación, Universidad de Granada, Campus de Cartuja, 18071 Granada, Spain
e-mail: jgodino@ugr.es

J.J. Ortiz
Universidad de Granada, Campus de Melilla, Santander, 1, 52071 Melilla, Spain
e-mail: jortiz@ugr.es;

R. Roa
Universidad de Granada, Campus de Cartuja, 18071 Granada, Spain
e-mail: rroa@ugr.es

M.R. Wilhelmi
Departamento de Matemáticas, Universidad Pública de Navarra,
Campus de Arrosadia, 31006 Pamplona, Spain
e-mail: miguelr.wilhelmi@unavarra.es

C. Batanero, G. Burrill, and C. Reading (eds.), *Teaching Statistics in School*   271
Mathematics-Challenges for Teaching and Teacher Education: A Joint ICMI/IASE Study,
DOI 10.1007/978-94-007-1131-0_27, © Springer Science+Business Media B.V. 2011

students' learning and that are required in organising the teachers' training designs and in assessing their efficacy. Although there is a general consensus that mathematics teachers should master the disciplinary content, there is no similar agreement about how such mastery should be achieved and how the discipline should be conceived. It is, however, recognised that mathematical or statistical knowledge alone does not assure professional competence and that other capabilities are required, including knowledge about how students learn, their conceptions, types of thinking, strategies, difficulties, and potential errors. Teachers should also be able to organise the teaching, design learning tasks, use adequate resources, and understand the factors that condition the teaching and learning processes (Ponte, 2008).

In this chapter, the notion of pedagogical content knowledge proposed by Shulman (1987), which has been extensively applied in the teaching of mathematics, and other models created for mathematics education, are briefly described. Then, attention is focussed on the models for statistical pedagogical knowledge presented at the Joint ICMI Study Conference. In Sect. 4 a model for teachers' pedagogical knowledge, which is based on a previous theoretical framework developed for mathematics cognition and instruction (Godino, Batanero, & Font, 2007), is presented. This model extends the components identified in the models described in Sects. 2 and 3 and can be adapted to the specific character of statistical knowledge (from both the epistemological and didactic points of view).

An implication of the analysis is the need to develop and assess teachers' competencies to carry out didactical analysis of their own practice, which takes into account the different components of pedagogical knowledge. A possible formative cycle that serves these purposes is briefly described in the final section.

2 Models for Mathematical Pedagogical Knowledge

A reason for the lack of impact of research into practice is that teachers, who are the main agents of change, are only viewed as simple components of the educational system, who automatically apply the information they receive. The complexity of teaching and the high level of initiative and autonomy required by the teachers are highlighted in the research on "teacher thinking" (beliefs, conceptions, and attitudes) and on teacher professional knowledge and competencies. An increasing number of authors interested in this theme point to the insufficiency of mathematical knowledge alone to achieve truly effective teaching outcomes (Shulman, 1987; Hill & Ball, 2004). Consequently, this research is producing models of teacher knowledge, in order to design educational plans and elaborate tools for assessing the efficacy of such actions. In this section, we present a synthesis of some models that were specifically developed for mathematics education.

Shulman (1987) identified seven categories of knowledge that underpin expert teaching: (a) content knowledge or knowledge about the discipline; (b) general pedagogical knowledge; (c) curriculum knowledge; (d) pedagogical content knowledge, or pedagogical knowledge specific for the discipline (PCK); (e) knowledge of learners and their characteristics; (f) knowledge of education contexts; and (g) knowledge of educational ends, purposes, and values. Ponte and Chapman (2006) emphasised PCK as an important component in the education of teachers. The categories of knowledge described by Shulman have played an important role in developing research programmes and curricular materials and are still valid, although the initial interpretations and the names given to them have changed over time. For example, Ball and her colleagues (Ball, Lubienski, & Mewborn, 2001; Hill, Ball, & Schilling, 2008) developed the notion of mathematical knowledge for teaching (MKT) in which they distinguished six main categories.

1. Common content knowledge (CCK): the mathematical knowledge teachers are responsible for developing in students.
2. Specialised content knowledge (SCK): the mathematical knowledge that is used in teaching, but not directly taught to students, for example, knowledge about why the algorithms for the arithmetic operations work.
3. Knowledge at the mathematical horizon: understanding the broader set of mathematical ideas to which a particular idea connects, for example, understanding some epistemological obstacles related to the historical development of probability.
4. Knowledge of content and students (KCS): the amalgamated knowledge that teachers possess about how students learn content.
5. Knowledge of content and teaching (KCT): the design of instruction, including how to choose examples and representations, and how to guide student discussions towards accurate mathematical ideas.
6. Knowledge of curriculum: ways to sequence and structure the development of a mathematical topic.

In fact components 4–6 are a decomposition of Shulman's PCK and comprise the competencies that are deeply embedded in the work of teaching knowing. For example, knowledge of what makes a topic difficult for students, ways in which learners tend to develop understanding of a particular idea, ways to sequence and structure the development of a mathematical topic, including representations likely to help students learn (Hill et al., 2008). As stated by Graeber and Tirosh (2008, p. 124), "the fact that many researchers do not offer a definition of PCK but rather attempt to characterise it with lists or examples is another indication that the concept is still somewhat ill defined".

In addition to MKT, several researchers are proposing other tools to conceptualise the knowledge needed in teaching mathematics. Schoenfeld and Kilpatrick (2008, p. 322) offer a provisional framework for *proficiency in teaching mathematics* consisting of the following set of dimensions: (a) Knowing school mathematics in depth and breadth; (b) knowing students as thinkers; (c) knowing students as learners; (d) crafting and managing learning environments; (e) developing classroom norms and supporting classroom discourse as part of "teaching for understanding"; (f) building relationships that support learning; and (g) reflecting on their own practice.

A number of questions still need to be explored in research in teacher education, including the role of beliefs and values in the development of PCK, whether different teaching/learning paradigms require different components of PCK, what are adequate methods for assessing PCK; and what are more global theoretical models for describing the teachers' knowledge, beliefs, and affects, such as, teachers' orientation, perspective, and identity (Philipp, 2007).

3 Models for Statistical Pedagogical Knowledge

Two key elements in the didactical analysis of teaching and learning processes are the epistemic (mathematical content) and cognitive (students' learning) components. In anthropological and semiotic perspectives, mathematics is considered as a human activity arising from people's practices when working with specific problem-solving situations. This point of view also takes into account the specificity of statistics (see Ottaviani & Gattuso, in this book), since the epistemic facet is specific for each particular content, and therefore, for the case of statistics. Moreover, there are specific statistics problems, representations, and procedures that are different from those found in geometry, physics, or algebra. Basic statistical problems are related to inference and decision-making under uncertainty (involving random variation) and involve specific statistical practices: randomisation, collecting sample data, tabulation and transnumeration, data reduction, and using statistical models (Wild & Pfannkuch, 1999). These practices lead to the emergence of specific representations (e.g., specific graphs and terms), concepts (e.g., distribution, significance, correlation), procedures (e.g., analysis of variance), properties (e.g., bias, efficiency, independence) and arguments (e.g., the central limit theorem is given with a probabilistic statement, simulation is sometimes used to justify a result). Hence there are specific statistical practices and specific statistical objects and processes related to statistics problems. Consequently, since there is a specific epistemology of statistics, we should also recognise a specific didactics of statistics, given that the epistemic facet interacts with all the other facets of teachers' knowledge (cognitive, instructional, and curricular knowledge). This justifies the effort made by several statistics educators, in particular Burgess (2008), Garfield and Ben-Zvi (2008), and Watson, Callingham, and Donne (2008) to adapt and develop PCK or MKT models for statistical education.

Burgess (2008) defined teacher knowledge to teach statistics for the case when this teaching is based on statistical investigations. As research on teacher knowledge to teach statistics is scant and recent, he based his approach on studies carried out in mathematics education. Burgess built a model for statistical pedagogical knowledge starting from Ball et al. (2001) and extending and adapting to statistics education, by including categories from the Wild and Pfannkuch's framework (1999) for statistical thinking. Hence, based on these two theoretical models, Burgess proposed a two-dimensional grid to analyse the statistical knowledge for teaching. In one dimension (mathematical knowledge for teaching) he considered four categories: common knowledge of content; specialised knowledge of content; knowledge of content and students; and knowledge of content and teaching. In the other dimension (statistical thinking in empirical inquiry) he included the following categories: four types of fundamental statistical thinking (need for data, transnumeration, variation, reasoning with models, integration of statistical, and contextual); two components in the statistics research process (investigative cycle, interrogative cycle); and dispositions towards statistics.

The grid was used by the author to describe the knowledge put in practice by two statistics teachers, and the knowledge those teachers failed to apply even when they had the opportunity, in the context of teaching experiences based on statistical investigations. Results allowed Burgess to build a profile for each teacher's knowledge. His study served to describe the components of teacher knowledge that emerged during the teaching of statistics investigations and how lack of appropriate knowledge created missed opportunities in relation to the teaching and learning of statistics.

Garfield and Ben-Zvi (2008) described their experiences in training teachers in statistics, which were based on the application of six instructional design principles from Cobb and McClain (2004):

1. Focus on developing central statistical ideas rather than on presenting a set of tools and procedures.
2. Use real and motivating data sets to engage students in making and testing conjectures.
3. Use classroom activities to support the development of students' reasoning.
4. Integrate the use of appropriate technological tools that allow students to test their conjectures, explore and analyse data, and develop their statistical reasoning.
5. Promote classroom discourse that includes statistical arguments and sustained exchanges that focus on significant statistical ideas.
6. Use assessment to learn what students know and to monitor the development of their statistical learning as well as to evaluate instructional plans and progress.

Garfield and Ben-Zvi (2008) used these principles to design and teach courses. They teach also these principles to the students explicitly as they prepared to become teachers of statistics. Consequently, these prospective teachers had the opportunity to experience the learning of statistics following an instructional model that allowed them to know and understand the didactical knowledge incorporated in the principles mentioned.

The Cobb and McClain's (2004) principles of instructional design, adopted by Garfield and Ben-Zvi for teacher training courses, can be interpreted as an implicit model for teacher didactical knowledge. The first principle (focus on developing central statistical ideas) involves the epistemic component. Garfield and Ben-Zvi selected the following key statistical ideas: data, distribution, variation, central tendency, randomness, co-variation, and sampling. The second principle is related to both the epistemic component (real data sets refer to statistical problems and related conjectures) and the affective component (students' motivation and commitment). The third principle calls on the instructional facet (classroom activities, exploration, discussion and argumentation, cooperative work) and the cognitive facet (development of students' reasoning). The fourth principle refers to tools and media. The fifth principle involves an interactional component: promoting classroom discourse that focuses on significant statistical ideas. Finally, the sixth principle highlights the role of assessment in teaching and learning.

Assessment and measurement are important tools in developing teachers' PCK, as highlighted by Watson et al. (2008). In their work presented at the Joint Study Conference, Watson et al. described and applied a questionnaire that was developed to assess the different components of Shulman's PCK (see also Calligham & Watson, this book). Their questionnaire, based on Watson (2001), also included some items measuring the teachers' beliefs about statistics and its teaching, and their confidence to teach particular statistical topics. Watson et al. viewed PCK as a general notion including the different categories initially proposed by Shulman, i.e., disciplinary content knowledge and pedagogical content knowledge related to students, curriculum, teaching: "this approach appears to treat PCK as the underlying and encompassing phrase to summarise Shulman's original intentions" (Watson et al., 2008, p. 1). Some items included by these authors in their questionnaire to assess teachers' PCK were based on the answers given by students to questions used in previous survey research carried out by Watson. "The major focus of PCK in items in this study is teachers' content knowledge, its reflection in knowledge of their students' content knowledge, and their PCK in using student responses to devise teaching intervention" (p. 1).

Although the models for PCK or MKT described in the previous paragraphs are useful for training teachers to teach statistics, their categories are still general and could be made more precise. It would be useful to develop models that provide detailed and further operative criteria that can be applied in designing procedures or materials directed to educating teachers. In the following section we describe a theoretical model that attempts to complement and expand those described in the previous sections. This model is applicable to both mathematics and statistics (consequently for preparing mathematics and statistics teachers).

4 Expanding the Analysis of Mathematical and Statistical Pedagogical Knowledge

In this section we describe a specific model, which is based on a theoretical integrative framework developed for research in mathematics education. The onto-semiotic approach (synthesised in Godino et al., 2007) combines three dimensions in mathematical knowledge and teaching: (a) the epistemological component, which is conceived from an anthropological and socio-cultural perspective; (b) the cognitive component, which is given a semiotic foundation; and (c) the instructional component, which is based on social constructivism. Mathematics is conceived as a human activity linked to solving certain types of problem-situation, whereas mathematical objects are viewed as emerging from the systems of practices carried out to solve these problems. The above assumptions are also applicable to statistics, and hence the categories of teachers' knowledge derived from the onto-semiotic approach also serve to characterise the statistical pedagogical knowledge. The different types of mathematics and statistics objects considered in this perspective are first clarified, then the different facets and levels considered in the mathematical or statistical pedagogical knowledge are described, and finally the idea of didactic suitability and its components are expanded.

4.1   Types of Mathematical and Statistical Objects

Different types of knowledge are put in practice when carrying out mathematical or statistical practices and when interpreting their results. For example, when comparing two distributions (statistical problem) some symbolic or graphical representations, concepts, propositions, and procedures are used to elaborate the argument needed to make a decision as regards to those distributions (such as justifying whether the differences in averages or spread for these distributions are statistically significant). In the example, the following types of mathematical objects, introduced in the onto-semiotic approach to describe the mathematical practices, are identified:

1. *Language*: terms, expressions, symbols, graphs used to represent the distributions, their parameters, or the operations carried out with them.
2. *Situations*: extra or intra-mathematical problems or applications, for example, comparing the two distributions or carrying out a statistical test for the differences in averages or spread.
3. *Concepts*: given by their definitions or descriptions (variable, distribution, parameter, average, standard deviation).
4. *Propositions*: properties or attributes of concepts (e.g., the sum of frequencies is equal to the number of cases; two distributions with very different means are different).
5. *Procedures*: operations, algorithms, techniques (computing the mean and standard deviations; computing the significance of differences).
6. *Arguments*: used to validate and explain the propositions and procedures (deductive or inductive reasoning).

By considering these six types of mathematical or statistical objects, the traditional distinction between conceptual and procedural knowledge, which is insufficient to describe all the objects that intervene and emerge in mathematical or statistical activity, is expanded. Problem-situations are the origin and reason of mathematical or statistical activity; language is needed to represent the other types of objects and is an instrument for action; arguments justify the procedures and propositions that relate different concepts. These and other theoretical tools, as well as a classification of mathematical processes, are described in detail in Godino et al. (2007).

4.2 Facets and Levels of Mathematical and Statistical Knowledge for Teaching

A statistics teacher needs a deep knowledge of statistics, which includes competence in understanding and applying the different types of objects described in Sect. 4.1 for the particular statistical content he or she is teaching. Moreover, the teacher needs a deep mathematical or statistical knowledge for teaching. Teaching and learning processes involve a group of students, the teacher, and some didactic resources, all of them interacting within an institutional context. Consequently the mathematical or statistical knowledge for teaching should also include the different facets or components that are necessary to study teaching and learning processes and that are synthesised in Fig. 27.1. Didactic research is producing a substantial amount of knowledge for each of these facets that teachers should acquire and apply to achieve efficient teaching.

A short description of the facets of the model is given below (see Godino et al., (2007) for a more complete description):

1. *Epistemic facet*: The intended and implemented institutional meaning for a given mathematical or statistical content, that is, the set of problems, procedures, concepts, properties, language, and arguments included in the teaching and its distribution over the teaching time.

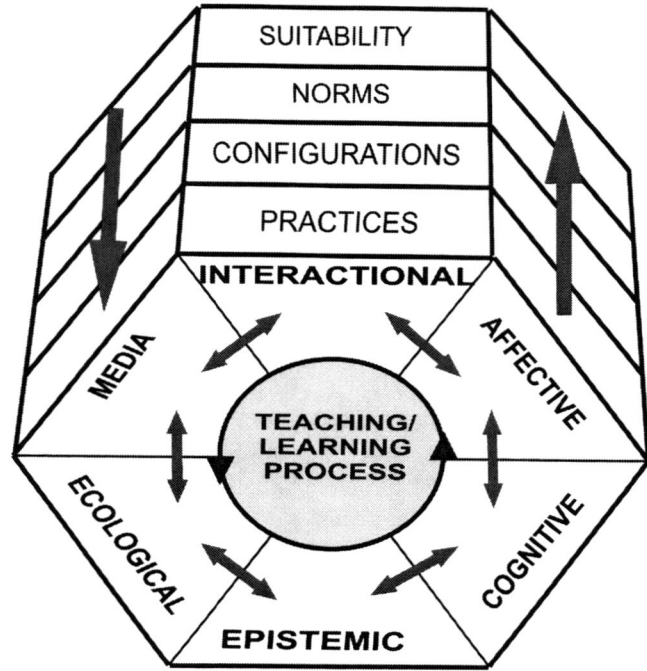

Fig. 27.1 Facets and levels of teachers' knowledge

2. *Cognitive facet*: Students' levels of development and understanding of the topic, and students' strategies, difficulties, and errors as regards the intended content (personal meaning).
3. *Affective facet*: Students' attitudes, emotions, and motivations regarding the content and the study process.
4. *Media facet*: Didactic and technological resources available for teaching and the possible ways to use and distribute these resources over time.
5. *Interactional facet*: Possible organisations of the classroom discourse and the interactions between the teacher and the students that help solve the students' difficulties and conflicts.
6. *Ecological facet*: Relationships of the topic with the official curriculum, other mathematical or statistical themes and with the social, political, and economical settings that support and condition the teaching and learning.

Teaching and learning processes can also be analysed from four different levels or points of view that provide additional categories for teachers' knowledge.

(a) *Mathematical–statistical or didactic practices*: Mathematical or statistical actions that students carry out to solve the problems posed, as well as the actions carried out by the teacher in order to promote learning and contextualise the content.
(b) *Configurations of mathematical or statistical objects and processes*: Mathematics objects (e.g., problems, procedures, concepts, properties, language, or arguments) and processes (e.g., generalisation, representation) that intervene and emerge in the aforementioned practices.
(c) *Norms*: Rules, habits, and conventions that condition and make possible the study process and affect each facet and their interactions.
(d) *Didactic suitability*: Objective criteria that serve to improve the teaching and learning and guide the evaluation of the teaching/learning process.

Teachers' progressive knowledge in each of these facets and levels for a specific content develops their understanding of the teaching complexity and their competence in finding possible causes for learning conflicts. Although in Fig. 27.1 the components and levels of teachers' knowledge are separated, in order to highlight their difference, in fact all of them interact. As an example, below, the interactions of didactic suitability with the facets 1–6 in the teachers' knowledge are analysed.

4.3 Didactic Suitability

Didactic suitability for a particular teaching and learning process should be evaluated for each of the six facets described in Sect. 4.2 as the teaching process may be suitable from the statistical point of view and not suitable, for example, from the affective point of view. Consequently six different types of suitability can be considered (Godino, Wilhelmi, & Bencomo, 2005):

1. *Epistemic suitability* measures the extent to which the implemented meaning (statistical content implemented in a classroom or course) represents adequately the intended meaning (the curricular guidelines for this course or classroom).
2. *Cognitive suitability* is the degree to which the implemented meaning is appropriate to the students' cognitive development. That is, the degree to which the implemented meaning is included in the students' *zone of proximal development*, and whether the students' learning (personal meaning achieved) is close to the intended meaning.
3. *Emotional suitability* describes the students' involvement (interest, motivation, attitudes) in the study process.
4. *Media suitability* reflects the availability and adequacy of material and temporal resources in the teaching process.
5. *Interactive suitability* is the extent to which the organisation of the teaching and the classroom discourse serve to identify and solve possible conflicts and difficulties that appear during the instructional process.
6. *Ecological suitability* is the extent to which the teaching process is in agreement with the school and society educational goals, and takes into account other possible social and cultural factors.

The different categories for teacher knowledge in the models described in Sects. 2 and 3 include, to a greater or lesser extent, the facets assumed in the onto-semiotic model. The levels of analysis crossing each facet in this last model involve a deepening in the analysis of the knowledge needed to design teacher education and to assess teacher knowledge. Moreover the idea of suitability and the different suitability criteria provide a guide to design, implement, and assess teacher professional development plans, and to support the teachers' reflection on their own practice.

5 Implications for Teachers' Education

Statistics teachers should develop competence to recognise the statistical objects and processes that intervene in the students' statistical practices, be aware of the norms that support and condition learning, affect, resources, and interactions in the classroom. Consequently, the education and assessment of teachers' professional knowledge should take into account the different facets and levels described in Sect. 4. The multi-dimensional and systemic nature of this knowledge also requires multiple strategies for developing and assessing this knowledge, such as those described in other chapters in this book.

A main challenge for teacher educators is finding suitable ways to articulate the teachers' learning of statistics and transmitting an epistemological vision of statistics in agreement with social constructivism, as well as developing teachers' statistics pedagogical knowledge. A possible tool is the formative cycle designed by Godino, Batanero, Roa, and Wilhelmi (2008), which was tried in an experience with prospective primary school teachers.

The formative cycle started with a statistical project that was completed by the prospective teachers in teams, following a socio-constructivist instructional design. Collecting data to complete the project led the future teachers to compare frequency distributions, and thus justify the introduction of statistical tables, graphs, and summaries. Another feature of this project was the multivariate approach to data analysis, which is also specific to statistics, as decision-making in random situations often requires taking into account, not just one variable, but a multiple approach. The project also provided the prospective teachers with a teaching model where the traditional *knowledge division* in textbooks (concepts versus procedures) was overcome and where statistical concepts and techniques were justified by a real problem, so that these concepts acquired a *situational meaning* for the teachers.

In a second stage, the project served to provoke didactical reflection on pedagogical content knowledge. After discussing the solution to the problem posed and the statistical conclusions for the research project, the prospective teachers were asked to analyse the different facets and suitability criteria described in Sect. 4 in the teaching/learning process they had lived in their own classroom. Many prospective teachers in the Godino et al.'s (2008) experience had difficulties in analysing the different components for pedagogical knowledge and in assessing the didactical suitability of the teaching process. This outcome was reasonable, given the scarce time devoted to preparing the teachers who took part in the experience and the complexity of pedagogical knowledge. However, the activity proved to be useful to introduce systematic reflection on the different facets affecting the teaching and learning of statistics. Moreover, responses by even the most advanced future teachers showed some underlying conceptions about teaching and learning mathematics that should be made explicit and confronted. It also provided a multivariate approach to didactical analysis by including the different dimensions that interact with the teaching and learning processes of statistics that were described in the previous sections.

To conclude we suggest the need to improve the models for the didactic knowledge required to teach statistics that take into account the specificity of statistics. Improving the statistics education of school teachers will also require significant changes in the initial teachers' preparation syllabus and assigning more time to teachers' statistics education.

Acknowledgements Research supported by the projects: SEJ2007-60110/EDUC. MCYT-FEDER and EDU2010-14947, MICIIN-FEDER.

References

Ball, D. L., Lubienski, S. T., & Mewborn, D. S. (2001). Research on teaching mathematics: The unsolved problem of teachers' mathematical knowledge. In V. Richardson (Ed.), *Handbook of research on teaching* (4th ed., pp. 433–456). Washington, DC: American Educational Research Association.

Batanero, C., Burrill, G., Reading, C., & Rossman, A. (Eds.). (2008). *Joint ICMI/IASE Study: Teaching Statistics in School Mathematics. Challenges for Teaching and Teacher Education. Proceedings of the ICMI Study 18 and 2008 IASE Round Table Conference.* Monterrey, Mexico: International Commission on Mathematical Instruction and International Association for Statistical Education. Online: www.stat.auckland.ac.nz/~iase/publications

Burgess, T. A. (2008). Teacher knowledge for teaching statistics through investigations. In C. Batanero, G. Burrill, C. Reading, & A. Rossman (2008).

Cobb, P., & McClain, K. (2004). Principles of instructional design for supporting the development of students' statistical reasoning. In D. Ben-Zvi & J. Garfield (Eds.), *The challenge of developing statistical literacy, reasoning, and thinking* (pp. 375–396). Dordrecht, The Netherlands: Kluwer.

Garfield, J., & Ben-Zvi, D. (2008). Preparing school teachers to develop students' statistical reasoning. In C. Batanero, G. Burrill, C. Reading, & A. Rossman (2008).

Godino, J. D., Batanero, C., & Font, V. (2007). The onto-semiotic approach to research in mathematics education. *ZDM. The International Journal on Mathematics Education, 39*(1–2), 127–135.

Godino, J. D., Batanero, C., Roa, R., & Wilhelmi, M. R. (2008). Assessing and developing pedagogical content and statistical knowledge of primary school teachers through project work. In C. Batanero, G. Burrill, C. Reading, & A. Rossman (2008).

Godino, J. D., Wilhelmi, M. R., & Bencomo, D. (2005). Suitability criteria of a mathematical instruction process: A teaching experience of the function notion. *Mediterranean Journal for Research in Mathematics Education, 4*(2), 1–26.

Graeber, A., & Tirosh, D. (2008). Pedagogical content knowledge. In P. Sullivan & T. Wood (Eds.), *Knowledge and beliefs in mathematics teaching and teaching development* (pp. 117–132). Rotterdam, The Netherlands: Sense Publishers.

Hill, H., & Ball, D. L. (2004). Learning mathematics for teaching: Results from California's mathematics professional development institutes. *Journal for Research in Mathematics Education, 35*(5), 330–351.

Hill, H. C., Ball, D. L., & Schilling, S. G. (2008). Unpacking pedagogical content knowledge: Conceptualizing and measuring teachers' topic-specific knowledge of students. *Journal for Research in Mathematics Education, 39*, 372–400.

Philipp, R. A. (2007). Mathematics teachers' beliefs and affect. In F. K. Lester (Ed.), *Second handbook of research on mathematics teaching and learning* (pp. 257–315). Charlotte, NC: Information Age Publishing and National Council of Teachers of Mathematics.

Ponte, J. P. (2008). Preparing teachers to meet the challenges of statistics education. In C. Batanero, G. Burrill, C. Reading, & A. Rossman (2008).

Ponte, J. P., & Chapman, O. (2006). Mathematics teachers' knowledge and practice. In A. Gutierrez & P. Boero (Eds.), *Handbook of research of the psychology of mathematics education: Past, present and future* (pp. 461–494). Rotterdam, The Netherlands: Sense Publishers.

Schoenfeld, A. H., & Kilpatrick, J. (2008). Towards a theory of proficiency in teaching mathematics. In D. Tirosh & T. Wood (Eds.), *Tools and processes in mathematics teacher education* (pp. 321–354). Rotterdam, The Netherlands: Sense Publishers.

Shulman, L. S. (1987). Knowledge and teaching: Foundations of the new reform. *Harvard Educational Review, 57*, 1–22.

Sowder, J. T. (2007). The mathematical education and development of teachers. In F. K. Lester (Ed.), *Second handbook of research on mathematics teaching and learning* (pp. 157–223). Charlotte, NC: Information Age Publishing and National Council of Teachers of Mathematics.

Watson, J. M. (2001). Profiling teachers' competence and confidence to teach particular mathematics topics: The case of chance and data. *Journal of Mathematics Teacher Education, 4*, 305–337.

Watson, J. M., Callingham, R. A., & Donne, J. M. (2008). Establishing PCK for teaching statistics. In C. Batanero, G. Burrill, C. Reading, & A. Rossman (2008).

Wild, C. J., & Pfannkuch, M. (1999). Statistical thinking in empirical enquiry. *International Statistical Review, 67*(3), 223–265.

Wood, T. (Ed.). (2008). *The international handbook of mathematics teacher education.* Rotterdam, The Netherlands: Sense Publishers.

Chapter 28
Measuring Levels of Statistical Pedagogical Content Knowledge

Rosemary Callingham and Jane Watson

Abstract The introduction of statistics and probability into the school curriculum has raised awareness of the expectations on teachers who have to teach it. A review of the related field of mathematics education indicates that teachers need more than content knowledge. They must also respond to their students' statistical understandings in ways that move students' current understanding to higher levels. Efforts to measure such statistical pedagogical content knowledge are still in their infancy. Findings from a large-scale Australian study are reported to exemplify these efforts, and the implications for future research are discussed.

1 Introduction

As early as 1988, many of the participants in the International Statistical Institute Round Table in Budapest (Hawkins, 1990) were discussing the lack of adequate preparation of school teachers for teaching statistics and the consequent deficiencies observed. Rubin and Rosebery (1990), for example, described anecdotally the difficulties displayed by classroom teachers that are indicators of the types of understanding that researchers are attempting to measure more formally today. Despite statistics being part of the curriculum in the later years of schooling (Holmes, 1980), the advent of statistics and probability in the broader school curriculum in the early 1990s, including the primary or elementary years (e.g., Australian Education Council, 1991; National Council of Teachers of Mathematics, 1989) raised awareness of issues surrounding the teaching of these topics.

R. Callingham (✉)
Faculty of Education, University of Tasmania, Locked Bag 1307,
Launceston, TAS, Australia 7250
e-mail: Rosemary.Callingham@utas.edu.au

J. Watson
University of Tasmania, Private bag 66, Hobart, TAS, Australia 7001
e-mail: Jane.Watson@utas.edu.au

C. Batanero, G. Burrill, and C. Reading (eds.), *Teaching Statistics in School Mathematics-Challenges for Teaching and Teacher Education: A Joint ICMI/IASE Study*, DOI 10.1007/978-94-007-1131-0_28, © Springer Science+Business Media B.V. 2011

In this chapter, the understanding of the nature of teachers' knowledge that is needed at the school level and emerging attempts to measure this knowledge are considered (chapters by Burgess; Godino, Ortiz, Roa, & Wilhelmi; Ponte; Pfannkuch & Ben-Zvi; and Makar & Fielding-Wells, in this book present other analyses of teachers' pedagogical and professional knowledge). First, research in the related field of mathematics education is examined as a way of informing discussion about statistics education; second the field of statistics education itself is canvassed, identifying issues in relation to teachers' knowledge. Finally aspects of pedagogical content knowledge for teaching statistics and recent efforts to measure this construct are described to exemplify current approaches, and the implications are discussed.

2 Background

The measurement of teachers' mathematical knowledge has predominantly focused on content knowledge (Hill, Sleep, Lewis, & Ball, 2007), especially arithmetic, with statistical knowledge limited to reading simple graphs. In a seminal paper, Shulman (1987) identified seven knowledge types needed by teachers to be competent in the modern classroom. This work has prompted a variety of attempts to describe, identify, and measure more precisely the nature of teachers' knowledge, in particular in the field of mathematics education. Although the measurement of teachers' mathematical knowledge is well established, often measured by relatively simplistic tests of mathematical knowledge (Hill et al., 2007), more contested has been the level of content knowledge required by teachers if they are to teach mathematics effectively. Ma (1999) suggested that elementary teachers do not need high levels of mathematical knowledge, but do require a "profound understanding of fundamental mathematics" (p. 22), including a deep grasp of the interconnections and relationships among different aspects of mathematical knowledge. It seems likely that such thinking could apply also to statistics.

A different approach to measuring teachers' mathematical knowledge is that taken by the Training and Development Agency for Schools (2009) in the United Kingdom. All teachers are required to take a test of "numeracy" before they are accredited to teach, including many items that could be considered statistical in nature, for example, reading box plots of data from testing programmes. Although addressing some aspects of Shulman's (1987) "knowledge of education contexts", the test remains content-focused.

In recent years, attention has shifted from content knowledge to the description and measurement of "Pedagogical Content Knowledge" (PCK), which Shulman (1987) conceptualised as "that special amalgam of content and pedagogy that is uniquely the province of teachers, their own special form of professional understanding" (p. 6). Shulman's ideas have been refined and developed over time. Hill, Rowan, and Ball (2005), for example, described Mathematical Knowledge for Teaching (MKT), which they defined as that "...mathematical knowledge used to carry out the work of teaching mathematics" (p. 373). Chick (2007) suggested that PCK may be inferred from the nature of the tasks set by teachers, and in particular their use of stimulus material that has a range of possibilities or "affordances" for use in the classroom. Such work suggests that different topics may require diverse aspects of PCK, and the creative and imaginative use of suitable artefacts may play a role in identifying PCK.

Recent work has focused on the formal measurement of aspects of PCK. Hill, Schilling, and Ball (2004) measured teachers' MKT using a multiple choice test based on mathematical content commonly found in elementary school courses, such as number concepts and operations, and pattern and algebra. No statistical knowledge was included. These instruments were then used to link teacher knowledge with students' outcomes (Hill et al., 2005). A recent international study (MT-21) considered both content knowledge and "Mathematics Pedagogy Knowledge" to evaluate teacher education programmes. The content survey included statistics, but the Mathematics Pedagogy Knowledge component was organised around three sub-scales: Curriculum, Teaching, and Students (Schmidt et al., 2007, p. 26). Of particular relevance to the discussion presented here in relation to the types of items employed is the Students scale, which used items that required teachers to respond to student answers and identify the errors made on a variety of topics.

2.1 The Situation in Statistics Education

Measuring teacher knowledge specifically for teaching statistics has a shorter history than for teaching mathematics. At the International Statistical Institute Roundtable in 1992, Begg (1993) proposed a research agenda that began to address such issues at the school level. This agenda focused on professional development and its effectiveness that implied, although did not address explicitly, some measure of teachers' statistical knowledge. Shaughnessy (1992) further indicated the lack of research in this area by placing teachers' conceptions of probability and statistics as one of his seven significant future research questions.

By 2001 however, issues in school statistics were recognised, and issues around teachers' capacity to teach statistical concepts were becoming part of the research agenda. One approach to identifying and measuring teachers' knowledge in statistics was that taken by Watson (2001) using a "profile". This instrument attempted to address multiple domains of teacher knowledge including self-efficacy or confidence in teaching statistical concepts, beliefs about the value and use of statistics, and pedagogical content knowledge using items in which teachers responded to questions based on student survey items. Groth (2007) provided a framework for teaching statistics at the high school level adapting the ideas of "common" and "specialised" knowledge of Hill et al. (2004) and acknowledged that the specialised knowledge area required a growing research base. This is particularly the case in relation to non-mathematical knowledge, which encompasses the pedagogical activities that take place in the classroom.

Watson, Callingham, and Donne (2008) reported on the development of a measure of pedagogical content knowledge in statistics. The instrument used was a pen-and-paper survey similar to Watson's (2001) profiling instrument but extended the items in which teachers were asked to give a response to student work. Using Rasch measurement techniques (Bond & Fox, 2007), the 12-item scale of statistical pedagogical content knowledge had good measurement qualities and provided measures of teachers' performance that could be interpreted in hierarchical levels. One aspect of this study was the relative difficulty teachers had in suggesting suitable "next steps" to move students' understanding forward, even when they could predict students' responses. This finding suggests that instruments identifying PCK may fall short if they only include items asking teachers to identify appropriate responses without any accompanying follow-up action.

In an attempt to refine the measures, Watson, Callingham, and Nathan (2009) used similar items in an interview situation. An initial analysis of the teachers' responses indicated four components of teacher knowledge, including content knowledge, knowledge of students as learners, and two aspects of PCK. One PCK component was characterised by teachers employing content-specific strategies to develop students' understanding; the second involved constructing a shift from a specific to a more general statistical context, such as making explicit connections between different aspects of statistical knowledge through connecting, for example, a data display with notions of probability. This latter category extended the understanding of PCK and embeds specific statistical teaching strategies within the domain of statistical pedagogical knowledge. The interview format, however, has some practical limitations in terms of its use with a large number of teachers, although it does provide insights into the nature of statistical PCK.

PCK for teaching statistics is undoubtedly a complex construct. The study reported in the next section represents an attempt to measure statistical PCK using an instrument sufficiently sensitive to track changes in teachers' understanding over time.

3 Teacher Knowledge for Teaching Statistics

The *StatSmart* project incorporated a professional learning programme over 3 years assisting teachers to appreciate the developmental processes that students go through in reaching statistical understanding (e.g., Watson, 2006). Teachers were also provided with resources including *Tinkerplots* software (Konold & Miller, 2005) and attended a 2-day conference each year at which activities and ideas were presented to help them devise learning activities suitable for their class levels. The project timeline included repeated monitoring of teachers and their students. Design details were discussed in Callingham and Watson (2007). The survey data reported here were collected from teachers after the first professional learning conference towards the end of the first year of the study, and again midway through the project.

The teacher group was composed of 45 very experienced teachers who mostly had limited tertiary-level mathematics study within which none explicitly mentioned studying statistics. They taught middle-year students (Grades 5–10) in a variety of settings or in high school, often including post-compulsory years. Although teaching in the different jurisdictions that comprise education in Australia (i.e., government, Catholic and independent schools) using different curriculum documents, their discussions suggested that they were more homogenous than the background data might imply. Completion of the survey was an expectation of the project and 42 teachers responded to the first survey. The second survey was answered by 26 teachers, of whom 18 had also responded to the first administration. The lower response rate was probably due to the time of year when the survey was sent out, some months after the second professional conference. The eight new teachers had very similar backgrounds to the others.

Only the 12 PCK items from the instrument (Watson, 2001) are discussed here. These items were of three types. The first group (four items) provided teachers with survey questions that had been given to students and asked them to anticipate appropriate and inappropriate responses that their own students might provide. A second group (four items) extended these initial questions by asking teachers to indicate how they might use the item stimulus in the classroom, using Chick's (2007) notion of affordances. An example is the item in the context of a newspaper headline on "odds" shown in Fig. 28.1, in which question a was of the first type (anticipating students' responses) and question b was of the second type (classroom affordance). The stimulus contexts for these items included an incorrect pie chart from a media article and a newspaper report about an association between car accidents and heart disease (Watson, 2006), as well as the "odds" headline in Fig. 28.1.

The third type of item, of which there were four, explicitly invited teachers to respond to particular students' answers. Real student answers from surveys were provided and teachers were asked to say how they would respond to the student. An example of two items employing proportional reasoning in the context of

---

North at 7-2, but we can still win match, says coach

1. What does "7-2" mean in this headline about the North against South football match? Give as much detail as you can.
2. From the numbers, who would be expected to win the game?

The above item was given to a group of students.

a. What kinds of responses would you expect from your students? Write down some appropriate and inappropriate responses.
b. How would/could you use this item in the classroom? For example, how would you intervene to address the inappropriate responses?

Fig. 28.1 An example of a PCK item asking teachers to anticipate students' responses

Box A and Box B are filled with red and blue marbles as follows. Each box is shaken. You want to get a blue marble, but you are only allowed to pick out one marble without looking. Which box should you choose?

a. Box A.

b. Box B.

c. It doesn't matter.

Please explain your answer.

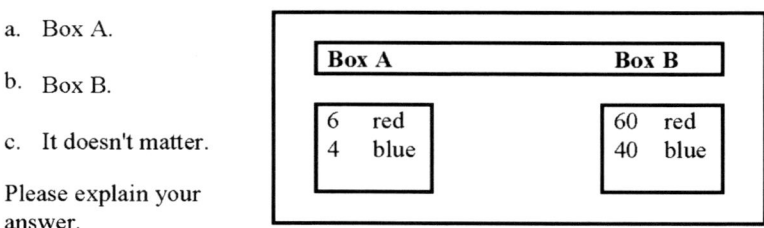

Consider each of the following answers and explanations to the above problem and explain how you would respond to each answer.

Student 1: (B), because you could get red or blue.

Student 2: (A), because there are only two more reds in A and 20 more in B.

Fig. 28.2 Example of a student response item for addressing teachers' statistical PCK

determining a probability is shown in Fig. 28.2. The other two PCK items were set in the context of a two-way table.

In the second teacher survey in 2008 some modifications were made to the original survey items in order to extend the information obtained into some different statistical content areas. The items on odds, pie chart, and the proportional reasoning item shown in Fig. 28.2 remained the same. The harder proportional reasoning item, using a two-way table, was dropped because many teachers avoided answering it. The association item was modified to make the teacher questions more explicit, reducing the number of individual questions. A new item addressing average replaced the two-way table question. These modifications provided a total of 11 items, of which 4 items addressed anticipated student responses, 3 items addressed classroom affordance, and 4 items asked teachers to respond to students' answers. Of these 11 items, 6 were common across both administrations of the profile (odds, pie chart, and the proportional reasoning task shown in Fig. 28.2) to provide a basis for scale anchoring using Rasch measurement (Bond & Fox, 2007). All items were scored using a rubric based on the increasing complexity of the response. The scores for each item varied from 0–2 to 0–5, where 0 in every instance indicated no response or a totally irrelevant response (e.g., Watson et al., 2008).

3.1 Analysis

In order to explore the construct of statistical PCK a number of analyses were carried out. First, a Principal Components factor analysis with Kaiser normalisation and varimax rotation was undertaken to examine the construct in relation to the four components identified by Watson et al. (2009).

The initial factor analysis using the 2007 data suggested four factors that together accounted for 67.6% of the variance. The first of these included all four items in which teachers were asked to respond to students' work. The second factor addressed anticipated student responses to the items on odds and a pie graph, and classroom use of odds (three items). The third included anticipated student responses to the association item (three items), and the fourth factor included suggested classroom use of two items that included graphing: pie graph and association. An identical analysis was run on the data from the second administration of the instrument in 2008. Only three factors were identified that accounted for 70.1% of the variance, and eight of the eleven items loaded onto more than one factor. The number of teachers ($n = 26$) was smaller and 18 of these were completing a second survey. It is possible that the teachers who had been in the project for some time were now interpreting the items in different ways, leading to the different results. The changes did suggest, however, that creating scales based on the initial factor analysis was unlikely to provide consistent results.

Second, the data were analysed using Rasch measurement (Bond & Fox, 2007) in which person ability and item difficulty are placed on the same interval measurement scale. The hierarchical nature of the scoring rubrics provided direction along a single variable of PCK, and the variable hence established had good measurement characteristics. Measures of person "ability" in logits, the unit of Rasch measurement, for all 42 teachers who responded to the 2007 administration were obtained. The second administration, in 2008, was similarly analysed but anchored to the item difficulty of the first so that all measures were against the same "ruler". In this way, direct comparisons of teacher performance could be made across both administrations of the instrument to determine change over time.

3.2 Interpretation of the Construct

The Rasch analysis process spreads the items along the extent of the variable, creating an hierarchical scale where the more difficult items appear at the top of the construct and the easiest items occur at the bottom. By considering the "gaps" between clusters of items and the demands of the items within clusters, four levels of PCK were identified. A description of the nature of teachers' responses at each level of the PCK hierarchy was obtained by a qualitative analysis of the item responses occurring at each level and is described in this section.

Aware level. Teachers are likely only to suggest a single appropriate or inappropriate student response to the items. They display little broader statistical understanding and do not make suitable suggestions for addressing students' understanding. For example, in relation to using the odds question in the classroom a typical Aware level response is, "Great for relating to football betting in real life; Hear this on the news and footy show".

Emerging level. Teachers use some statistical knowledge to suggest several *either* correct *or* incorrect responses for students. Two PCK items, related to odds and a graph of association, expect the ability to suggest *both* correct and incorrect student responses, for example, in relation to the odds question: "No idea; Not sure; North have a good chance to win; for every $2 placed you receive $7 back or for every 9 matches North would win 7". Generic rather than content-specific strategies are suggested for classroom intervention, which implies good teaching but not necessarily in the context of statistics.

Competent level. The items with the highest codes at this level reflect more traditional and familiar topics in the school statistics curriculum: graphs and probability. Some statistically appropriate interventions are suggested but only in the context of familiar classroom activities. For example, for the proportional reasoning task, "Question: Is it all luck? Would I be more likely to get a red? Why? In both boxes? How could I show my chance of getting a blue in numbers?"

Accomplished level. The highest code for most PCK items appears at this level, demanding the suggestion of *both* correct and incorrect student responses and an integration of appropriate statistical content with student-centred intervention strategies. For example, in using the lung disease two-way table item in the classroom, a response showing an integrated view of statistical PCK is "Research cause and effect relationship for the diseases. Develop understanding of dependent, independent variables, control and how difficult it is to isolate other factors – need to 'read between the statistics.' What other information is needed? If you eat carrots you will die!"

As teachers progress up the scale, they demonstrate more complex and inter-related responses, showing both understanding of appropriate statistical knowledge and of their students.

3.3   Teacher Change

To consider teacher change, a paired sample *t*-test was conducted on the mean teacher ability measures obtained from Rasch measurement from 2007 ($\bar{x} = 0.47$, s.d. $= 0.78$) and 2008 ($\bar{x} = 1.15$, s.d. $= 1.44$) for the 18 teachers who completed both surveys. This test indicated a significant increase in the mean value of teacher ability (df $= 17$, $t = 2.26$, $p = 0.040$) with an effect size of 0.59, indicating that the scale could provide a measure of teacher change.

The four hierarchical levels of statistical PCK identified, however, provide a criterion-referenced approach to considering teacher knowledge that can go beyond considering mean scores. By comparing teacher ability measures obtained through the Rasch measurement approach against the items on the same scale, the number of teachers in each level of the hierarchy can be compared. The proportion of teachers in each level of the PCK construct in 2007 was ($n = 42$): Aware 14% (6); Emerging 28% (12); Competent 50% (21); and Accomplished 7% (3). In 2008, the corresponding figures were ($n = 26$): Aware 15% (4); Emerging 15% (4); Competent 46% (12); and Accomplished 23% (6). Teachers in the Aware level in 2008 were all teachers who had entered the project late. Of particular interest is the increased number of teachers in the Accomplished level, from three in 2007 to six in 2008. Using this profile approach has the potential to be able to provide more targeted intervention with teachers, by matching professional learning activities to teachers' demonstrated level of PCK.

4   Implications for Teacher Training and Research

The *StatSmart* study represents a new direction in PCK measurement by using items in which teachers have to respond in ways that draw on their classroom practice, which are scored according to the quality of that response. The results suggest that in addition to understanding the statistical content, at high levels of PCK teachers need to be able to anticipate the range of students' likely answers to questions and to respond in ways that will further develop students' understanding. Although this profiling approach may lack the level of detail that might be obtained through interviews and observations, it has the advantage that it can be used for large-scale projects.

The relatively small pool of teachers, and their long experience, limits the inferences that might be drawn in general about teachers' statistical PCK. It is possible that with a larger pool of teachers more levels might be identified because a larger and more representative sample would provide more information about items at the top and particularly the bottom of the scale. Further refining the measure of pedagogical content knowledge would be useful for evaluating professional learning or pre-service teacher education programmes, for example.

Developing measures of teachers' statistical pedagogical content knowledge is an important continuing goal of statistics education. The next stage in research related to measuring teachers' statistical pedagogical knowledge will be to examine the association of teachers' statistical PCK and student learning outcomes. Such research will then prepare the ground for evaluating professional learning programmes for teachers in terms of *change* in teachers' statistical PCK and *change* in their students' learning outcomes.

Acknowledgements This study was supported by an Australian Research Council, Grant No. LP0669106, with Linkage Partners the Australian Bureau of Statistics, Key Curriculum Press, and the Baker Centre for School Mathematics at Prince Alfred College, Adelaide.

References

Australian Education Council. (1991). *A national statement on mathematics for Australian schools*. Carlton, VIC: Author.

Begg, A. (1993). Establishing a research agenda for statistics education. In L. Pereira-Mendoza (Ed.), *Introducing data analysis in the schools: Who should teach it and how?* (pp. 212–218). Voorburg, The Netherlands: International Statistical Institute.

Bond, T. G., & Fox, C. M. (2007). *Applying the Rasch model: Fundamental measurement in the human sciences* (2nd ed.). Mahwah, NJ: Lawrence Erlbaum.

Callingham, R., & Watson J. M. (2007, December). *Overcoming research design issues using Rasch measurement: The StatSmart project*. Paper presented at the Australian Association for Research in Education International Educational Research Conference 2007, Fremantle, Australia. Online: www.aare.edu.au/07pap/cal07042.pdf

Chick, H. L. (2007). Teaching and learning by example. In J. Watson & K. Beswick (Eds.), *Mathematics: Essential research, essential practice. Proceedings of the 30th Annual Conference of the Mathematics Education Research Group of Australasia* (Vol. 1, pp. 3–21). Sydney: MERGA.

Groth, R. E. (2007). Toward a conceptualisation of statistical knowledge for teaching. *Journal for Research in Mathematics Education, 38*, 427–437.

Hawkins, A. (Ed.). (1990). *Training teachers to teach statistics: Proceedings of the International Statistical Institute Round Table Conference*. Voorburg, The Netherlands: International Statistical Institute.

Hill, H. C., Rowan, R., & Ball, D. L. (2005). Effects of teachers' mathematical knowledge for teaching on student achievement. *American Educational Research Journal, 42*(2), 371–406.

Hill, H. C., Schilling, S. G., & Ball, D. L. (2004). Developing measures of teachers' mathematics for teaching. *Elementary School Journal, 105*, 11–30.

Hill, H. C., Sleep, L., Lewis, J. M., & Ball, D. L. (2007). Assessing teachers mathematical knowledge: What knowledge matters and what evidence counts? In F. K. Lester Jr. (Ed.), *Second handbook of research on mathematics teaching and learning* (pp. 111–156). Reston, VA: National Council of Teachers of Mathematics.

Holmes, P. (1980). *Teaching statistics 11–16*. Slough, UK: Schools Council and Foulsham Educational.

Konold, C., & Miller, C. D. (2005). *TinkerPlots: Dynamic data exploration [computer software]*. Emeryville, CA: Key Curriculum Press.

Ma, L. (1999). *Knowing and teaching elementary mathematics: Teachers' understanding of fundamental mathematics in China and the United States*. Mahwah, NJ: Lawrence Erlbaum.

National Council of Teachers of Mathematics. (1989). *Curriculum and evaluation standards for school mathematics*. Reston, VA: Author.

Rubin, A., & Rosebery, A. S. (1990). Teachers' misunderstandings in statistical reasoning: Evidence from a field test of innovative materials. In A. Hawkins (Ed.), *Training teachers to teach statistics: Proceedings of the International Statistical Institute Round Table Conference* (pp. 72–89). Voorburg, The Netherlands: International Statistical Institute.

Schmidt, W. H., Tatto, M. T., Bankov, K., Blömeke, S., Cedillo, T., Gogan, L., et al. (2007). *The preparation gap: Teacher education for middle school mathematics in six countries* (MT21 report). East Lansing, MI: Center for Research in Mathematics and Science Education, Michigan State University.

Shaughnessy, J. M. (1992). Research in probability and statistics: Reflections and directions. In D. A. Grouws (Ed.), *Handbook of research on mathematics teaching and learning* (pp. 465–494). New York: National Council of Teachers of Mathematics & MacMillan.

Shulman, L. S. (1987). Knowledge and teaching: Foundations of the new reform. *Harvard Educational Review, 57*, 1–22.

Training and Development Agency for Schools. (2009). *Numeracy practice materials*. Online: www.tda.gov.uk/skillstests/numeracy/practicematerials.aspx

Watson, J. M. (2001). Profiling teachers' competence and confidence to teach particular mathematics topics: The case of chance and data. *Journal of Mathematics Teacher Education, 4*, 305–337.

Watson, J. M. (2006). *Statistical literacy at school: Growth and goals*. Mahwah, NJ: Lawrence Erlbaum.

Watson, J. M., Callingham, R., & Donne, J. (2008). Establishing pedagogical content knowledge for teaching statistics. In C. Batanero, G. Burrill, C. Reading, & A. Rossman (Eds.), *Joint ICMI/IASE Study: Teaching Statistics in School Mathematics. Challenges for Teaching and*

Teacher Education. Proceedings of the ICMI Study 18 and 2008 IASE Round Table Conference. Monterrey, Mexico: International Commission on Mathematical Instruction and International Association for Statistical Education. Online: www.stat.auckland.ac.nz/~iase/publications

Watson, J. M., Callingham, R., & Nathan, E. (2009). Probing teachers' pedagogical content knowledge in statistics: "How will Tom get to school tomorrow?" In R. Hunter, B. Bicknell & T. Burgess (Eds.), *Crossing divides. Proceedings of the 32nd Annual Conference of the Mathematics Education Research Group of Australasia* (Vol. 2, pp. 563–570). Palmerston North, NZ: MERGA.

Part IV
Challenges and Experiences in Teacher Education

Carmen Batanero

It was made clear in the analyses presented in the previous parts that changing the teaching of statistics in schools will depend on the extent to which teachers are educated to adequately face the teaching of statistics at school level. The complex nature of teacher's knowledge described in Part III also suggest that attention should be paid not only to developing teachers' statistical knowledge and thinking, but also to improving their pedagogical content knowledge and to promote adequate conceptions and beliefs about statistics education.

Topic 3 (*Analysing current practices in teacher education regarding the teaching of statistics*) and Topic 4 (*Empowering teachers to teach statistics: A look into the future*) in the Joint Study Conference were devoted to analysing how statistics teachers are trained in different countries and how to improve this education. Along the discussions in the related working groups the need to use meaningful activities in supporting teachers' learning was emphasised, since didactical research and experiences presented at the Conference suggested that new knowledge is not automatically gained by teachers through their participation in professional development courses.

The purpose of this part is to expand the ideas discussed in Topics 3 and 4 in the conference, and to analyse the criteria that courses for educating teachers to teach statistics should fulfil. At the same time, the chapters in this part provide examples of successful experience in teachers' education in statistics education that were presented or discussed in the Joint Study Conference.

The nine chapters in this part deal with different but complementary themes. In the first chapter in this part, Joao Pedro da Ponte expands his reflections from the opening plenary session about the conditions that teacher education in statistics education have to meet to assure that the preparation of teachers may have real impact on classroom practice. The author suggests that teacher education should be related to the practice of teaching, the school and the profession, provide challenge

C. Batanero (✉)
Departamento de Didáctica de la Matemática, Universidad de Granada,
Facultad de Ciencias de la Educación, Campus de Cartuja, 18071 Granada, Spain
e-mail: batanero@ugr.es

and support, as well as recognise and empower teachers. The author also presents suggestions for combining these major elements in teacher education programmes and analyses two examples in Portugal of how these ideas can be put into practice in workshops for teachers.

As was highlighted in the analysis of different curricula in Part I and in the conference Topic 1, the main emphasis of statistics teaching should be on developing students' statistical literacy and reasoning, two main components that should also be taken into account in preparing the teachers. These next two chapters in this part address these two components.

Today, many national and international agencies make statistical information available on the Internet. Consequently, decision-making is increasingly based on evidence, and citizens need basic statistical skills to manage in today's society. Ridgway, Nicholson and McCusker conclude that statistical literacy is more than applying statistics mechanically; it is the ability to read and critically interpret data and use statistics as evidence in arguments in everyday and professional contexts. The authors describe some innovative curricula and experiences supported by statistical agencies with a focus on developing students' and teachers' statistical literacy. In particular they discuss the possibilities that new technologies offer to teachers to find effective ways to use multivariate data and graphs to develop statistical literacy in their students.

To complement these ideas, Pfannkuch and Ben-Zvi suggest that courses directed to teachers should be conceived as a "statistical reasoning learning environment" in which teachers develop a deep and meaningful understanding of statistics that later is transmitted to their students. The focus of these courses should be on fundamental statistical ideas, while at the same time the teachers should experience the complete statistical investigation cycle PPDAC (Problem, Planning, Data, Analysis, Conclusion) that have been discussed extensively in different chapters of the book. Two additional components in courses directed to teachers that are analysed by Pfannkuch and Ben-Zvi are formative assessment and learning to understand students' reasoning.

The next three chapters in Part IV analyse other inter-related themes largely discussed at the Joint Study Conference in relation to the education of teachers: the benefits and challenges of using real data in the education of students and teachers (Chap. 32), the need to develop teachers' ability to teach statistics through investigations (Chap. 33) and the knowledge teachers need to effectively use technology in the teaching of statistics (Chap. 34).

As in Parts I and II, the recommendation in training teachers to teach statistics is for a data-orientated education, where teachers design their own statistical project, collect and analyse their own data, and draw conclusions that are based on these data. Hall analyses the challenges and pedagogical issues faced by teachers when using real data, the different status of data in mathematics and statistics classrooms; and the potential that real data offer for learning, motivation and interdisciplinarity. She discusses the advantages of using primary versus secondary sources of data and offers suggestions for obtaining and using real data. The chapter concludes with an analysis of the use of real data in supporting teachers' learning and presents an example in Canada, where CensusAtSchool data is used in development workshops for elementary teachers.

Projects and investigations are common in other school subjects, like science and social studies, but are rarely used in a mathematics classroom where statistics is usually taught. Makar and Fielding-Wells suggest the need to give more relevance to the statistical investigation cycle in the training of teachers, in order to develop their statistical thinking abilities, and a deep understanding of statistics. Their chapter highlights ways for teacher educators to support teachers' learning to teach statistical inquiry. Results of two longitudinal studies are used to formulate recommendations to develop teachers' proficiency in this area.

Working with real data and projects often require the use of technology. Preparing teachers to use technology is not an easy task, as many factors influence teachers' decisions about how to use these tools in teaching statistics. These factors are analysed by Lee and Hollebrands, who also offer a theoretical model of what they term "technological pedagogical statistical knowledge" (TPSK) or the knowledge needed to effectively use technology to teach statistics. The chapter is complemented with a discussion of issues to consider in developing this knowledge in teachers and examples of teacher education courses focused on increasing teachers' TPSK.

The next two chapters analyse particular methods that were successful in educating statistics teachers. Groth and Xu analyse two experiences where case analysis was used to promote professional development in teachers in two very different settings: A case discussion among a group of prospective secondary mathematics teachers in the Mid-Atlantic United States and another case discussion among a group of in service high school teachers in China. These examples are used by the authors to show how case analysis can help teachers to increase their general pedagogical knowledge, statistical pedagogical knowledge, and statistical knowledge. The authors also use these examples to discuss the types of classroom discourse that statistics teacher educators should take into account to facilitate case discussions among the teachers.

Quick changes in curriculum pose the problem of providing continuous education for an increasing number of teachers. Meletiou and Serradó argue that distance education offers flexibility and convenience in solving financial and logistic difficulties of engaging teachers in face-to-face professional development courses. The authors discuss the main pedagogical issues and challenges underlying distance education in general, and teacher education in particular. They also describe four examples of programmes that have utilised distance education to offer teacher training in statistics education.

Chapter 36 was developed from two papers presented in the Conference Topic 6 (*Building collaboration between mathematics and statistics educators in teacher education*). The last chapter in this part is another example of experiences presented and discussed in this Topic.

An increasing number of national statistical offices around the world are today recognising the importance of promoting statistical literacy at school level. Support from these institutions for statistics education includes making their data easily accessible, and promoting educational projects to develop teaching materials or educating school teachers. In the last chapter, North and Scheiber describe some successful examples of collaboration between statistical offices and schools or educational authorities to provide support to train teachers around the world. These examples include programmes from national statistical offices in Canada, Portugal, New Zealand and South Africa, as well as two international projects, the CensusAtSchool or the International Statistical Literacy Project.

In summary although chapters included in this part derived from papers presented in different topics in the conference, there is an underlying theme, the preparation of statistics teachers. These chapters provide valuable insights and recommendations for teacher educators and educational authorities in charge of educating teachers.

Chapter 29
Preparing Teachers to Meet the Challenges of Statistics Education

João Pedro da Ponte

Abstract This chapter addresses how teacher education in statistics education may have real impact on classroom practice. Teacher knowledge about statistics and teaching statistics as well as teachers' practices are discussed with emphasis on tasks and classroom communication, contrasting direct teaching to teaching for exploratory learning. Then, teacher education strategies with emphasis on problematising, interacting, and resourcing are discussed. Finally, the design of teacher education programmes is considered, stressing the need to relate teacher education to teaching practice, the school and the professional group, as well as to provide challenge and support, and to recognise and empower teachers. Examples from in- and pre-service teacher education are presented.

1 Introduction

The most serious problem in teacher education is its minimal impact on teaching practice (Smith, 2001). The aim of this chapter is to discuss how this may be changed in the field of statistics education. Such discussion is based on a perspective about the content and nature of teacher knowledge in statistics and teaching statistics and its relation to teaching practices. The chapter also addresses issues on teacher education strategies and designs, and suggests several key ideas regarding the teacher education setting and the role of the teacher educator, illustrating them with two examples, one from in-service and another from pre-service teacher education. Finally, some implications for teacher education and research are provided.

J.P. da Ponte (✉)
Instituto de Educação, Universidade de Lisboa,
Alameda da Universidade, 1649-013 Lisbon, Portugal
e-mail: jpponte@ie.ul.pt

C. Batanero, G. Burrill, and C. Reading (eds.), *Teaching Statistics in School*
Mathematics-Challenges for Teaching and Teacher Education: A Joint ICMI/IASE Study,
DOI 10.1007/978-94-007-1131-0_29, © Springer Science+Business Media B.V. 2011

2   Teacher Knowledge About Statistics and Teaching Statistics

Teacher education in statistics may follow many routes. It depends, for example, on the perspective that is assumed for teaching statistics. In fact, it makes a difference when such teaching is centred in: (a) mastering concepts and procedures, computing statistics measures, and representing data in routine exercises; (b) data handling, collecting, representing, and interpreting readymade data provided by the teacher, the textbook or the Internet; or (c) doing statistical investigations, which involve a full cycle of posing questions, collecting, analysing, interpreting, and critiquing data and arguments.

First, consider content. Curricula emphasise statistical ideas in different ways. Some give prominence to basic statistical summaries (mean, median, mode, standard deviation, quartile…), others to statistical representations (graphs, tables, diagrams…), and some strive to address important ideas in statistics (such as centre, variability and distribution and other notions discussed by Burrill & Biehler, in this book) or to connect statistics and probability (Chaput, Girard, & Henry, 2008; this book; Borovcnik, this book). Curricula also differ largely in the ways they address the process of statistical investigation. For example, how much is discussed about formulating questions that seek generalisation? And what about designing for differences in data collection (using random selection in sampling designs and randomisation in experimental designs)? Does data analysis address variability just within a group or also between groups? Does it describe and quantify sampling error? Does it quantify association and fitting of models of association? With what depth and breadth of examples does it consider the interpretation of results of a statistical investigation? And what are the main targets for primary and secondary education? Whereas it is not difficult to make a list of statistical topics and concepts that teachers need to know to teach statistics (see Chaps. 20–25), the extent to which teachers should know and how they may learn these topics is still a matter that needs further research.

Another important element of teachers' statistical education concerns the professional knowledge required for teaching of statistics. A large number of studies has been carried out in the framework of "pedagogical content knowledge" (PCK), a notion proposed by Shulman (1986). The key idea is that "pedagogy" and "content" combine in a special way in teachers' professional knowledge. Recently, Hill and Ball (2004) suggested the concept of "specialised knowledge of content" (SKC), as a particular way for teachers to master subject matter, that supports their activity in planning and handling classes and in assessing students' knowledge, strategies, and difficulties. PCK and SKC are appealing notions, as they resonate with the experience and concerns of teachers and teacher educators for whom both content and pedagogy are important. These notions, however, involve some ambiguity concerning the nature of such knowledge – whether it is declarative knowledge or it is essentially practical; the problem is that most researchers that use these notions tend to view them as declarative knowledge, and not as action-oriented professional craft knowledge (Ponte & Chapman, 2006).

Professional knowledge required for teaching statistics may be regarded as standing on three main poles: (a) knowledge of students, including their learning

Table 29.1 Professional knowledge related to teaching practice

Planning	Curriculum objectives
	Classroom structure (introduction/exploration/discussion)
	Tasks
	Materials
	Organising students' work
	Management of time
	Assessment
Conducting	Introducing tasks and negotiating the work and norms (contract)
	Handling classroom communication
	Negotiation of statistical meanings
	Making decisions according to the flow of the lesson
Reflecting	Were the curriculum objectives met?
	Did the students learn what was sought?
	Was the planning appropriate?
	Were the classroom events handled properly?

processes, thinking strategies, difficulties, interests, and culture; (b) knowledge of the curriculum, including purposes, levels of development, connections with mathematical topics and with other subjects; and (c) knowledge of teaching practice (Table 29.1), including the planning for instruction, conducting the classroom activity, and reflecting on teaching practice.

Professional knowledge may be regarded as a blend including a set of facts and principles (declarative knowledge), knowing how (process knowledge), and knowing how to be specific about the teaching activity (Schön, 1983). This view has important implications for the assessment of such knowledge by paper and pencil instruments (see, e.g., Watson, Callingham, & Donne, 2008), since these measures capture just some aspects of teachers' declarative knowledge, missing its practice-oriented aspects.

This view also has important implications for teacher education since this knowledge is informed by theory and by practice. Advocates of situated knowing will argue that professional knowledge is highly connected to the specific institution and context where the teacher works (Putnam & Borko, 1997). In the next section two key elements of teaching practice (tasks and classroom communication and teaching styles) are analysed. The issue of how teacher education settings may support teachers in developing their professional knowledge and teaching practices will be discussed in Sects. 4 and 5.

3 Teaching Practice

3.1 Tasks and Classroom Communication

Statistics learning largely depends on the activity that students carry out in the classroom and such activity largely depends on the tasks proposed by the teacher. Therefore, tasks are essential elements in framing teachers' practices. The exercise is the most characteristic task in many other school subjects, including mathematics (Christiansen & Walther, 1986) and also statistics. But, besides exercises, students need to get involved in doing other kinds of tasks such as investigations, explorations, and projects (see MacGillivray & Pereira Mendoza, in this book). It should be noted that it is impossible to classify a task in absolute terms, since its nature is always relative to the person who does it.

Another important element of teaching practice is handling classroom communication. In many classrooms, the teacher dominates the discourse, either providing explanations and examples or posing successive questions and giving feedback. Such classes tend to follow the sequence IRF – the teacher "initiates" a question, a student "responds", and the teacher provides immediate "feedback", accepting or rejecting the response. However, there are other patterns of classroom discourse, based, for example, on inquiry questions (Wood, 1994).

The students may be encouraged to share ideas with their colleagues, working individually, in pairs, in groups, or as a whole class. Classroom discussions are important for negotiation of meanings (Bishop & Goffree, 1986). In such discussions, different representations are contrasted, the conventional representations are analysed, and the proper use of statistical language is fixed. This is also when the main ideas related to the task are clarified, formalised, and institutionalised as accepted knowledge in the classroom community. During group work, the way the teacher interacts with the students is also very important. If the teacher does not respond to the students' questions, he or she may lose the motivation to keep working, whereas if the teacher provides the answer, he or she cancels out most of the possible learning benefit for the students. Therefore, teachers have to regularly deal with difficult dilemmas in conducting classroom communication.

3.2   Two Styles of Teaching Practices

The analysis of different kinds of tasks, roles, and communication patterns suggests a contrast of two main styles of teaching statistics practices that may be found in classrooms in different grade levels, sometimes used by the same teachers at different moments:

- In *Direct teaching*: (a) tasks are reduced to standard exercises and the situations are artificial; (b) for each task there is a strategy and a correct answer; (c) students receive "explanations" and the teacher shows "examples" so that students learn "how to do things"; (d) the teacher poses questions and provides immediate feedback and the student asks "clarification" questions; and (e) teachers and textbooks are the authorities.
- In *Exploratory learning*: (a) there is a variety of tasks, including explorations, investigations, problems, projects, and exercises; (b) the situations are realistic and often, there are several strategies to deal with a task; (c) students receive tasks to discover strategies to solve them while the teacher asks the student to explain and justify his/her reasoning, so that the student is also an authority;

(e) students are encouraged to discuss with colleagues (working in groups or pairs); and (f) there are frequent discussions with the whole class and meanings are negotiated in the classroom.

Classroom practice depends on the teacher but also depends on the students and, at a different level, on several external factors. Some conditions make it very difficult for a teacher to move from direct teaching to exploratory learning. For example, in a statistics class with explorations, investigations, and projects it is impossible to predict all the ideas and questions that students may raise and, therefore, such a class is much more complex to manage than a class based on the exposition of content and completion of exercises. In addition, many students do not have previous experience in working on projects and investigations and need special support. Notwithstanding such difficulties and limitations, this work is essential in a statistics class that aims to develop students' statistical understanding, ability to investigate and carry out projects, and reasoning ability (see MacGillivray & Pereira Mendoza, in this book).

If the students are to experience significant statistics learning, they need to work for an extended time on a field of problems, leading to important statistical ideas, such as for example, correlation – at least for several classes. During such an activity, they have the opportunity to grasp the non-trivial aspects of the new knowledge, connect them to previous knowledge, and develop new representations and strategies. To carry out such teaching, teachers need a positive personal relation with statistical explorations and investigations and statistical reasoning as well as a capacity to design or select and enact such tasks and teaching units. These attitudes and competencies need to be a central focus in teacher education (Chick & Pierce, 2008) which is discussed in the next sections.

4   The Central Problem in Teacher Education

There is widespread consensus that teacher education is an essential element for quality teaching in any subject, including statistics. Teacher education is often seen as the major key to improve education, and sometimes it is even seen as the only key that is worthwhile to consider. In consequence, very high expectations are usually put on teacher education.

However, teacher education is also often subject to strong criticisms. Despite huge investments in money, time, resources, and personal involvement, progress seems to be thin. Smith (2001), for example, made an extensive critique of many features of common mathematics teacher education programmes and most of them apply equally well to statistics teacher education. She indicated that those programmes frequently focus on a particular issue and this corresponds to seeing teaching as a technical and routine activity, encouraging "fix ups" in the borders of practice instead of a global re-examination of such practice. The author also critiqued the fact that teacher education activities are often of very short duration and condensed in time and often not related to the teaching content. Furthermore, teacher education tends to follow an "academic" model, with a predefined curriculum and structured activities based on a paradigm of transmission of knowledge. As she indicated, one of the most serious weaknesses is that such teacher education programmes do not include support for practical implementation of new ideas, assuming that such ideas will be easily handled by teachers when they return to their classrooms. As a consequence, these teacher education courses have little impact on classroom practice.

Research and development in statistics education has produced a high quantity of knowledge and resources. However, providing teachers with this knowledge and these resources in an intensive and structured way, without taking into account what teachers need and the conditions in which they work, simply does not work. Teachers are able to learn and change their practices, when they follow a natural professional development path. Therefore, teacher education for in-service teachers must support their professional development process, recognising that such process is framed by interacting and sharing experiences with other teachers and requires articulation of the interests, needs, and resources of teachers and their professional contexts. Teacher education for pre-service teachers, in addition, requires ingenious ways of working on theoretical ideas in contexts that replicate or simulate professional practice. So, this leaves the problem of how to integrate knowledge on statistics education and knowledge on teachers' professional development and this is the central problem of teacher education design. Below, ideas and experiences on this issue are described and further suggestions are included elsewhere in this book (see Chaps. 30–37).

5   Teacher Education: Strategies and Designs

Teacher education strategies that take into account teachers' learning processes may be organised in three major groups that emphasise problematising, interacting, and resourcing. Strategies that emphasise problematising give prominence to conceptual aspects that the teacher him/herself needs to sort out, in relation to his or her practice. This includes the tradition of framing reflection as problem solving or reflection in practice (Schön, 1983) as well as action-research and other activities that involve observing and inquiring practice (Zeichner & Noffke, 2001). Strategies that give prominence to interacting, draw on notions of networking, collaboration and learning communities (Jaworski, 2005), suggesting that a key element of teacher professional development is the mediating power of the collective group. Strategies that emphasise resourcing teachers are based on the notion that mediation processes that bring about change may be supported by different kinds of artefacts, such as purposely designed teaching materials or technology-based teacher education environments (Lampert & Ball, 1998). Resourcing teachers with dense conceptual information and materials that are not closely related to their actual classroom practice is a common approach in teacher education but it is far from being an efficient one.

Of course, it is possible to combine these different emphases in several ways. A very powerful strategy that promotes such combination is working through projects, an activity with a long tradition in, and that has become a major strategy for teaching statistics (McGilliwray & Pereira-Mendoza, this book). A project may be seen as a set of activities that aim to reach a goal, is based on the agency of the actors, and connects individual and collective processes. A project moves forward precariously between success and failure. Carrying it out requires unity between conception and execution, taking into account the singularity of the situation, moving permanently between theory and practice, and managing complexity and uncertainty.

Teacher education is therefore an activity that requires careful design. Settings for teacher education need to have teachers involved in acting and reflecting on their actions and to have them involved in collective activities as well as in assuming their own agency and autonomy. Such settings may differ in time scale – from one day, to one trimestre, one year, or several years – and in the value ascribed to teacher agency – from the lower level as a consumer of ideas and materials, to the higher level as an active participant in negotiating and decision-making concerning the activities of the course. These settings may also differ in the way they frame the relationship between theory and practice, either focusing on theory, on practice, or on the relationship between theory and practice. Another very important dimension in which teacher education settings may differ is their relation to the school and other institutional arrangements, for example, whether teacher education is part of school activity or not and whether it is part of the collective work of all teachers who teach statistics and thus sustained by the school's structures and resources or whether it is just a responsibility of each individual teacher.

In designing teacher education programmes, it is necessary to take into consideration the many factors that influence teacher education processes such as the conceptions, knowledge, and resources of teacher educators and of teacher education institutions, the official regulations and educational policy, and the school and the social context. The conceptions, needs, and interests of teachers also influence those programmes – and perhaps should influence them in a stronger and more explicit way – and the same happens – or should happen – with the perspectives and results of research on teacher education.

At the heart of the teacher education process is the developing teacher with a direct contribution of the teacher educator and of teacher education setting, and also important influences of the teacher professional context and perhaps other factors. This process involves assessment of the needs and interests of the participating teachers done both by themselves and by the teacher educator, as well as the negotiation of a working "contract", the planning and undertaking of the activities and their relation to professional practice, and the closing of the activities leading to new challenges.

6   Key Ideas for Teacher Education in Statistics Education

Below driving heuristics to combine these major elements of the design of teacher education programmes are presented as four key ideas:

1. *Teacher education should be related to professional practice*. Teacher education may be related to professional practice in a number of different ways (Smith, 2001).

Teacher education may be based on practice, seeking to recognise the existing problems in the situations that the teacher experiences and framing their solution in the light of theory. Teacher education may be situated in practice, using the materials that represent the teaching activity (students' work, mathematical/ statistical tasks, classroom episodes) as opportunities for critique and investigation, leading teachers to develop knowledge by analysing real situations. And, finally, teacher education may be based on teachers' own practice, as teachers collect data from their practice and reflect about them with the support of the teacher educator and of other participating teachers.

2. *Teacher education should be based at the school and the professional group.* Teacher education needs to support the development of new practices in the professional context and address directly any problems arising from such changes. This means that the starting point is diagnosis of students' real difficulties. It also means carrying out intervention and professional development projects, establishing verifiable objectives and working collaboratively. And, finally, this means learning within the group with teachers' planning together and regularly exchanging experiences. The classroom and the subject group (or its formal subgroups) become essential acting spaces.

3. *Teacher education should provide challenge and support.* Teacher educators need to productively combine elements of challenge and elements of support. They need to provide examples that reflect important statistical ideas and, at the same time, sound teaching practice. Such examples must be based on worthwhile tasks, contribute to improve the classroom discourse, and help to create learning environments that encourage reasoning, thereby expecting teachers to assume intellectual risks. Teacher educators also need to create disequilibrium in teachers, challenging their conceptions about statistics, about who can do statistics and what it means to be successful in statistics. This necessarily involves moments of discomfort, but there is no other way to learn and make progress (Smith, 2001). And finally, teacher educators need to encourage teachers' collaboration, defining common goals, combining individual and common objectives, and negotiating ways of working together. In some moments challenge may be stronger, while in other moments support must prevail, always taking into account that the needs of one teacher may be quite different from those of another.

4. *Teacher education should recognise and empower teachers.* Teacher education must be on going throughout teachers' careers. It needs to take into account the teachers' contexts, that is, the students, the schools' culture, current practices, resources and regulations, the educational system guidelines and calendar, as well as the support for innovation from educational administrators. In addition, teacher education needs to use teachers' knowledge and competency about students, curriculum, schools, and communities, and also to use external contributions, for example, from statisticians, statistical societies, agencies, and projects (such as the ALEA project in Portugal, //alea-estp.ine.pt/; see also North & Scheiber, in this book) or university consultants. Besides, teacher education needs to consider issues of sustainability and cohesiveness, and, therefore, devise a set of experiences that amplify each other and contribute towards a general long-term integrative movement.

7 Practice-Based Teacher Education Programmes in Statistics

In this section two examples of how the ideas of the previous section may be put in practice will be briefly mentioned. The first example concerns in-service teacher education in Portugal, related to the introduction of a curriculum for basic education (grades 1–9) that emphasises data handing and statistical investigations, and, in all topics, teaching for exploratory learning. Such teacher education is carried out in courses lasting for 3–4 months and with a face-to-face working time of 25 h (teachers may work another 25 h autonomously). The courses are organised by grade level bands, and also by theme, one of them being Data Handling (the others are Numbers/Algebra and Geometry). In 25 h of work it is not possible to cover all the relevant issues related to statistics and statistics education and, therefore, just a few ideas are addressed. Each course is attended by up to 20 teachers and most of the work is done in small groups, usually comprising three or four members. The teacher educators for these courses are recruited from teachers experienced in teaching statistics, who undertook a specific preparation. The courses are organised in six sessions of about 4 h each with the following content:

1. Introduction to the new curriculum, especially comprising data handing; analysis of selected aspects of the curriculum documents; and work on sample tasks aligned with the curriculum orientations.
2. Design of a lesson on data handling using tasks aligned with the new curriculum.
3. Design of strategies and instruments for data collection regarding students' work in a lesson using such tasks.
4. Actual use of the tasks designed and collection of data in one or more classes taught by some teachers in the group.
5. Reflection about the class, analysis of the data collected, and organisation of a presentation by each small group to all the teachers attending the course.
6. Actual presentation of the work carried out and discussion by the other participating teachers.

To carry out these activities the teachers have face-to face-sessions but also meet together in their free time. These courses include some elements outlined in the previous sections. The teacher education process is related to teachers' professional practice as actual tasks and samples of students' work are used to exemplify the new curriculum orientations. In addition, new tasks are produced and used in the teachers' own classrooms, and become the objects of reflection and discussion. There is an important element of collaboration since the work is carried out by groups of teachers – planning tasks, constructing materials, observing, analysing data, and constructing a report. There are also some elements of teachers researching their professional practice, although on a very simple level, with the design of instruments for data collection and analysing and reporting results.

On the other hand, the links to actual school settings depends on the number of teachers from a single school that attend each course. In a course like this teachers cannot learn a full course on statistics or on teaching statistics. But, overall, the course leaders reported that the activities were largely successful, with many teachers showing a high interest in continuing to try such curriculum perspectives in the classroom. It is also gratifying to learn that some experienced teachers indicated that what they do in the classroom is very similar to what the new curriculum recommends.

The second example comes from pre-service teacher education at the University of Lisbon, and in fact, it is not much different from the former. First, prospective teachers need some orientation regarding the main ideas of the curriculum, they analyse curriculum documents and textbooks and also work on sample tasks that illustrate important curriculum ideas. Next, working in small groups (three or four), and with or without the support of a cooperating teacher from a local school, they design one or two related lessons on a statistics topic and some instruments to collect data through students' work. Afterwards, the lesson(s) are taught, either by the prospective teacher or the cooperating teacher (depending on the arrangement), and data is collected.

The later stages of the work include reflection on the lesson(s), analysis of the data collected, and organisation of a presentation to the whole class. The teacher educator strives to organise things such that in the final discussion all major issues regarding statistics concepts and teaching statistics are adequately addressed. Again, the success of this activity depends largely on the completion of previous study of statistics in the university programme. It only will be fruitful if the prospective teachers have in the future many other opportunities of involvement with statistics and statistics education. But the authors' evaluation is that this has proved helpful in conveying to prospective teachers the main ideas about the purposes and processes of teaching statistics at school level.

8   Implications for Teacher Education and Research

As discussed in this chapter, teacher education in statistics faces several challenges. As a better perspective on the goals of statistics education, the processes of curriculum development in statistics, and the problems related to students' learning is achieved, the traditional processes of "teacher training" based on transmission models are recognised to be very ineffective in developing teachers' professional knowledge and in promoting changes in teaching practice. New designs for teacher education in statistics that take into account the specific processes of teacher development and are adjusted to the needs, resources, and culture of each particular school in each particular country are needed. In the same way, more research on the actual activity and results of courses designed in this way – both at pre- and in-service levels – is needed so that these courses can be improved and such experiences are made available to teacher educators around the world.

References

Batanero, C., Burrill, G., Reading, C., & Rossman, A. (Eds.). (2008). *Joint ICMI/IASE Study: Teaching Statistics in School Mathematics. Challenges for Teaching and Teacher Education. Proceedings of the ICMI Study 18 and 2008 IASE Round Table Conference.* Monterrey, Mexico: International Commission on Mathematical Instruction and International Association for Statistical Education. Online: www.stat.auckland.ac.nz/~iase/publications

Bishop, A., & Goffree, F. (1986). Classroom organization and dynamics. In B. Christiansen, A. G. Howson, & M. Otte (Eds.), *Perspectives on mathematics education* (pp. 309–365). Dordrecht: Reidel.

Chaput, B., Girard, J. C., & Henry, M. (2008). Modeling and simulation in statistics education. In C. Batanero, G. Burrill, C. Reading, & A. Rossman (2008).

Chick, H. L., & Pierce, R. U. (2008). Teaching statistics at the primary school level: Beliefs, affordances, and pedagogical content knowledge. In C. Batanero, G. Burrill, C. Reading, & A. Rossman (2008).

Christiansen, B., & Walther, G. (1986). Task and activity. In B. Christiansen, A. G. Howson, & M. Otte (Eds.), *Perspectives on mathematics education* (pp. 243–307). Dordrecht: Reidel.

Hill, H. C., & Ball, D. L. (2004). Learning mathematics for teaching: Results from California's mathematics professional development institutes. *Journal for Research in Mathematics Education, 35*(5), 330–351.

Jaworski, B. (2005). Learning communities in mathematics: Creating an inquiry community between teachers and didacticians. *Research in Mathematics Education, 7,* 101–120.

Lampert, M., & Ball, D. L. (1998). *Teaching, multimedia, and mathematics.* New York: Teachers College Press.

Ponte, J. P., & Chapman, O. (2006). Mathematics teachers' knowledge and practices. In A. Gutierrez & P. Boero (Eds.), *Handbook of research on the psychology of mathematics education: Past, present and future* (pp. 461–494). Roterdham: Sense Publishers.

Putnam, R. T., & Borko, H. (1997). Teacher learning: Implications of new views of cognition. In B. J. Bridlde, T. L. Good, & I. F. Goodson (Eds.), *International handbook of teachers and teaching* (Vol. 2, pp. 1223–1296). Dordrecht: Kluwer.

Schön, D. A. (1983). *The reflective practitioner: How professionals think in action.* New York: Basic Books.

Shulman, L. S. (1986). Those who understand: Knowledge growth in teaching. *Educational Researcher, 15*(2), 4–14.

Smith, M. S. (2001). *Practice-based professional development for teachers of mathematics.* Reston, VA: National Council of Teachers of Mathematics.

Watson, J. M., Callingham, R. A., & Donne, J. M. (2008). Establishing PCK for teaching statistics. In C. Batanero, G. Burrill, C. Reading, & A. Rossman (2008).

Wood, T. (1994). Patterns of interaction and the culture of the mathematics classroom. In S. Lerman (Ed.), *Cultural perspectives on the mathematics classroom* (pp. 149–168). Dordrecht: Kluwer.

Zeichner, K., & Noffke, S. (2001). Practitioner research. In V. Richardson (Ed.), *Handbook of research on teaching* (pp. 298–330). Washington, DC: American Educational Research Association.

Chapter 30
Developing Statistical Literacy in Students and Teachers

Jim Ridgway, James Nicholson, and Sean McCusker

Abstract While statistical literacy is gaining much more recognition as something that all citizens need in order to function fully in modern society, there is much less agreement as to exactly what is meant by the term. This chapter discusses what statistical literacy is, why it is important for children at school and for teachers, and the need for our understanding to evolve to keep pace with developments worldwide. It explores the potential of new curricula introduced in South Africa and New Zealand, and the work being done in many different countries by statistical agencies to support teachers' statistical literacy. A case study where naïve students and teachers develop skills by engaging with complex evidence on a topic of real social import is also described.

1 Introduction

Statistical literacy has long been an issue of concern for an important minority, but has become much more prominent in the first decade of the new millennium. The explosion in information accessibility means that statistical literacy, for the first time, is widely viewed as an essential life skill for a fully functioning citizen. In many countries there is a drive towards evidence-informed decision-making. Governments and other producers of large-scale data sets are increasingly concerned about the capacity of user groups to understand and make intelligent use of their outputs. For example, the Global Project (www.oecd.org/progress) is actively looking at ways to develop new indicators of societal progress that are better reflections of the quality of life and sustainability than measures such as gross domestic product (GDP) alone.

J. Ridgway (✉), J. Nicholson, and S. McCusker
SMART Centre, School of Education, University of Durham, Leazes Road,
Durham City, DH1 1TA, UK
e-mail: Jim.Ridgway@durham.ac.uk; j.r.nicholson@durham.ac.uk;
Sean.McCusker@durham.ac.uk

C. Batanero, G. Burrill, and C. Reading (eds.), *Teaching Statistics in School*
Mathematics-Challenges for Teaching and Teacher Education: A Joint ICMI/IASE Study,
DOI 10.1007/978-94-007-1131-0_30, © Springer Science+Business Media B.V. 2011

The curriculum in schools tends to be slow to respond to changes in the outside world. However, there are exceptions: South Africa in 2005 and New Zealand in 2008 introduced radically new curricula that place statistical literacy at the core of the statistics component of mathematics. In many other countries, statistical offices and agencies are providing resources that can be used to support the introduction of statistical literacy in schools. However, without wide-reaching professional development at pre-service or in-service stages for teachers, such resources are unlikely to have the impact they warrant. Moreover, new technologies have changed the scale of data collection, the analyses that are possible, and public access to data. New technologies mean that multivariate data is ubiquitous via open access databases. The challenge for teachers is to find effective ways to use such large-scale data sets. This challenge applies equally to teachers of mathematics who explicitly teach statistical concepts and to teachers in other disciplines who need to understand evidence-based arguments.

Moreover, there is no consensus as to exactly what constitutes statistical literacy. In this chapter some conceptions of statistical literacy are described and some issues surrounding it in the school context are explored. An experience intended to develop teachers' and students' statistical literacy is analysed and some implications for training teachers are explored.

2 What Is Statistical Literacy?

In 1999 the Statistical Reasoning, Thinking and Literacy (SRTL) forums started and are held biennially. These three aspects of statistics are inter-related, but in the first five SRTL forums there has been a much stronger emphasis on statistical reasoning and thinking than on statistical literacy. The *Statistics Education Research Journal* provides a repository of research-based knowledge about statistics education, including many aspects of statistical reasoning and thinking – but only two articles refer directly to statistical literacy in the title (Watson & Callingham, 2003; Carmichael, Callingham, Watson, & Hay, 2009).

One can trace the early development of statistical reasoning from the nineteenth century, when the industrial revolution brought widespread problems of disease, housing, sanitation and working conditions, stimulating an era of social enquiry. Current concerns over global issues like climate change, poverty and acquired immune deficiency syndrome (AIDS) together with the technology revolution are combining to create the conditions for a similar revolution in statistical literacy. Various efforts have been made to describe the requirements for people to be able to participate fully in society. These have included the capacity to interpret statistical information presented graphically, in tables or in words, and to have some critical appreciation of whether certain conclusions can be justified by the information available. If statistical literacy is to be introduced into the school curriculum, there needs to be an understanding of the nature of the statistical concepts required, and of the logical dependencies between statistical inference, and decisions and actions in real situations.

Statistical literacy encompasses a number of ideas. Wallman (1993) emphasised the ability to value, understand and evaluate statistical evidence that influences our daily lives; Gal (2002) emphasised the ability to interpret, evaluate and communicate statistical evidence, and provided an initial model that posits that statistically literate behaviour comprises both a knowledge component (comprised of five cognitive elements: literacy skills, statistical knowledge, mathematical knowledge, context knowledge and critical questions) and a dispositional element (comprised of critical stance, beliefs and attitudes). Gal and Murray (in press) expand on that model to provide a conceptual analysis of the task demands presented by statistics agencies in web-based environments (now the dominant dissemination medium for statistical agencies).

There is increasing sophistication in the ways that data are presented and analysed in the media when addressing complex problems facing society, and our conception of what statistical literacy is must evolve to match these developments. Few policy questions have no statistical component, be it in government or in commercial enterprise, and knowledge of statistics is an essential skill because data, variation, and chance are omnipresent (Moore, 1998). Moreover, since complex data are pervasive within different domains, we believe that teachers across different disciplines need statistical literacy, as defined by Gal and by Wallman, to integrate evidence-based arguments into their classroom practices.

3   Statistics Producers and Statistical Literacy

Key agencies in the development of statistical literacy include government statistics offices, non-governmental agencies and sites supporting Web 2.0 activities, as well as providers of print and video media. Politicians and policy makers are beginning to understand that aggregating data across a heterogeneous population may give an inaccurate picture of every subgroup, and at the very least it is likely to lead to less than optimal strategies for the allocation of scarce resources. The capacity now exists to collect data in sufficient quantity to allow analysis by a number of categorisations, for example age, sex, ethnicity, or social background. However, traditional analysis methods rely on models and assumptions that may not be appropriate, and the outputs of the analysis can be difficult to interpret, in that they generate little intuitive understanding of the story behind the data.

Some agencies perceive the need for educational practitioners to engage with this wider community, and there are encouraging signs that this is starting to happen. There is a growing awareness of the need to develop new indicators that better measure the important aspects of a society, and to be able to communicate effectively with data consumers about the information available in these indicators. For example, the Istanbul Declaration (Organisation for Economic Co-operation and Development [OECD], 2007) promotes an evidence-based approach to social progress. It recommends involving citizens in the definition of 'progress' and appropriate indicators, and advocates 'appropriate investment in building statistical capacity'.

Some national statistics offices are developing ways to make their data more useful and better understood. For example, Statistics Canada (www.statcan.ca/start. html) works directly with educators to develop resources (see Townsend, 2007). Schools have free access to their entire database of social and economic statistics, as well as to census information. So students have access to a huge amount of detailed data, with an intuitive interrogation interface (E-STAT) allowing them to produce their own tables of summary data, and there are audio podcasts which offer insights into what can be done in projects with the data. The Australian Bureau of Statistics (www.abs.gov.au/) uses SuperTable to make all their data easily accessible to all users. Like E-STAT, SuperTable provides a powerful analytical tool for interrogating data. In the United Kingdom the Office of National Statistics (ONS) has produced activities and lessons plans based on real, large datasets that were specially prepared for immediate use in classrooms (www.stats4schools.gov.uk/). The ONS has set up a Data Visualisation Centre where innovative interactive displays with complex data are accessible, including dynamic population pyramids, and a personal inflation index calculator. CommuterView allows the user to explore patterns of commuter behaviour around any population centre. The power of geographical information systems to incorporate spatial dimensions into statistical information is demonstrated clearly. The CensusAtSchool project (www.censusatschool.org.uk/) offers access to interesting national data along with similar survey data from a number of international partners for comparison.

Those working in the area of official statistics are making important contributions to statistical literacy, and finding ways of disseminating information about their activities more widely (see also North & Scheiber, this book). The International Statistical Literacy Project published a book (Sánchez, 2008) summarising the efforts and outlining the resources produced by government statistical offices in Portugal, New Zealand, Canada, Finland, Italy and Australia, and in 2011 the *Journal of the International Association for Official Statistics* publishes a special issue on statistical literacy.

Mittag and his collaborators (e.g., Grunewald & Mittag, 2006) have developed interfaces that allow some user interaction with OECD data; an important feature of this work is a facility to connect to on-line tutorials about statistical concepts. Gapminder (www.gapminder.org) is an excellent tool for the exploration of multivariate relationships, and has been adopted by the United Nations Statistics Division for the display of the Millennium Development Goals. The entire content of OECD's *Factbook* is available on-line in Gapminder. Sites promoting Web 2.0 activities such as Many Eyes (services.alphaworks.ibm.com/manyeyes/) offer a variety of interfaces, and some encourage users to upload and discuss data. Some government statistics offices (and OECD) are working to make their data available in such forums. Web 2.0 technology affords opportunities for communicating with young people in ways familiar to them. Dynamic visualisation of data has some important champions. For example, Giovannini (2007) argues that providers of statistics have to exploit all available technologies to help users make sense of data and Radermacher (2007) argues for 'intelligent graphics' that are data-driven, allow user interaction, and make use of animation.

To take advantage of this variety and amount of information we need to build our ability to interpret and display multivariate data effectively; this work is still at an early stage of development. There is now a window of opportunity for the statistics education community to collaborate actively with data providers to help provide effective multivariate displays that can significantly improve the quality of their websites and interactive publications, and the quality of Web 2.0 offerings.

4   Statistical Literacy and Curricular Issues

Statistical concepts in most school curriculums are rooted in the 1920s. In the United Kingdom at least, in the teaching of statistics at school-level there is a focus on the mastery of technique, and rather little on interpretation of results (Ridgway, Nicholson, & McCusker, 2007a). The techniques themselves focus on the analysis of univariate and bivariate data, make implicit assumptions about linearity, and are largely useless in dealing with any data sets students might encounter in their lives outside school. This situation is (at best) almost useless for developing statistical literacy, and at worst is damaging and potentially dangerous. If students know nothing about interaction or Simpson's paradox (e.g., where a conclusion about the relative efficacy of two treatments, or the direction of a trend line, can be reversed when a third variable is considered), they are likely to draw conclusions about multivariate data that are completely wrong, and may make important decisions using false conclusions.

A number of authors have argued (e.g., Chance, 2002; Nicholson, Ridgway, & McCusker, 2006; Ridgway, Nicholson, & McCusker, 2007b) that reasoning with data is pervasive within society but that the current statistics curriculum does not equip tomorrow's citizens with the skills to engage in appropriate statistical reasoning. Curriculum areas such as geography, citizenship, and sociology all deal with complex contexts where multiple factors impact on situations, and where relevant, real, data are available. These subjects currently make little use of relevant quantitative information because it is perceived that students will have great difficulties in making sense of multivariate data.

4.1   Implementing New Statistics Curricula

While poor curriculum materials can seriously impair the health of statistical literacy amongst teachers and pupils, there are examples of innovations in national statistics curricula that have the potential to make considerable progress in this important area.

New Zealand introduced a revised mathematics curriculum in 2008 that includes three strands within statistics, namely Statistical Investigation, Interpreting Statistical Reports, and Exploring Probability, and these are developed right through the curriculum, beginning with the introduction of basic principles in early primary education and progressing to ambitious project work at the end of secondary school. Students meet multivariate data early on in this process, and developing new ways of delivering such innovative content has been one of the challenges for teachers implementing the new curriculum. Arnold (2008) describes a pilot study where a group of teachers formed a professional learning community to identify needs in implementing the changes, and participated in workshops designed to address those needs.

South Africa introduced statistics into its school curriculum for the first time in 2005 (North & Zewotir, 2006). Starting from nothing, they were able to place an emphasis on statistical reasoning rather than focusing on the computational aspects of statistics. The foundation and intermediate phases of this curriculum focus on acquiring skills, and use contexts relating to human rights, and other social, economic, and environmental issues, and at the same time develop an awareness of broader issues we would identify as aspects of statistical literacy. In the senior phase, these techniques are used to investigate and solve problems in significant contexts, using relevant contemporary issues such as AIDS, crime, abuse, and environmental concerns (Wessels, 2008). The South African Statistical Association (Stats SA) produces a lot of topical resources (www.stat.auckland.ac.nz/~iase/islp/sa) that teachers can access which support the ambitions of the new curriculum. Crucially the implementation of the new curriculum includes an extensive, long-term, programme aimed at developing teachers. North and Scheiber (2008) describe the co-operation between Stats SA and the Department of Education that started with a group of teachers attending the Sixth International Conference on Teaching Statistics (ICOTS) in Cape Town in 2002, continuing through the 57th Session of the International Statistical Institute in Durban in 2009, where the ISIbalo project allowed a large number of teachers and students to engage with experts in statistical education from across the world to continue to develop their expertise and range of knowledge. See North and Scheiber, in this book, for more details of the role played by some statistical offices in helping develop statistics education.

4.2   Assessing People's Statistical Literacy

Assessment is a major driver in education systems: the assessment of statistical thinking in the United Kingdom is rather shallow, and the assessment of statistical literacy is essentially non-existent, so there are few incentives for teachers in the United Kingdom school education system to spend time devising authentic tasks based on real data (see Garfield & Franklin, in this book for a fuller discussion of assessment issues). Indeed, in many assessment items in the high stakes examinations taken by all students at the end of compulsory education (General Certificate of Secondary Education) in the United Kingdom, data were used with no context attached, or with no use made of the context at all. Moreover, there were instances in which using contextual reasoning would lead to an incorrect answer ('incorrect' in the sense of not agreeing with the published solution of the examiner) (Nicholson, Ridgway, & McCusker, 2009).

Watson and Callingham (2003) have done some work on developing a questionnaire to assess statistical literacy and a hierarchy of statistical literacy. However, there are two areas of interest that they do not address, namely reasoning with more complex data and the use of Information and Communications Technology (ICT) to reason from evidence. Both of these areas are increasingly important because of the impact new technologies are having on the way information is handled.

Schield has looked at people's capacity for critical thinking using data (e.g. Schield 2000, 2006) and at Simpson's paradox and the role of confounding factors (e.g., Schield & Burnham, 2003). The fundamental distinction between designed experiments and observational studies is critical in understanding what can (or should) be inferred from a data analysis, and what can or should not be inferred, yet these areas are under-represented in the literature on statistical literacy. There is additionally a large literature on the problems that students and adults have with simple concepts, such as interpreting static two dimensional graphs, and tabular information (e.g., Batanero, Godino, Vallecillos, Green, & Holmes, 1994; see also Chaps. 20–26). One might predict that working with multivariate data would be impossible for people with no statistical training. However, empirical explorations (e.g., Ridgway, McCusker, & Nicholson, 2006) show that computer-based three dimensional variable tasks are no more difficult for 12–14 year olds than are two dimensional paper-based tasks. Du Feu (2005) has shown that much younger children can work meaningfully with multivariate data displays that they have created in the form of tactile graphs built from LEGO® bricks.

4.3 Improving Teachers' Statistical Literacy

Teachers often come from non-quantitative backgrounds, and so do not feel confident using quantitative methods. Statistics is about making sense of numerical information in context, and the confidence that teachers have in their capacity to convey this to students can be greatly enhanced by well-documented resources. Paradoxically, it may be easier to work in more cognitively complex contexts, with real data, where interesting stories are found, than to work with materials which use no context for data (Ridgway et al., 2006).

Biehler (2008) identifies tensions in the training of mathematics teachers who will teach statistics within mathematics which stem from the differences in the two disciplines (see Burrill & Biehler, this book). Pre-service training of teachers has to accommodate classroom practice, general theories of learning, and much more, so there is very limited scope for developing subject knowledge and pedagogic content knowledge associated with statistical literacy. We urgently need to find ways of providing effective in-service training through engaging teachers in meaningful online collaborations in this area. The creation of dynamic interfaces that facilitate interaction with multivariate data offers exciting opportunities for statistics education for both students and teachers, but also poses a number of challenges to our understandings as educators. In the next section a case study using appropriate materials is described.

5 Using New Technologies to Enhance Statistical Literacy: A Case Study

In the previous sections we have suggested how new curricula can be implemented with appropriate support for teachers when there is a genuine collaboration between experts in statistics and expert classroom teachers and educators. A curriculum development project that explored the use of multivariate data, and the potential new technology offers in developing teachers' confidence in aspects of statistical literacy is briefly described below. A fuller description can be found in Ridgway, Nicholson, and McCusker (2008), where teachers identified data interpretation as the biggest problem area within the statistics curriculum, felt that pupils struggle with it, and found it difficult to help pupils.

Eight schools took part in a study that explored 12–14 year-old students' ability to reason with multivariate data (students covered the full range of academic ability). Here, we use a single data set on rates of infection in sexually transmitted infections (STI) (Fig. 30.1), to exemplify some general findings. The data represent population data from sexual health clinics published by the United Kingdom Health Protection Agency (the interactive data display can be accessed through

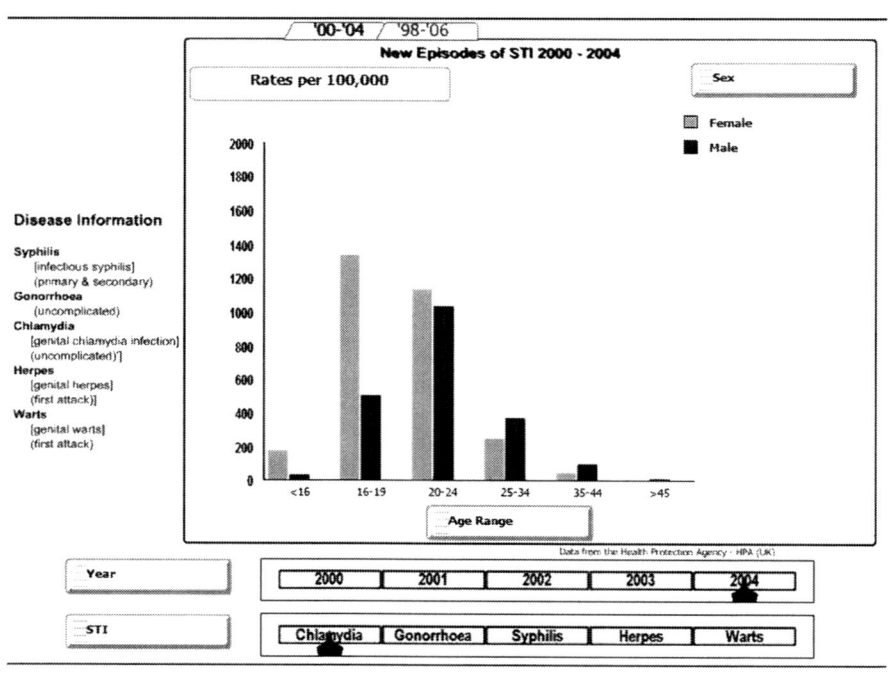

Fig. 30.1 Rates of new Chlamydia infections in 2004

www.dur.ac.uk/smart.centre). The dependent variable is the rate (per 100,000) of new infections of a number of STIs between 2000 and 2004 for males and females in a number of age groups. The interface allows the user to drag and drop the position of the four explanatory variables. In Fig. 30.1, *Age Range* is shown on the *x*-axis, with separate displays for males and females. Specific values on two other variables (*Year* and *STI*) are selected via sliders (in the example, the *STI* is Chlamydia, and the *Year* is 2004). The sliders can be used to change values dynamically. For example, trends over time can be observed by sliding along the *Year* variable.

The relationships within these data are more complex than anything pupils have previously encountered, so building confidence in their ability to describe relationships is important. Students quickly got used to swapping the positions of the variable labels for *Sex*, *Age Range,* and *Year*, and used these explorations to build up a clear picture of the relationships in the data. Changes in the incidence of Chlamydia over time (explored by dragging the *Year* slider back and forth) drew out some interesting differences (in the peak incidence of *STIs* in different age groups of males and females) and some interesting similarities (the scale of the problem increases dramatically for both groups over a short period of time). Other conjectures about some aspects of the data could be explored by reconfiguring the display. In the configuration shown in Fig. 30.1, moving the slider over *Year* gives the impression that there has been a steady increase in rates for all age groups of girls between 2000 and 2004. When the *Sex* and *Year* variable positions are interchanged so that all the data about Chlamydia infections in girls in successive years can be seen together, the impression can be confirmed – growth in the incidence of Chlamydia is roughly linear for all age groups.

An important remark is that statistics and graphs will show relationships between variables, but do not provide explanations as to why they exist. Some causal statements are obviously wrong when the data is inconsistent with what would happen if the reasoning were true; but being consistent with the data is not sufficient to assure causality. These materials provide many opportunities for teachers to explore alternative plausible explanations beyond cause and effect for relationships that are evident in the data, and to identify scenarios that are inconsistent with the story told by the data. This is an area where teachers have previously reported feeling very insecure, and they could feel more comfortable with it when using these materials precisely because the data are complex and the context rich enough to support multiple plausible causal explanations.

Classroom observations (Ridgway et al., 2008) provided evidence that 12–14 year olds across the attainment range can engage with and understand complex messages in multivariate data. This gives grounds for optimism about the possibility of developing statistical literacy in schools, and in the adult population. Teachers with weak backgrounds in mathematics (and little knowledge of statistics) were able to use these materials effectively in class. To be sure, they did not use the language of 'variables' and 'interactions' but were able to facilitate engagement and good pupil insights into the data. In later interviews, teachers reported that pupils were able to make sense of the relationships in the data, including describing interactions between factors and, crucially, teachers felt much more comfortable in dealing with the relationships in the data in these materials than they do normally when discussing quantitative data.

The level of communication of statistical ideas was reported to be much higher than in normal mathematics lessons. Students who, before hand, had been much less forthcoming in talking about the data they met in traditional mathematics courses were prepared to talk about these real, important, contexts.

6   Next Steps in Developing Statistical Literacy

Fundamental change to classroom practice in any curriculum requires careful management. If there is a major restructuring of the curriculum (as in South Africa and in New Zealand) then official documents, support materials, assessment items and teacher education initiatives are likely to provide a coherent framework within which that change can be accomplished.

In many countries, key agencies such as statistics offices are keen to promote statistical literacy, and can provide contemporaneous resources to support teachers in using real data in the classroom, but are often unable to provide pedagogical training to accompany those resources. In countries where the curriculum does not embrace the fundamental ideas of statistical literacy, the use of such resources is liable to fail since teachers have little incentive to embrace statistical literacy.

Reasoning with multivariate data is a key aspect of statistical literacy that has received little attention in most curricula. The changing nature of information gathering, analysis and communication means that there is an imperative for change in our conception of statistical literacy. Small-scale classroom trials give us clear evidence that pupils can reason successfully with multivariate data, and that teachers in mathematics and in other disciplines can handle this reasoning. This is an arena which offers the prospect of developing statistical literacy via realistic data sets on important topics, that can be explored by pupils and teachers in new ways.

References

Arnold, P. (2008). Developing new statistical content knowledge with secondary school mathematics teachers. In C. Batanero, G. Burrill, C. Reading, & A. Rossman (2008).

Batanero, C., Burrill, G., Reading, C., & Rossman, A. (Eds.). (2008). *Joint ICMI/IASE Study: Teaching Statistics in School Mathematics. Challenges for Teaching and Teacher Education. Proceedings of the ICMI Study 18 and 2008 IASE Round Table Conference.* Monterrey, Mexico: International Commission on Mathematical Instruction and International Association for Statistical Education. Online: www.stat.auckland.ac.nz/~iase/publications

Batanero, C., Godino, J. D., Vallecillos, A., Green, D., & Holmes, P. (1994). Errors and difficulties in understanding elementary statistical concepts. *International Journal of Mathematical Education in Science and Technology, 25*(4), 527–547.

Biehler, R. (2008). From statistical literacy to fundamental ideas in mathematics: How can we bridge the tension in order to support teachers of statistics. In C. Batanero, G. Burrill, C. Reading, & A. Rossman (2008).

Carmichael, C., Callingham, R., Watson, J., & Hay, I. (2009). Factors influencing the development of middle school students' interest in statistical literacy. *Statistics Education Research Journal, 8*(1), 62–81. Online: www.stat.auckland.ac.nz/~iase/serj/

Chance, B. (2002). Components of statistical thinking and implications for instruction and assessment. *Journal of Statistics Education, 10*(3), 1–18. Online: www.amstat.org/publications/jse/

du Feu, C. (2005). Bluebells and bias, stitchwort and statistics. *Teaching Statistics, 27*(2), 34–36.

Gal, I. (2002). Adult statistical literacy: Meanings, components, responsibilities. *International Statistical Review, 70*(1), 1–25.

Gal, I. & Murray S.T. (2011) Diversity in users' statistical literacy and information needs: Institutional and educational implications. *Statistical Journal of the International Association for Official Statistics, 27*(3,4), 185–196.

Giovannini, E. (2007). Dynamic graphics: Turning key indicators into knowledge. *Proceedings of the 56th Session of the International Statistical Institute, Lisbon, Portugal.* Voorburg, The Netherlands: International Statistical Institute. Online: www.stat.auckland.ac.nz/~iase/publications

Grunewald, W., & Mittag, H. J. (2006). The use of advanced visualisation tools for communicating European data on earnings to the citizen. In A. Rossman & B. Chance (Eds.), *Proceedings of the Seventh International Conference on Teaching Statistics.* Salvador, Bahia, Brazil: International Statistical Institute and International Association for Statistical Education. Online: www.stat.auckland.ac.nz/~iase/publications

Moore, D. S. (1998). Statistics among the liberal arts. *Journal of the American Statistical Association, 93*(444), 1253–1259.

Nicholson, J., Ridgway, J., & McCusker, S. (2006). Reasoning with data – time for a rethink? *Teaching Statistics, 28*(1), 2–9.

Nicholson, J., Ridgway, J., & McCusker, S. (2009). One small step for a pupil – one giant leap for citizens. In P. Murphy (Ed.), *Proceedings of the International Association for Statistics Education Satellite Conference, Next Steps in Statistics Education.* Durban SA: International Association for Statistical Education. Online: www.stat.auckland.ac.nz/~iase/publications

North, D., & Scheiber, J. (2008). Introducing statistics at school level in South Africa: The crucial role played by the national statistics office in training in-service teachers. In C. Batanero, G. Burrill, C. Reading, & A. Rossman (2008).

North, D., & Zewotir, T. (2006). Introducing statistics at school level in South Africa. In A. Rossman & B. Chance (Eds.), *Proceedings of the Seventh International Conference on Teaching Statistics.* Salvador, Bahia, Brazil: International Statistical Institute and International Association for Statistical Education. Online: www.stat.auckland.ac.nz/~iase/publications

OECD (2007). *Istanbul declaration.* Online: www.oecd.org/dataoecd/14/46/38883774.pdf

Radermacher, W. (2007). Make sense of the world by having fun with statistics. *Proceedings of the International Statistical Institute 56th Session.* Lisbon, Portugal: International Statistical Institute. Online: www.stat.auckland.ac.nz/~iase/publications

Ridgway, J., McCusker, S., & Nicholson, J. (2006). *Reasoning with evidence – Development of a scale.* Paper presented at the 32nd Annual Conference of the International Association for Educational Assessment, Singapore. Online: www.iaea2006.seab.gov.sg/

Ridgway, J., Nicholson, J. R., & McCusker, S. (2007a). Teaching statistics – despite its applications. *Teaching Statistics, 29*(2), 44–48.

Ridgway, J., Nicholson, J. R., & McCusker, S. (2007b). Reasoning with evidence – new opportunities for assessment, curriculum and public understanding. *Proceedings of the International Statistical Institute 56th Session.* Lisbon, Portugal: International Statistical Institute. Online: www.stat.auckland.ac.nz/~iase/publications

Ridgway, J., Nicholson, J.R., & McCusker, S. (2008). Reconceptualising 'statistics' and 'education'. In C. Batanero, G. Burrill, C. Reading, & A. Rossman (2008).

Sánchez, J. (Ed.). (2008). Government statistical offices and statistical literacy. Los Angeles, CA: International Statistical Literacy Project. Online: www.stat.auckland.ac.nz/~iase/islp/

Schield, M. (2000). Statistical literacy: Describing and comparing rates and percentages. *Proceedings of the 2000 American Statistical Association Section on Statistical Education* (pp. 76–81). Alexandria, VA: American Statistical Association. Online: www.StatLit.org/pdf/2000SchieldASA.pdf

Schield, M. (2006). Presenting confounding and standardization graphically. *STATS Magazine*, Fall 2006. 14–18. Alexandria, VA: American Statistical Association.

Schield, M., & Burnham, T. (2003). Confounder-induced spuriosity and reversal: Algebraic conditions for binary data using a non-interactive model. *Proceedings of the 2003 American Statistical Association Section on Statistical Education* (pp. 3690–3697). Alexandria, VA: American Statistical Association. Online: www.StatLit.org/pdf/2003SchieldBurnhamASA.pdf

Townsend, M. (2007). Statistics Canada's learning resources: A key channel for educators. *Proceedings of the International Statistical Institute 56th Session*. Lisbon, Portugal: International Statistical Institute. Online: www.stat.auckland.ac.nz/~iase/publications

Wallman, K. K. (1993). Enhancing statistical literacy: Enriching our society. *Journal of the American Statistical Society, 88*, 1–8.

Watson, J. M., & Callingham, R. (2003). Statistical literacy: A complex hierarchical construct. *Statistics Education Research Journal, 2*(2), 3–46. Online: www.stat.auckland.ac.nz/serj/

Wessels, H. (2008). Statistics in the South African school curriculum: Content, assessment and teacher training. In C. Batanero, G. Burrill, C. Reading, & A. Rossman (2008).

Chapter 31
Developing Teachers' Statistical Thinking

Maxine Pfannkuch and Dani Ben-Zvi

Abstract In this chapter learning experiences that teachers need in order to develop their ability to think and reason statistically are described. It is argued that teacher courses should be designed around five major themes: developing understanding of key statistical concepts; developing the ability to explore and learn from data; developing statistical argumentation; using formative assessment; and learning to understand students' reasoning.

1 Introduction

The recent plethora of research in statistics education as well as articles by some prominent statisticians on the necessity to reform teaching approaches in statistics have led to a paradigm shift in the conceptualisation of statistics teaching. This paradigm shift evolved mainly from technology developments and the identification and promulgation of the characteristics of statistical thinking and the "big ideas" underpinning statistics (e.g., Moore, 1990; Wild & Pfannkuch, 1999). The explication and exploration of these ideas by researchers have contributed to teaching approaches that emphasise Exploratory Data Analysis (EDA), attention to building students' conceptual understandings and curricula that aim to develop students' reasoning,

M. Pfannkuch (✉)
Department of Statistics, The University of Auckland, Private Bag 92019,
Auckland 1142, New Zealand
e-mail: m.pfannkuch@auckland.ac.nz

D. Ben-Zvi
Faculty of Education The University of Haifa, Mount Carmel, Haifa, 31905 Israel
e-mail: dbenzvi@univ.haifa.ac.il

C. Batanero, G. Burrill, and C. Reading (eds.), *Teaching Statistics in School Mathematics-Challenges for Teaching and Teacher Education: A Joint ICMI/IASE Study*, DOI 10.1007/978-94-007-1131-0_31, © Springer Science+Business Media B.V. 2011

thinking and literacy (Ben-Zvi & Garfield, 2004; Garfield & Ben-Zvi, 2008). Furthermore, innovative software such as *TinkerPlots* (Konold & Miller, 2005) and *Fathom* (Key Curriculum Press, 2006) are cementing in technology as an integral part of teaching, learning and reasoning processes. In addition, statistics and statistical reasoning are becoming part of the mainstream school curriculum in many countries.

Consequently, many teachers, whose experience is mainly grounded in descriptive statistics, are challenged by recent approaches and guidelines for statistics teaching and learning (e.g., Franklin et al., 2005) and curricula for students (e.g., Ministry of Education, 2007). Such huge shifts in teaching approaches and thinking about the nature, role and purpose of statistics require teacher educators to design and implement courses that will develop teachers' statistical thinking.

Based on the assumption that to develop an effective teacher of statistics a course devoted entirely to statistics education is essential, this chapter proposes some fundamental learning experiences for teachers in order to develop their statistical thinking. It is also assumed that teachers need to acquire not only statistical knowledge but also professional knowledge to teach statistics, and that an effective and positive learning environment can develop in teachers a deep and meaningful understanding of statistics and ability to think and reason statistically.

2 Fundamental Course Components

Based on the work of Shulman (1986) and Ball, Thames, and Phelps (2008) it is posited that a course should have three main goals. The first goal is to develop and improve teachers' understanding of statistics, since it is generally acknowledged that they lack statistical knowledge, good statistical thinking and points of view that are now required by new curricula (e.g., de Oliveira et al., 2008). The second goal is to enable teachers to understand the prior knowledge, beliefs and reasoning prevalent in their students, the value in listening carefully to their students' emerging reasoning processes and how to build and scaffold students' conceptions. The third goal is to facilitate teachers' understanding of how curricula, technology and sequences of instructional activities build students' concepts across the year levels.

There are currently some courses around the world that are focused on the preparation of teachers of statistics at different educational levels; some of them help teachers develop their statistical thinking. For example, Garfield and Ben-Zvi (2009) described courses that are intended to help pre-service and in-service teachers understand and appreciate the importance of statistics as well as develop them into competent and effective teachers. They view the classroom as a "statistical reasoning learning environment" in which teachers develop a deep and meaningful understanding of statistics and help students to develop their ability to think and reason statistically. Other teacher training courses use different theoretical stances but essentially provide similar learning environments for students to develop their statistical thinking. For example, Godino, Batanero, Roa, and Wilhelmi (2008) described how they designed activities for teachers by paying attention to six components of pedagogical content knowledge, which assist in exposing and developing students' statistical reasoning. Sanchez and Blancarte (2008), on the other hand, used Wild and Pfannkuch's (1999) statistical thinking model as their centrepiece to design statistical experiences for their teachers.

Whatever the theoretical stance these various courses use, some basic principles of instructional design are evident and are similar to those described by Cobb and McClain (2004). In cognisance of these principles and goals proposed for teacher education, five major interrelated themes are suggested for a course, each of which is discussed below:

1. Developing the understanding of key statistical concepts;
2. Developing the ability to explore and learn from data;
3. Developing statistical argumentation;
4. Using formative assessment; and
5. Understanding students' reasoning.

2.1 Developing Understanding of Key Statistical Concepts

To help teachers understand statistics at a deep conceptual level, the focus of the course should be on developing understanding of the more relevant ideas of statistics and the interconnections among them rather than presenting statistics as a set of tools and procedures. The development of these ideas should not only be within a curriculum level but across curriculum levels in order for teachers to appreciate how concepts can be built up over time. These big ideas include but are not limited to *data, patterns in data, variability, distribution and inference* (see Burrill & Biehler, this book as well as different chapters in Part III that describe research on teachers' understanding of some of these ideas). Teacher educators should involve teachers in activities to develop these concepts and the interrelationship among them and provide research readings on the fundamental ideas, some of which are discussed below.

Data. At the very heart of statistics is recognising that the main purpose of collecting and investigating data is to learn more about real situations and that data-based evidence is needed for making decisions and evaluating information. However, the soundness of such learning and judgements about real situations depends on understanding data generation, data analysis and the types of conclusions that can be drawn from data. Data generation includes understanding, for example, how to capture and translate notions such as *prompt service* into data that can be collected, how to capture measurements that are relevant to the problem situation and how to design methods of collecting data that avoid bias and measurement error. Some examples of research on students' reasoning about data generation has been conducted by Konold and Higgins (2003) and Lehrer and Romberg (1996). See also Hall, in this book for a discussion of the role of data in teaching statistics and training teachers.

Patterns in data. Looking for patterns in data is akin to attempting to unlock the story hidden within the data. Reasoning from data plots requires the ability to notice, decode, assess and judge the messages inherent in the plot (see also González, Espinel, & Ainley, this book). Part of the decoding will include finding of centres and looking for correlation/regression. To interpret messages in data teachers need to understand, for example, the idea of a centre of a distribution as a "signal in a noisy process", that the relationship between two quantitative variables may vary in a predictable way, and that a perceived pattern may be due to random variation. Engel, Sedlmeier, and Worn (2008) believe that if teachers perceive data as a mixture of signal and noise they will acquire an important statistical thinking skill (see also Engel & Sedlmeier, this book). But seeking patterns in data also involves learning that different types of representation, including re-categorisation of data, reveal different aspects of the story within the data and from these representations a story can be synthesised (Shaughnessy & Pfannkuch, 2002). Furthermore, the interpretation of patterns in data requires recognition of the importance of contextual knowledge; "what patterns mean, and indeed, whether they mean anything at all, depends on context" (Cobb, 2007, p. 338). The building of students' concepts of centre, spread and association should be highlighted, with teachers being cognisant of some examples of the extended research in these topics, for example, the research of Mokros and Russell (1995) on the idea of centre and representativeness, Moritz (2004) on reasoning about covariation and Konold and Pollatsek (2002) on signal and noise. Other research in these topics are summarised in Chaps. 21 and 25.

Variability. Variability is omnipresent throughout the statistical enquiry cycle and is fundamental to statistical thinking (Moore, 1990). In fact, "statistical problem solving and decision-making depend on understanding, explaining, and quantifying the variability in the data" (Franklin et al., 2005, p. 6). Teachers need to learn about how to deal with many sources of variation inherent in data and how to use variation ideas to design investigations. Variation ideas include measurement variability, natural variability, induced variability, sampling variability and chance variability from sampling or from random assignment to groups in experiments (Franklin et al., 2005). Apart from considering and understanding variation at a meta-level teachers need to know how to interpret and teach the different variability measures used in statistics such as range, interquartile range, standard deviation and correlation coefficient. Recent research on students' understanding of variability includes Ben-Zvi and Garfield (2004), Shaughnessy (2007), Canada (2008) and Sanchez and Garcia (2008) (see also Sánchez, Borim, & Coutinho, this book).

Distribution. According to Wild (2006) distributions are "a fundamental component of statistical reasoning" (p. 22), since "statisticians look at variation through a lens which is distribution" (p. 11). Learning to reason from distributions includes the following understandings: A set of data may be examined and explored as an entity (a distribution) rather than as a set of individual cases; a graph of these (quantitative) data can be summarised in terms of shape, centre or trend, and spread; visually examining distributions is an important and necessary part of data analysis; distributions may be formed from sets of individual data values, sets of possible values for a random variable or from summary statistics such as means (e.g., sampling distribution of sample means) and probability distributions allow us to make traditional formal statistical inferences (see Reading & Canada, this book).

From the student perspective teachers need to learn how students will intuitively represent data and how they as teachers can scaffold students to consider other representations, which will allow them to learn more from the data. However, this is not always easy for the students (see González, Espinel, & Ainley, this book). For example, Konold and Higgins (2003) demonstrate how students can be scaffolded from their intuitive individual case bar plots to dot plots, while Bakker and Gravemeijer (2004) explain how students can make the transition from dot plot to box plot. Some readings that can be given to teachers to understand their students' reasoning about distributions would be about shifting students' intuitive focus on individual cases to a consideration of the global distribution (e.g., Ben-Zvi & Arcavi, 2001).

Inference. "Statistical inference moves beyond the data in hand to draw conclusions about some wider universe, taking into account that variation is everywhere and the conclusions are uncertain" (Moore, 2007, p. xxviii). That is, a "big idea" of statistics is for people to understand that much of statistical work involves taking random samples and using them to make estimates or decisions about the populations or processes from which they are drawn or generated (see Harradine, Batanero, & Rossman, this book). Recent research suggests that developing informal types of inferential reasoning across the curriculum might be essential for building students' concepts about statistical inference (e.g., Pratt & Ainley, 2008).

Concepts such as samples and sampling underpin inference. Groth and Bergner (2005) demonstrate that teachers have impoverished metaphors for the concept of a sample and suggest that teacher educators need to design learning pathways that enhance teachers' knowledge of samples. They also need to take into account students' conceptual development of sampling reasoning, which includes notions of a sample, sampling methods, sample size and randomness (Watson, 2006). Linked to the notion of a sample are the sample distribution, population and population distribution, all of which should be understood.

Teachers also need to experience visually sampling variability of sample distributions with the same and different sample sizes for both qualitative and quantitative data. Wild, Pfannkuch, Regan, and Horton (2011) have recently developed some ways of building students' concepts of sampling variability and proposed some informal decision rules for making a claim when comparing two groups using box plots. These rules progressively get more sophisticated across grade levels and work towards an understanding of statistical inference including simple versions of significance testing and confidence intervals. In response to Cobb's (2007) call for statistics teachers to place the logic of inference rather than the normal distribution at the heart of introductory and secondary school courses, Rossman and Chance (see statweb.calpoly.edu/csi), for example, are promoting curricula involving resampling methods such as randomisation and permutation tests as inference techniques rather than normal-based probability models.

In summary, developing understanding of at least the five key statistical concepts – data, patterns in data, distribution, variability and inference – form the foundation for improving teachers' statistical thinking. The concepts are best developed within an interactive inquiry-based learning environment.

2.2   Developing the Ability to Explore and Learn from Data

Since data are at the heart of statistical work, courses should use real (or realistic) and motivating data as the focus for statistical learning. Teachers can be challenged to explore and learn from data in ways similar to the ways their students will explore data (Godino et al., 2008). Experiencing the whole empirical enquiry cycle from understanding the contextual situation, formulating problems, defining variables, determining methods of measurement, designing methods of data collection, collecting data, and so forth, is a fundamental learning experience (Sanchez & Blancarte, 2008; Garfield & Ben-Zvi, 2009; Hall, this book). Both sample survey and experimental processes of enquiry should be understood.

Intrinsic to the process of enquiry is experiencing the exploration of multivariate data sets. Learning to be a data detective by wondering whether some factors might explain differences between two groups or whether there is a relationship between two variables is part of learning the game of statistics. The emphasis should be on teachers posing their own questions about the data, interrogating the data and learning new information about the real world from the data. Use of technology such as *TinkerPlots* can quickly enculturate teachers into unlocking stories in data (Rubin & Hammerman, 2006). An important aspect of teacher courses should be exposing them to a range of technological tools (e.g., graphics calculators, simulation software, Web applets) that can be used to explore and simulate data, test conjectures by analysing data, model situations and help develop understanding of abstract concepts. Teachers, however, should be challenged to discuss how and why such technology enhances students' learning (see Pratt, Davies, & Connor, this book).

By experiencing school-type investigations and sharing with other teachers the problems they encounter and what they learn from the investigations, teachers may appreciate how their students might interact with similar investigations. Also they should become aware of good sorts of data that can be accessed on the Internet, such as the databases of the *CensusAtSchool* project (www.censusatschool.com), and datasets of the Consortium for the Advancement of Undergraduate Statistics Education (*CAUSE*; www.causeweb.org/); see also Hall, in this book.

Learning to explore and learn from data within an enquiry cycle and experiencing activities that aim to build concepts underpinning statistical thought (e.g., least squares regression concept) are vital for improving teachers' reasoning from data. Presenting findings, including justifying inferences and conclusions, to other teachers is also part of the statistics learning process, particularly in developing skills in statistical argumentation.

2.3 Developing Statistical Argumentation

Argumentation in general refers to discourse for persuasion, logical proof, and evidence-based belief, and more generally, discussion in which disagreements and reasoning are presented and critiqued (Kirschner, Buckingham Shum, & Carr, 2003). Contemporary argumentation research has shifted in focus, calling for a move from formal grammatical structures towards socio-cultural perspectives (Andriessen, 2006). In these new forms of argument, the goal is to engage others in generative inquiry – seeking collaborative construction of knowledge and inviting critique. Statistical inference is closely related to these argumentative activities. Deriving logical conclusions from data – whether formally or informally – is accompanied by the need to provide persuasive arguments based on data analysis in an inquiry-based environment. Ben-Zvi (2006) has found that argumentation is a natural tool for students' articulation of informal statistical inferences.

Abelson (1995) proposes two essential dimensions to informal argumentation: (a) the act or process of deriving logical conclusions from data (inference), and (b) providing persuasive arguments based on the data analysis (rhetoric and narrative). However, the skills of argumentation are typically not part of statistics taught in school. Instead, the teaching of statistics ignores the argumentative, give-and-take nature of statistical claims. As a consequence, students tend to develop their own characteristic misperceptions of statistics.

Therefore, another important goal for teacher courses is to develop an appreciation for the value of classroom statistical discourse and argumentation. This is different from teachers asking questions and students responding. The kind of discourse promoted is dialogue where students learn to question each other, respond to each other's questions as well as defend their answers and data-based arguments. Cobb and McClain (2004) describe the characteristics of effective classroom discourse in which statistical arguments explain why the way in which the data have been organised gives rise to insights into the phenomenon under investigation; students engage in sustained exchanges that focus on significant statistical ideas. Teacher educators need to model how to create a classroom climate where teachers feel safe expressing their views, even if they are tentative. Allowing questions that begin with "what do you think", "why do you think", or "what would happen if" can lead to good class discussions. Such an approach develops an argumentation culture in which teachers expect inferences to be evaluated, motivating them to make linkages explicit.

Teachers should be aware of studies in multiple disciplines that report the widespread difficulty students have with argument, including confusing personal beliefs with evidence, inadequate sampling, jumping to conclusions, difficulties expressing uncertainties in data, reinterpreting data with causal explanations that contradict scientific principles and not connecting questions and evidence (Driver, Newton, & Osborne, 2000). In statistics, evidence must be considered in ways that are valid statistically – attending to sampling, variability, suitable measures in comparing groups and context. Argumentation, however, can be improved with substantial experience and support (Driver et al., 2000), but requires going beyond simply explaining thinking to focus on shared intellectual substance, sustained dialogue, chains of reasoning and a classroom culture of inquiry. Essential to supporting and developing teachers' argumentation is giving them ongoing feedback during the learning process, that is, formative assessment.

2.4   Formative Assessment

Assessment is an integral part of teaching statistics at all levels (see Garfield & Franklin, this book). Over the past decades calls for alternative forms of assessment have led to increased use of assessment to provide feedback to students and to improve student learning, rather than merely to provide summative measures of achievement. Some countries such as the United States of America have standardised assessments, which rely on multi-choice tests that primarily assess computational skills and factual knowledge. Therefore, such teachers need to become knowledgeable about alternative methods of assessment that provide formative information useful for guiding students' learning. For example, Garfield and Ben-Zvi (2009) had teachers learn about student projects as a form of authentic assessment. Other forms of alternative assessment were also used in this research to assess students' statistical literacy (e.g., critique a graph in a newspaper) or reasoning (e.g., write a meaningful short essay). Teachers were referred to assessment resources such as the *ARTIST* Website (https://app.gen.umn.edu/artist).

   In teacher education courses, the case needs to be clearly made that students will value what the teacher assesses. Therefore, assessments need to be aligned with learning goals and teachers should be assessed in their course using alternative methods. For example, teachers can be asked to work in a collaborative group to develop a class lesson plan, find and analyse a good data set and present to the class a Web resource that they think would be a good tool to promote student learning and explain why. Teachers can also be asked to work in a group to design an activity for students aimed at developing reasoning about one central statistical idea. Teachers can implement the activity in their class, collect assessment data and share and discuss it during one of the course lessons.

2.5   Understanding Students' Reasoning

The learning trajectories that teachers plan for their students need to be built on a sound understanding of the types of intuitions, prior knowledge and conceptions that they may expect their students to exhibit. By reading research on students' statistics conceptions teachers can begin to appreciate and to identify how students might be reasoning and what intuitions they bring into the classroom. Using the readings as a starting point, teachers can also be challenged to argue why and how an activity enhances and develops students' statistical reasoning. Another method is to involve teachers in research projects that focus on studying and analysing students' emerging statistical reasoning and argumentation skills (Garfield & Ben-Zvi, 2009). Alternatively teachers can deliberately analyse their own students' reasoning from assessment tasks, interviews and monitoring student feedback while teaching.

By critically reflecting on and taking notice of how their students are thinking, together with reading research, teachers can begin to build up a coherent body of knowledge on students' understanding of statistical concepts and how they can scaffold naïve conceptions to more normative ones. Without this knowledge, teachers will continue to blindly follow the designated curriculum and use resources that have no regard to how students think. The outcome is that students' thinking is not developed, a finding that Watson (2006) concluded from her many longitudinal studies of students' statistical reasoning. Her main hypothesis for such a situation was the method of instruction. Whilst the method of instruction is a major factor, overcoming such a challenge will involve both improving teachers' understanding of students' thinking and improving teachers' thinking. Therefore, teacher educators need not only to build teachers' statistical concepts but also to make teachers aware of how students' conceptual understanding may develop. Research described in different chapters in this book has found that teachers' thinking is not much different from students and hence the onus is on teacher educators to link, for the teachers, how they are thinking to the ways in which students think.

3   Implications for Research and Training the Teachers

In this chapter, learning experiences that teachers need in order to develop their ability to think and reason statistically were proposed. We described a comprehensive learning environment that is designed around five major themes: developing understanding of key statistical concepts; developing the ability to explore and learn from data; developing statistical argumentation; using formative assessment and learning to understand students' reasoning.

The main implication of this proposal to develop teachers' statistical thinking and conceptual understandings is that there also needs to be a paradigm shift in teacher education. A statistics course developed along the lines suggested is essential if teachers are to be enculturated into a new conceptualisation of statistics teaching. Teachers need to be immersed in a statistical reasoning and learning environment in which a spirit of enquiry is at its heart. Learning statistical reasoning through engaging in EDA, using technology, practising argumentation and reflecting on research will assist teachers in not only developing their statistical thinking but also their students' thinking. However, further research and research instruments are needed for ascertaining whether the type of course proposed in this chapter is effective in developing teachers' statistical thinking.

References

Abelson, R. (1995). *Statistics as principled argument*. Hillsdale, NJ: Lawrence Erlbaum Associates.
Andriessen, J. (2006). Arguing to learn. In R. K. Sawyer (Ed.), *The Cambridge handbook of the learning sciences* (pp. 443–459). New York: Cambridge University Press.
Bakker, A., & Gravemeijer, K. (2004). Learning to reason about distribution. In D. Ben-Zvi & J. Garfield (Eds.), *The challenge of developing statistical literacy, reasoning, and thinking* (pp. 147–168). Dordrecht, The Netherlands: Kluwer.
Ball, D., Thames, M., & Phelps, G. (2008). Content knowledge for teaching. What makes it special? *Journal of Teacher Education, 59*(5), 389–407.
Batanero, C., Burrill, G., Reading, C., & Rossman, A. (Eds.). (2008). *Joint ICMI/IASE Study: Teaching Statistics in School Mathematics. Challenges for Teaching and Teacher Education. Proceedings of the ICMI Study 18 and 2008 IASE Round Table Conference.* Monterrey, Mexico: International Commission on Mathematical Instruction and International Association for Statistical Education. Online: www.stat.auckland.ac.nz/~iase/publications
Ben-Zvi, D. (2006). Scaffolding students' informal inference and argumentation. In A. Rossman & B. Chance (Eds.), *Proceedings of the Seventh International Conference on Teaching Statistics*. Salvador, Bahia, Brazil: International Statistical Institute and International Association for Statistical Education. Online: www.stat.auckland.ac.nz/~iase/publications
Ben-Zvi, D., & Arcavi, A. (2001). Junior high school students' construction of global views of data and data representations. *Educational Studies in Mathematics, 45*, 35–65.
Ben-Zvi, D., & Garfield, J. (Eds.). (2004). *The challenge of developing statistical literacy, reasoning and thinking*. Dordrecht, The Netherlands: Kluwer.
Canada, D. (2008). Variability in a sampling context: Enhancing elementary preservice teachers' conceptions. In C. Batanero, G. Burrill, C. Reading, & A. Rossman (2008).
Cobb, G. (2007). One possible frame for thinking about experiential learning. *International Statistical Review, 75*(3), 336–347.
Cobb, P., & McClain, K. (2004). Principles of instructional design for supporting the development of students' statistical reasoning. In D. Ben-Zvi & J. Garfield (Eds.), *The challenge of developing statistical literacy, reasoning, and thinking* (pp. 375–396). Dordrecht, The Netherlands: Kluwer.
De Oliveira, A., Paranaiba, P., Kataoka, V., de Souza, A., Fernandes, F., & de Oliveira, M. (2008). Statistics teaching for prospective teachers of mathematics: Reflections and perspectives. In C. Batanero, G. Burrill, C. Reading, & A. Rossman (2008).
Driver, R., Newton, P., & Osborne, J. (2000). Establishing the norms of scientific argumentation in classrooms. *Science & Education, 84*, 287–312.
Engel, J., Sedlmeier, P., & Worn, C. (2008). Modeling scatterplot data and the signal-noise metaphor: Towards statistical literacy for pre-service teachers. In C. Batanero, G. Burrill, C. Reading, & A. Rossman (2008).
Franklin, C., Kader, G., Mewborn, D., Moreno, J., Peck, R., Perry, M., & Scheaffer, R. (2005). *Guidelines for assessment and instruction in statistics education (GAISE) report: A pre-k-12 curriculum framework*. Alexandria, VA: American Statistical Association. Online: www.amstat.org/Education/gaise/
Garfield, J., & Ben-Zvi, D. (2008). *Developing students' statistical reasoning: Connecting research and teaching practice*. Dordrecht, The Netherlands: Springer.
Garfield, J., & Ben-Zvi, D. (2009). Helping students develop statistical reasoning: Implementing a Statistical Reasoning Learning Environment. *Teaching Statistics, 31*(3), 72–77.
Godino, J. D., Batanero, C., Roa, R., & Wilhelmi, M. (2008). Assessing and developing pedagogical content and statistical knowledge of primary school teachers through project work. In C. Batanero, G. Burrill, C. Reading, & A. Rossman (2008).
Groth, R., & Bergner, J. (2005). Pre-service elementary school teachers' metaphors for the concept of statistical sample. *Statistics Education Research Journal, 4*(2), 27–42.

Key Curriculum Press. (2006). *Fathom dynamic data™ software*. Emeryville, CA: Key Curriculum Press.

Kirschner, P. A., Buckingham Shum, S. J., & Carr, C. S. (Eds.). (2003). *Visualizing argumentation: Software tools for collaborative and educational sense-making*. London: Springer.

Konold, C., & Higgins, T. (2003). Reasoning about data. In J. Kilpatrick, W. G. Martin, & D. Schifter (Eds.), *A research companion to principles and standards for school mathematics* (pp. 193–214). Reston, VA: National Council of Teachers of Mathematics.

Konold, C., & Miller, C. D. (2005). *TinkerPlots: Dynamic data explorations*. Emeryville, CA: Key Curriculum Press.

Konold, C., & Pollatsek, A. (2002). Data analysis as the search for signals in noisy processes. *Journal for Research in Mathematics Education, 33*(4), 259–289.

Lehrer, R., & Romberg, T. (1996). Exploring children's data modelling. *Cognition and Instruction, 14*(1), 69–108.

Ministry of Education. (2007). *The New Zealand curriculum*. Wellington, New Zealand: Learning Media Limited.

Mokros, J., & Russell, S. (1995). Children's concepts of average and representativeness. *Journal for Research in Mathematics Education, 26*(1), 20–39.

Moore, D. (1990). Uncertainty. In L. Steen (Ed.), *On the shoulders of giants: New approaches to numeracy* (pp. 95–137). Washington, DC: National Academy Press.

Moore, D. (2007). *The basic practice of statistics* (4th ed.). New York: W.H. Freeman.

Moritz, J. (2004). Reasoning about covariation. In D. Ben-Zvi & J. Garfield (Eds.), *The challenge of developing statistical literacy, reasoning, and thinking* (pp. 227–256). Dordrecht, The Netherlands: Kluwer.

Pratt, D., & Ainley, J. (2008). Introducing the special issue on informal inferential reasoning. *Statistics Education Research Journal, 7*(2), 3–4. Online: www.stat.auckland.ac.nz/serj/

Rubin, A., & Hammerman, J. (2006). Understanding data through new software representations. In G. Burrill (Ed.), *Thinking and reasoning with data and chance: 2006 NCTM yearbook* (pp. 241–256). Reston, VA: National Council of Teachers of Mathematics.

Sanchez, E., & Blancarte, A. (2008). Statistical thinking as a fundamental topic in training the teachers. In C. Batanero, G. Burrill, C. Reading, & A. Rossman (2008).

Sanchez, E., & Garcia, J. (2008). Acquisition of notions of statistical variation by in-service teachers. In C. Batanero, G. Burrill, C. Reading, & A. Rossman (2008).

Shaughnessy, M. (2007). Research on statistics learning and reasoning. In F. Lester (Ed.), *Second handbook of research on the teaching and learning of mathematics* (Vol. 2, pp. 957–1009). Charlotte, NC: Information Age Publishers and National Council of Teachers of Mathematics.

Shaughnessy, M., & Pfannkuch, M. (2002). How faithful is Old Faithful? Statistical thinking: A story of variation and prediction. *Mathematics Teacher, 95*(4), 252–259.

Shulman, L. S. (1986). Those who understand: Knowledge growth in teaching. *Educational Researcher, 15*, 4–14.

Watson, J. M. (2006). *Statistical literacy at school: Growth and goals*. Mahwah, NJ: Lawrence Erlbaum.

Wild, C. J. (2006). The concept of distribution. *Statistics Education Research Journal, 5*(2), 10–26. Online: www.stat.auckland.ac.nz/serj/

Wild, C. J., & Pfannkuch, M. (1999). Statistical thinking in empirical enquiry (with discussion). *International Statistical Review, 67*(3), 223–265.

Wild, C. J., Pfannkuch, M., Regan, M., & Horton, N. (2011). Towards more accessible conceptions of statistical inference. *Journal of the Royal Statistical Society. Series A, 174*(2), 247–295.

Chapter 32
Engaging Teachers and Students with Real Data: Benefits and Challenges

Jennifer Hall

Abstract Using real data in statistics education provides significant benefits for both teachers and students. In this chapter, considerations faced by teachers when using real data are explored with regard to student engagement and learning, as well as potential pedagogical issues. To address these issues, suggestions for obtaining and using real data are offered, as is the description of a successful example of a primary data collection project – Census at School. The chapter concludes by considering the use of real data in the training of teachers, and extends the example of Census at School to explore Canadian professional development workshops for elementary teachers.

1 Introduction

In many countries, statistics has only recently been introduced into the elementary and secondary school curriculum. For instance, in the United States, Franklin (2000) suggests that the inclusion of statistics within the K-12 curriculum was instigated by the National Council of Teachers of Mathematics' *Curriculum and Evaluation Standards for School Mathematics* (1989). Discussions about successful methods for teaching statistics and preparing teachers to teach statistics at the school level do not have a long-standing history, with the exception of a few countries (Shaughnessy, 2007).

In part due to this relative lack of discussion on training teachers in statistics education at the school level, findings from mathematics education research are often extrapolated to statistics. However, as argued in the Gattuso and Ottaviani chapter in this book, although mathematics and statistics share many similarities

J. Hall (✉)
University of Ottawa, 145 Jean-Jacques Lussier Private Ottawa, Ontario K1N 6N5, Canada
e-mail: jennifer.e.hall@uottawa.ca

C. Batanero, G. Burrill, and C. Reading (eds.), *Teaching Statistics in School Mathematics-Challenges for Teaching and Teacher Education: A Joint ICMI/IASE Study*, DOI 10.1007/978-94-007-1131-0_32, © Springer Science+Business Media B.V. 2011

as academic disciplines, they are not completely analogous. While mathematics, particularly at advanced levels, is sometimes viewed as deterministic and hierarchical in nature (Makar & Confrey, 2003), statistics is not accurately or fully portrayed by such a description. Rather, statistics differs from mathematics in that it is highly context-specific (Shaughnessy, 2007); there are many possible interpretations and outcomes rather than one answer. Furthermore, the applications of statistics tend to be wider-ranging than those of mathematics, as statistics can be incorporated into a greater variety of subject areas. Indeed, Cobb and Moore (1997) posit that statistics "exists not for itself but rather to offer other fields of study a coherent set of ideas and tools for dealing with data" (p. 801).

However, teachers often teach statistics in a similar manner to mathematics, which is thus not well-suited to the unique nature of statistics (Makar & Confrey, 2003; Meletiou-Mavrotheris & Stylianou, 2003; Meletiou-Mavrotheris, Paparistodemou, & Stylianou, 2009). Some authors (Lajoie & Romberg, 1998; Begg & Edwards, 1999; Shaughnessy, 2007) suggest that this may be attributed to teachers' lack of both content (subject area) knowledge and pedagogical content knowledge, a special domain of content knowledge that "embodies the aspects of content most germane to its teachability … the ways of representing and formulating the subject that make it comprehensible to others" (Shulman, 1986, p. 9). Furthermore, elementary teachers have been found to tend to teach statistics in a very rule-bound, disconnected manner due to their inexperience and discomfort with the subject area (Chick & Pierce, 2008). As a consequence of experiencing statistics taught in a decontextualised manner, students tend to become disengaged and lack conceptual understanding (Connor, Davies, & Holmes, 2006).

In this chapter, it is argued that the use of real and meaningful data in statistics education can help both teachers and students obtain fundamental statistical understandings and skills, and increase their engagement with the subject area. First, benefits and challenges are considered regarding the use of real data in the teaching and learning of statistics at the K-12 level. The choice of primary data versus secondary data is analysed, and links to reliable websites of secondary data are provided. Then, an example of a primary data collection programme for elementary and secondary students, Census at School, is described. Next, the discussion turns to how real data may be used in the training of teachers, and the Canadian example of Census at School professional development workshops for elementary teachers is explored.

2   Using Real Data: Considerations

Many benefits and challenges exist that are unique to teaching and learning statistics with real data, including student engagement, knowledge gained by students when working with real data, interdisciplinarity, and precautions in using real data. Although the considerations are outlined with regard to teaching students, most issues can also be extrapolated to teachers' own learning about statistics. Teachers are "also learners and their understanding of content and pedagogy is powerfully influenced by their own experiences as students" (Stoddart, Connell, Stofflett, & Peck, 1993, p. 229).

2.1 Student Engagement

Engaging students in statistics is one of the most important benefits of using real data. Too often, data presented in textbooks are decontextualised and seen as irrelevant to students' lives (Connor, 2002). Real data, particularly those that are both collected and analysed by students, take on a whole new dimension of interest and relevance. However, not all real data have the same appeal to students, so it is essential to use data that are relevant to students' interests. If students cannot connect to the dataset and problem being investigated, the possible engagement is lost and there is little more benefit obtained than there would be using synthetic data. McNab, Moss, Woodruff, and Nason (2006) suggest obtaining input from students when selecting a dataset or problem to investigate, in order to increase students' interest in statistical exploration. When data are about the students themselves, they tend to be excited to undertake statistical investigations to learn more about themselves and their peers (Turner, 2006; Wong, 2006; Catley, 2007). Furthermore, when statistical explorations are relevant, students become engaged, gain deep statistical understandings, and feel more positive about the subject area (Bingham, 2010).

2.2 What Students Learn When Working with Real Data

When students participate in primary data collection and analysis, they are exposed to "real" statistical issues that are rarely encountered with prepared synthetic datasets from textbooks. These issues include, but are not limited to, dealing with different types of data (qualitative, quantitative, ordinal, etc.), defining variables and categories of classification, dealing with reliability and validity issues in measurement, designing questionnaires or experiments, screening data, and dealing with outliers.

Designing questionnaires, experiments, and other means of data collection is an important facet of the statistical investigation cycle, which consists of problem-plan-data-analysis-conclusion (Wild & Pfannkuch, 1999). However, when students use provided data from textbooks, they typically only participate in the last two steps of the cycle – analysing data and making conclusions.

Students may collect their own data using pre-made data collection instruments (e.g., existing questionnaires), but their statistical understandings and skills become greatly enhanced when they are involved in the initial planning phases of a project, particularly the design of a data collection instrument or experiment. For instance, Connor et al. (2006) had elementary students collect real data through an experiment in which they investigated "how the accuracy of rolling marbles varies with age, throwing hand, gender, and distance from the target" (p. 190). The students were asked to make predictions about the outcome of the experiment prior to conducting it. Not only were the students engaged in undertaking the experiment, they were also very insightful about how the experimental design and context could affect the data.

If students collect data from their classmates, the sample size may be too small to illustrate certain statistical concepts, such as the normal distribution. However, this problem can be used as a springboard into discussions about sample size, distributions, and other related issues. Students can hypothesise about differences that may occur with the data in a larger sample size, and compare their class findings to a larger sample, such as a national-level dataset. The key benefit of having students use their own class set of primary data or secondary data regarding a familiar topic is that they can more easily undertake hypothesis testing and data exploration because they have a firm understanding of the dataset, and they are better able to logically explain their findings. For example, Turner (2006) had elementary students collect their own class data via the United Kingdom's CensusAtSchool questionnaire and analyse it using computer software. The students were able to quickly distinguish via computer graphical displays when they had made errors in data entry. The students felt confident about challenging what they perceived as problems in the data and the data handling methods since they had a firm understanding of the dataset itself, having collected data about their own lives.

2.3 Interdisciplinarity: The Example of Social Justice

Another benefit of using real data is the opportunity for interdisciplinary, or cross-curricular, learning opportunities (Nicholson, Ridgway, & McCusker, 2006; Ridgway, Nicholson, & McCusker, 2007). Using real data related to other subject areas (e.g., environmental studies, geography, social studies, health, history) allows students to not only garner an understanding of current (or historical) issues in that subject area, but also requires the use of statistical techniques to analyse the data and draw conclusions.

An example is the topic of social justice. Lesser (2007) argues that "tools used to identify statistical group differences can help people recognise, analyse or address social inequalities" (p. 2). Statistical investigations into social justice issues can cause students to question and challenge existing societal norms and structures. Certainly, explorations of social justice issues can only be done with real data; typically, secondary data are used in such investigations.

Some commonly explored topics include wealth distribution, racial profiling, and gentrification, as well as relationships between health conditions and socioeconomic status and between incarceration rates and ethnicity. For instance, Makar and Confrey (2004, 2007) worked with pre-service teachers to have them determine if a large-scale testing system was fair and valid for all students by making comparisons across student subgroups (e.g., gender, racial, economic). Makar and Confrey (2007) found that "strong personal engagement with the equity topic was a potent motivator for them to use more sophisticated statistical tools in their inquiry" (p. 489). Similarly, Campos, Wodewotzki, Jacobini, and Lombardo (2010) found that college and university students became highly engaged in statistics projects that focused on critical education topics, as they felt personally invested in the social problems analysed.

In teaching statistics for social justice (TSSJ), consideration should be made of the politicised nature of definitions for group membership (e.g., the "poverty line"), one of many issues related to the uncertainty and context-specificity of statistics. As with any real data, it is imperative to select a topic that will interest and engage students. Teachers may use community-level data or data related to a current societal issue to meet this goal. Using real data in TSSJ allows for cross-curricular teaching and learning opportunities (e.g., students may write letters to public officials or create petitions based on their statistical findings) and has the possibility to effect real change in society.

2.4   Precautions in Using Real Data

A common issue raised about using real data, particularly primary data, is that it is too time-consuming. Although collecting and using primary data does take a significant amount of time, particularly if students undergo the entire statistical investigation cycle of problem-plan-data-analysis-conclusion (Wild & Pfannkuch, 1999), it allows students to learn many skills and fully experience the related nature of statistical processes. As such, a large portion of the statistics curriculum may be covered, making this practice in fact very efficient (Catley, 2007).

Secondary data can be rapidly and efficiently obtained from various online sources, such as national statistical agencies' websites. However, not all students and teachers have access to computers with Internet access. Even in schools with computer labs, it is often challenging to be able to book a sufficient amount of time to complete the data collection from websites. Furthermore, some schools may not have printing facilities that allow for students to work with hard copies of the data, if their computer time is minimal. The process of obtaining online data sometimes involves following a lengthy step-by-step procedure, which can lead to student disengagement and frustration before they even see the data. Additionally, data that are obtained online may not be provided in formats that easily download into data analysis software programs, which can cause further frustration. Therefore, in order to maintain student interest in statistical investigations, it is imperative to use websites that are student-friendly in their interface and provide data in a manner that is easy to access and use. Some examples of such websites are provided in the following section.

3   Obtaining Real Data

Although teachers may be interested in using real data in their statistics teaching, they may find it challenging to obtain real data that are relevant to the statistical topic they are teaching. As such, they may abandon the idea of using real data in favour of using data provided in textbooks. To help counter this problem, this section offers suggestions for obtaining real data, both primary and secondary, and provides examples of reliable websites for obtaining secondary data. The section concludes with the description of Census at School, a successful international primary data collection project, in the Canadian context.

3.1   Primary Data Versus Secondary Data

When teachers design tasks for teaching statistics, they need to consider whether primary data or secondary data are more appropriate, as well as which type of data (e.g., continuous, discrete, numeric, categorical) is required. Primary data are particularly useful when a teacher wants students to learn about data collection and data management techniques, such as how to create a questionnaire and conduct data screening. Primary data could also be collected by means of experiments (science or otherwise), which would further allow for an opportunity for cross-curricular learning.

If a large dataset is required, secondary data are usually a more appropriate and efficient choice. With access to the Internet being far more prevalent than ever before, it is now simpler and more user-friendly for students to retrieve online data from their homes or schools. Students take an active role in retrieving online data, as opposed to a more passive and tedious role in retrieving textbook data, the latter of which tends to disinterest students (Connor et al., 2006). In many countries, national and provincial/state agencies (e.g., public health offices, ministries of education) provide data online to the public at no cost. Furthermore, data available online are frequently updated, so they are much more relevant than the data provided in textbooks, which become outdated shortly after the textbook is published and long before it is removed from student usage.

3.2   Online Data Resources

There are many reliable websites that allow students and teachers access to large-scale datasets of interest. For example, the *Journal of Statistics Education* website hosts a data archive (www.amstat.org/publications/jse/) that includes data files and related articles about such varied topics as automotive sales, births, television, smoking, sports, the lottery, and politics. Similarly, the Data and Story

Library (DASL) offers a large archive of data files and related stories (lib.stat. cmu.edu/DASL/) that can be helpful in demonstrating basic statistical practices. Furthermore, most countries' national statistical offices provide some datasets that are freely accessible to students and teachers. International organisations also provide statistical data on topical issues that are of interest to students and teachers. For instance, the United Nations Statistical Databases website (unstats.un.org/) offers a vast array of international statistics on such topics as population demographics, disabilities, service trade, and oil. Related, the United Nations Educational, Scientific and Cultural Organisation (UNESCO) Institute for Statistics website (www.uis. unesco.org) provides international longitudinal data on education, science and technology, culture, literacy, and communication. Other international organisations, such as the World Trade Organisation (stat.wto.org/) and the World Health Organisation (www.who.int/), also offer vast statistical databases that are easily accessible by students and teachers.

3.3 The Example of Census at School in Canada

A successful international example of a primary data collection programme for elementary and secondary school students is Census at School, which is conducted in Australia, Canada, Ireland, Japan, New Zealand, South Africa, the United Kingdom, and the United States of America (www.censusatschool.com). Although there is international co-operation and there are many similarities among the countries that participate, each country operates its Census at School programme differently.

For instance, in Canada, this programme consists of an in-class online survey (available at www.censusatschool.ca in English or www.recensementecole.ca in French) wherein students anonymously enter data about their lives. There are two versions of the questionnaire, one for Grade 4–8 students (ages 8–13) and one for Grade 9–12 students (ages 14–18), in order to have age-appropriate questions. The questionnaire covers a broad range of school subjects, such as mathematics, health, environmental studies, and social studies, and is thus cross-curricular. The questionnaire also features questions on out-of-school topics, such as eye colour and methods of communication with friends. The questionnaire results in both discrete and continuous numeric data and discrete categorical data through such questions as the number of languages spoken, favourite school subject, and recycling practices, plus measurements of arm span, height, and wrist circumference. Students enter their data anonymously into the national database, and the class dataset subsequently becomes available to download into a spreadsheet or statistical software program.

Further to the benefits offered by the primary data collection aspect of Census at School, the programme also offers a wealth of secondary data, so that comparisons may be drawn between datasets. For instance, the Canadian Census at School website offers annual summary results (percentages or means), both by province/ territory and by the country as a whole. Students may compare their class dataset to the provincial/territorial or national data, and in doing so, will also use important number sense skills, such as calculations involving fractions and percentages. Furthermore, large Census at School datasets may be retrieved from the other participating countries through a random data selector (rds.censusatschool.org.uk/). A teacher who has conducted the Census at School questionnaire with his/her class may retrieve a dataset from another participating country, and then have the students draw comparisons between the two datasets. Making international comparisons using real data is a rich learning opportunity for students, particularly in making statistical inferences about two populations and learning about students from around the world.

4   Training Teachers in the Use of Real Data

As has been shown, using real data with students has the potential to offer significant benefits, such as increased student engagement and opportunities for rich statistical learning, although some cautions have been made about potential obstacles. In this section, the related topic of training teachers in the use of real data is discussed, again highlighting some benefits and challenges. The section begins with a brief review of studies in which teachers were trained through the use of real data and concludes with a discussion of the Canadian example of Census at School elementary teacher professional development workshops, which involve the collection and analysis of primary data.

4.1   Related Studies

Many researchers have recognised that the value of using real data in statistics education goes beyond teaching K-12 students; it is also highly beneficial in helping teachers to obtain statistical understandings and an appreciation of statistics as a discipline. For example, Giambalvo and Gattuso (2008) conducted research with pre-service teachers who created lessons using real data about postal services. Although the pre-service teachers did not collect the data themselves, they were very engaged and saw the applicability of the data. The researchers concluded that the pre-service teachers realised the importance of statistics, particularly statistical reasoning, through working with real data. Similarly, Visnovska and Cobb (2010) worked with middle-school mathematics teachers in professional development workshops that spanned 5 years. Over time, the teachers developed increasingly sophisticated understandings of the value of using significant and relevant problem contexts in teaching statistics, both in terms of improving their own instructional practices and engaging students.

In another research project, Chick and Pierce (2008) asked pre-service teachers to create elementary-level statistics lesson plans using real data on local water storage levels. Although the lesson plans were very hands-on and active, the statistical thinking required by the lessons was very shallow in nature. Chick and Pierce suggest that this may be linked to the pre-service teachers' lack of pedagogical content knowledge and ambivalent feelings towards statistics. Although real data has great potential to engage both teachers and students, lessons involving its use are not guaranteed to have depth in terms of statistical concepts and content. When training teachers about using real data, one must ensure that they are trained in how to use it in a meaningful, statistically rich manner, which is linked to their pedagogical content knowledge.

4.2   Canadian Census at School Professional Development Workshops

Statistics Canada employs regional education outreach representatives around the country to conduct free professional development workshops about Statistics Canada's educational resources with both practising and pre-service teachers. Since its inception in 1996, Statistics Canada's education outreach programme has trained more than 30,000 teachers, approximately 10% of Canada's current K-12 teaching population. One workshop involves elementary teachers participating in the Census at School survey as though they were a class of students, and then analysing their class dataset using TinkerPlots™ software (Konold & Miller, 2005).

During the workshops, the Statistics Canada representative shares tips and techniques for conducting the Census at School project, in order to help teachers successfully undertake the project with their classes. Although explicit pointers are given, teacher workshop participants make some of the mistakes that children make, but these mistakes can turn into valuable learning opportunities. For instance, the questionnaire asks participants to measure their arm span. Measuring tapes are taped together at the 100-cm mark, but participants (and children) often forget to add 100 cm to the number they read (e.g., an arm span of 55 cm instead of 155 cm would be recorded). When the class dataset is displayed in graphical form, these erroneously recorded data points are easily spotted by participants. This can lead to discussions not only about the importance of checking data for accuracy before entering it into a database, but also about outliers. Discussions about when outliers are feasible responses as opposed to simply data entry errors can be enlightening learning opportunities, the likes of which do not typically occur with prepared datasets.

Importantly, the workshops purposely do not include any type of pre- or post-test to assess teachers' statistical knowledge. Rather, the workshops are intended to provide a stress-free, non-threatening, fun experience with statistical exploration for teachers who may have previously had negative experiences with the subject area. However, through participating in a workshop in the role of a student, teachers increase their own understandings of data collection, statistical analysis, and statistical software programs. In relation to this issue, Makar and Confrey (2004) suggest that teachers need to personally experience learning through innovative practices before they can implement them with students. Prior research by Hall (2008) on Statistics Canada's teacher workshops aligns with this suggestion. Workshop participants felt that they gained an understanding of statistical topics and technology, plus had fun with statistics; for many teachers, this was their first positive experience with the subject area. When teachers experience rewarding statistical explorations using real data, they can easily envision how their students would benefit from such an experience, and are more apt to implement these strategies in their classrooms.

5   Implications for Teaching and Training Teachers

In this chapter, the benefits and challenges of using real data in the teaching and learning of statistics have been discussed, both in the context of classroom teaching and teacher training. Real data can engage students and teachers, help them to develop statistical understandings, and provide interdisciplinary learning opportunities, such as the investigation of social justice issues. Although some issues exist in using real data, such as challenges locating relevant datasets and issues surrounding time constraints, they are outweighed by the substantial benefits that using real data offers. If our goal as statistics educators is to help students become statistically literate citizens who can understand and assess the plethora of statistical information that exists in everyday life, we need to teach statistics in a contextualised, relevant manner, which includes the use of real data. As pithily stated by Watkins, Scheaffer, and Cobb (2004, p. vii), "If you have pretend data, you can only pretend to analyse it".

References

Batanero, C., Burrill, G., Reading, C., & Rossman, A. (Eds.). (2008). *Joint ICMI/IASE Study: Teaching Statistics in School Mathematics – Challenges for Teaching and Teacher Education. Proceedings of the ICMI Study 18 and 2008 IASE Round Table Conference.* Monterrey, Mexico: International Commission on Mathematical Instruction and International Association for Statistical Education. Online: www.stat.auckland.ac.nz/~iase/publications

Begg, A., & Edwards, R. (1999). Teachers' ideas about teaching statistics. Paper presented at the Joint Conference of the Australian Association for Research in Education and the New Zealand Association for Research in Education. Melbourne, Australia. Online: www.aare.edu.au/99pap/

Bingham, A. D. (2010). Student attitudes toward real-world projects in an introductory statistics course. In C. Reading (Ed.), *Proceedings of the Eighth International Conference on Teaching Statistics.* Ljubljana, Slovenia: International Statistical Institute and International Association for Statistical Education. Online: www.stat.auckland.ac.nz/~iase/publications

Campos, C. R., Wodewotzki, M. L. L., Jacobini, O. R., & Lombardo, D. F. (2010). Statistics education in the context of the critical education: Teaching projects. In C. Reading (Ed.), *Proceedings of the Eighth International Conference on Teaching Statistics.* Ljubljana, Slovenia: International Statistical Institute and International Association for Statistical Education. Online: www.stat.auckland.ac.nz/~iase/publications

Catley, A. (2007). Statistical fun. *Mathematics Teaching Incorporating Micromath, 201*, 19–22.

Chick, H. L., & Pierce, R. U. (2008). Teaching statistics at the primary school level: Beliefs, affordances, and pedagogical content knowledge. In C. Batanero, G. Burrill, C. Reading, & A. Rossman (2008).

Cobb, G. W., & Moore, D. S. (1997). Mathematics, statistics, and teaching. *The American Mathematical Monthly, 104*(9), 801–823.

Connor, D. (2002). CensusAtSchool 2000: Creation to collation to classroom. In B. Phillips (Ed.), *Proceedings of Sixth International Conference on Teaching of Statistics*. Cape Town, South Africa: International Statistical Institute and International Association for Statistical Education. Online: www.stat.auckland.ac.nz/~iase/publications

Connor, D., Davies, N., & Holmes, P. (2006). Using real data and technology to develop statistical thinking. In G. Burrill (Ed.), *Thinking and reasoning with data and chance: 2006 NCTM Yearbook* (pp. 185–194). Reston, VA: National Council of Teachers of Mathematics.

Franklin, C. (2000, October). Are our teachers prepared to provide instruction in statistics at the K-12 levels? *Mathematics Education Dialogues, 10*. Online: www.nctm.org/resources/content. aspx?id=1776

Giambalvo, O., & Gattuso, L. (2008). Teachers training in a realistic context. In C. Batanero, G. Burrill, C. Reading, & A. Rossman (2008).

Hall, J. (2008). Using Census at School and TinkerPlots™ to support Ontario elementary teachers' statistics teaching and learning. In C. Batanero, G. Burrill, C. Reading, & A. Rossman (2008).

Konold, C., & Miller, C. D. (2005). *TinkerPlots: Dynamic data exploration*. Emeryville, CA: Key Curriculum Press.

Lajoie, S., & Romberg, T. (1998). Identifying an agenda for statistics instruction and assessment in K-12. In S. Lajoie (Ed.), *Reflections on statistics: Learning, teaching, and assessment in Grades K-12* (pp. xi–xxi). Mahwah, NJ: Lawrence Erlbaum

Lesser, L. (2007). Critical values and transforming data: Teaching statistics with social justice. *Journal of Statistics Education, 15*(1), 1–21. Online: www.amstat.org/publications/JSE/

Makar, K., & Confrey, J. (2003). Clumps, chunks, and spread out: Secondary preservice teachers' reasoning about variation. In C. Lee (Ed.), *Proceedings of the Third International Research Forum on Statistical Reasoning, Thinking and Literacy* [CD-ROM]. Mount Carmel, MI: Eastern Michigan University.

Makar, K., & Confrey, J. (2004). Modeling fairness in student achievement in mathematics using statistical software by preservice secondary teachers. In H. W. Henn & W. Blum (Eds.), *ICMI Study 14: Applications and modelling in mathematics education (Pre-conference volume)* (pp. 175–180). Dortmund, Germany: Universität Dortmund.

Makar, K., & Confrey, J. (2007). Moving the context of modelling to the forefront: Preservice teachers' investigations of equity in testing. In W. Blum, P. L. Galbraith, H. W. Henn, & M. Niss (Eds.), *Modelling and applications in mathematics education: The 14th ICMI study* (pp. 485–490). New York: Springer.

McNab, S. L., Moss, J., Woodruff, E., & Nason, R. A. (2006). "We were nicer, but we weren't fairer!" Mathematical modeling exploring "fairness" in data management. In G. Burrill (Ed.), *Thinking and reasoning with data and chance: 2006 NCTM Yearbook* (pp. 171–184). Reston, VA: National Council of Teachers of Mathematics.

Meletiou-Mavrotheris, M., Paparistodemou, E., & Stylianou, D. (2009). Enhancing statistics instruction in elementary schools: Integrating technology in professional development. *The Montana Mathematics Enthusiast, 6*(1 & 2), 57–78.

Meletiou-Mavrotheris, M., & Stylianou, D. (2003). On the formalist view of mathematics: Impact on statistics instruction and learning. In A. Mariotti (Ed.), *Proceedings of the Third Conference of the European Society for Research in Mathematics Education*. Bellaria, Italy: ERME. Online: www.dm.unipi.it/~didattica/CERME3/proceedings/

National Council of Teachers of Mathematics. (1989). *Curriculum and evaluation standards for school mathematics*. Reston, VA: Author.

Nicholson, J., Ridgway, J., & McCusker, S. (2006). Reasoning with data – time for a rethink? *Teaching Statistics, 28*(1), 2–9.

Ridgway, J., Nicholson, J., & McCusker, S. (2007). Teaching statistics – despite its applications. *Teaching Statistics, 29*(2), 44–48.

Shaughnessy, J. M. (2007). Research on statistics learning and reasoning. In F. Lester (Ed.), *Second handbook of research on mathematics teaching and learning* (pp. 957–1010). Greenwich, CT: Information Age Publishing and National Council of Teachers of Mathematics.

Shulman, L. S. (1986). Those who understand: Knowledge growth in teaching. *Educational Researcher, 15*(2), 4–14.

Stoddart, T., Connell, M., Stofflett, R., & Peck, D. (1993). Reconstructing elementary teacher candidates' understanding of mathematics and science content. *Teaching and Teacher Education, 9*(3), 229–241.

Turner, C. (2006). Height, foot length and threat to woodland: Positive learning from pupil relevant data. *Teaching Statistics, 28*(1), 22–25.

Visnovska, J., & Cobb, P. (2010). Supporting shifts in teachers' views and uses of problem contexts in teaching statistics. In C. Reading (Ed.), *Proceedings of the Eighth International Conference on Teaching Statistics.* Ljubljana, Slovenia: International Statistical Institute and International Association for Statistical Education. Online: www.stat.auckland.ac.nz/~iase/publications

Watkins, A., Scheaffer, R., & Cobb, G. (2004). *Statistics in action: Understanding a world of data.* Emeryville, CA: Key Curriculum Press.

Wild, C., & Pfannkuch, M. (1999). Statistical thinking in empirical inquiry. *International Statistical Review, 67*(3), 223–265.

Wong, I. (2006). Using CensusAtSchool data to motivate students. *Australian Mathematics Teacher, 62*(1), 38–40.

Chapter 33
Teaching Teachers to Teach Statistical Investigations

Katie Makar and Jill Fielding-Wells

Abstract Despite its importance for the discipline, the statistical investigation cycle is given little attention in schools. Teachers face unique challenges in teaching statistical inquiry, with elements unfamiliar to many mathematics classrooms: Coping with uncertainty, encouraging debate and competing interpretations, and supporting student collaboration. This chapter highlights ways for teacher educators to support teachers' learning to teach statistical inquiry. Results of two longitudinal studies are used to formulate recommendations to develop teachers' proficiency in this area.

1 Introduction

Wild (1994) defined statistics as an inquiry process "concerned with finding out about the real world by collecting, and then making sense of, data" (p. 164). Despite calls for more emphasis on the investigative process (Moore, 1997), the focus in school statistics continues to be on calculations, procedures, and graphs (Sorto, 2006). Some countries now include statistical inquiry or investigations (implicitly or explicitly) in their national curriculum or curriculum standards (e.g., see National Council of Teachers of Mathematics [NCTM], 2000; Davies, 2007; New Zealand Ministry of Education, 2007; Australian Curriculum, Assessment and Reporting Authority, 2009), but it is uncertain the extent to which schools in these countries have successfully implemented statistical investigations.

Little research has focused on the whole inquiry process (Lavigne & Lajoie, 2007). Instead, much of the research in statistics education has centred on data analysis, predominantly with well-defined problems in which many of the difficult decisions have been obscured.

K. Makar (✉) and J. Fielding-Wells
School of Education, The University of Queensland, Social Sciences Building, Brisbane 4072, Australia
e-mail: k.makar@uq.edu.au; j.wells2@uq.edu.au

C. Batanero, G. Burrill, and C. Reading (eds.), *Teaching Statistics in School Mathematics-Challenges for Teaching and Teacher Education: A Joint ICMI/IASE Study*, DOI 10.1007/978-94-007-1131-0_33, © Springer Science+Business Media B.V. 2011

This chapter overviews key understandings that teacher educators need to know in order to develop teachers' proficiency in teaching statistical inquiry. We focus in particular on the investigation cycle (see Pfannkuch & Ben-Zvi, this book, for further discussion of developing teachers' statistical thinking and MacGillivray & Pereira Mendoza, this book for further discussion of teaching statistics through projects and investigations). An example of a statistical investigation in a middle school classroom is used to illustrate each step of the investigative cycle. Challenges encountered in teaching investigations are discussed to alert teacher educators to key areas to focus their work with teachers. Two longitudinal studies are used to highlight ways that researchers have approached supporting teachers to develop this proficiency.

2 What Is a Statistical Investigation?

Wild and Pfannkuch's (1999) landmark paper described four dimensions of thinking used in statistical inquiry of authentic problems: Phases of the investigative process [Problem, Plan, Data, Analysis, and Conclusion (PPDAC)], types of thinking used, ongoing and iterative mental questioning (interrogating), and dispositions required. Problems in school statistics are frequently well-structured, where the planning, data collection, analysis, and conclusion are streamlined and unproblematic. Most problems in life, however, are ill-structured since the problem definition or solution pathways have a number of ambiguities that need to be addressed (King & Kitchener, 1994). They are sometimes unresolved with conflicting evidence, requiring students to consider potential causes of the problem, and generate multiple ideas on how to address it (Walker & Leary, 2008). Ill-structured problems often require discussion to negotiate which characteristics of a phenomenon can be measured to address the problem under investigation. For example, if the question being addressed is, "Which brand of bubble gum is the best?", then students will need to debate qualities valued in bubble gum that might qualify as "best" and identify possible measures to capture these qualities. Statistical inquiry situates statistical investigations within these complex settings. Below we give a brief example of a statistical investigation embedded in a middle school science unit designed to develop students' understanding of forces used in flight.

Problem. The driving question for this investigation was "What is the best design for a loopy aircraft?" At the beginning of the project, students constructed a loopy aircraft (made by affixing a paper loop to each end of a plastic straw). In performing initial test flights, they immersed themselves in the problem and negotiated how to define "best" to address the driving question. They defined "best" as the aircraft able to fly the greatest distance. Students also deepened their understanding of the science of flight through additional study.

Plan. In the second investigation step, students considered factors that could be altered on the loopy aircraft. They defined three variations (each) for the width, length, and placement of loops on the straw, requiring construction of 27 aircraft.

The students developed measurement protocols and a sampling design that would reduce unexplained variability and anomalies.

Data Collection. Once data collection was planned, students recorded flight distances of 5 flights for each of the 27 aircraft constructed. In reviewing the data, the students recognised a need to "clean" the data as measurements had been recorded with a mixture of meters and centimetres.

Data Analysis. When faced with 135 data points, students initially made superficial judgments about the best design (such as the single plane flying the greatest distance). In discussing this issue, they decided to add the five flight distances together to create a single measure for each aircraft to moderate the effect of anomalies. The teacher used this opportunity to introduce the mean. Students used *TinkerPlots* (Konold & Miller, 2005) to generate distributions of the mean distances to compare flights for each variable; for example, the distribution of distances of the nine planes with narrow wings could be compared to those with medium and wide wings.

Conclusions. Students used means to draw conclusions about the most advantageous width, length, and wing placement. Because the optimal width, length, and placement differed from the aircraft that flew the farthest, one student wondered whether they had ignored potential interactions between variables (a concept students might encounter in a second university statistics course). This raised the opportunity to initiate a new, more sophisticated PPDAC cycle with design modifications to further improve distance. At the end of the unit, students wrote a final report articulating their findings, conclusions, and justifications.

3   Assisting Teachers with Statistical Investigations

By breaking down the PPDAC cycle into its five phases, we draw on the literature to identify key issues to consider in teaching statistical investigations.

Problem. The problem-posing phase is essential in the investigation cycle as the investigation question acts as the initial "hook" and driving focus for the investigation. Research suggests that a critical aspect of this phase is for teachers to learn "to use the driving question to orchestrate a project" (Marx et al., 1994, p. 535). Questions developed for statistical investigations need to be:

*   *Interesting, challenging, and relevant* (Groves & Doig, 2004). The flight unit was motivating to students as it tapped into their interest in paper planes and was of a low-stakes competitive nature. The level of cognitive engagement was challenging, but attainable for middle school students.
*   *Statistical in nature* (Makar & Confrey, 2002; Arnold, 2008a). Questions need to be answered through gathering and interpreting data. In the flight unit, the students collected data to justify and defend their choice of "best" aircraft. The data also offered sufficient complexity to generate interesting results.

- *Ill-structured and ambiguous* (Borasi, 1992). Depth of investigation can be achieved by using questions which are ill-defined as they enable negotiation by students. For example, "What is the best design for a loopy aircraft?" raised issues such as whether "best" was an aircraft that flew the farthest, the most accurately, or spent the longest time airborne, each of which would lead the investigation into separate statistical areas.

There is significant scope here to assist teachers with the skills needed to develop problems worthy of investigation, or to guide students to do so (Allmond & Makar, 2010). One particularly critical aspect is the depth of knowledge and experience that teachers need to plan and conduct statistical investigations that develop rich statistical understandings (Arnold, 2008b).

Planning. School statistics often results in students being presented with data manufactured to demonstrate a pre-determined result. Teachers and students therefore have little experience with the reasoning and decision-making needed for planning data collection, recording methods, and appropriate statistical analyses. Research suggests that statistical content knowledge is viewed by teachers as the largest threat to perceptions of their competence (Hills, 2007) and thus suggests that with better statistical knowledge and experience, teachers could give students more support in planning investigations and making essential methodological decisions (Fielding-Wells, 2010).

Data Collection. Investigations enable students to choose their method of data collection, providing them with authentic feedback of their planning decisions that will generate deeper knowledge of methodology and efficiency (Krajcik et al., 1998). Therefore, it is important to help teachers recognise the deeper learning opportunities that arise when students are allowed to face problems and deliberate ways to resolve issues that arise during data collection.

Data Analysis. In most classrooms, data analysis is the entry point of student learning in statistics. Data presented to students have usually been cleaned and carefully selected to illustrate the lesson purpose, while in reality, data are more complex. Issues that arise in managing complex data can springboard discussions about treatment of outliers, errors, and unanticipated results. Allowing students to represent their own data also encourages the process of changing data representations to reveal alternate insights (Cobb, 1999). While time constraints are a valid concern for teachers, allowing students to identify errors and inefficiencies and negotiate alternate activities develops deeper understandings, perseverance, and efficiency.

Conclusion. The ability to communicate and critique statistical processes is necessary for the development of statistical literacy (Gal, 2002). In the conclusion phase, students interpret their results, reflect on the process, and draw critical inferences. At this time, teachers need to draw upon skilful questioning techniques and understanding of statistical analyses in order to facilitate students' reasoning in connecting their conclusions to the question under investigation and the evidence they have collected.

Other Issues. Through participation in the entire PPDAC cycle, students deepen understandings of the complexity of statistical processes. However, teachers need to allow students to make their own mistakes and support students in managing many of the challenges that arise. In doing so, they develop students' resilience and motivation as they implement plans and actions which result in improved understanding of their world. As a result, there is necessarily less order in the classroom that teachers often find counter-intuitive to teaching, with the noise and shifts in approaches to monitoring student behaviour confronting to existing classroom norms.

4   Challenges in Teaching Statistical Investigations

Developing students' inquiry skills requires teachers to engage students in statistical inquiry and support them through the investigative cycle. Those involved in teacher education and professional development must understand the nature of these challenges in order to support and validate teachers' experiences in learning to teach statistical inquiry. For example, teachers are typically frustrated with initial attempts at implementation (Anderson, 2002). "There is a danger that … initial difficulties with implementation and disappointment with student performance can lead to a premature rejection of [these] new pedagogies" (Krajcik et al., 1998, p. 341). Researchers have identified a number of issues to consider when supporting teachers to teach inquiry – in statistics, mathematics, and the sciences:

- *Envisioning inquiry.* Teachers often have difficulty envisioning inquiry in the classroom. Providing resources for teachers to use with students, allowing time to plan and learn collaboratively with other teachers, and creating opportunities to observe students in learning inquiry (R. Anderson, 2002; J. Anderson, 2005) will support them to develop this vision.
- *New teaching practices.* Teachers take on new roles when teaching inquiry, often requiring unfamiliar skills (Crawford, 2000; Arnold, 2008b). These roles highlight the diversity and complexity of teaching inquiry, reflecting shifts in the nature of teacher–student interactions. For example, the teacher takes on the role of motivator, modeller of inquiry practice, collaborator, and mentor. Teaching practices are extended with teachers becoming learners and innovators. Teacher educators can alert teachers to these new practices and discuss their implications.
- *Managing uncertainties.* Teachers need support to manage ambiguities and limitations in applying mathematical ideas. This will help them develop an ability in students to recognise the tentativeness of results, dependence on context, and that outcomes can be continually improved (Borasi, 1992). Experience as learners in conducting statistical investigations gives teachers direct experience in managing uncertainties (Makar, 2010).
- *Validation and support.* Emotional support and validation are needed for teachers to cope with new teaching practices, competing time and curricular pressures, and frustration (Marx et al., 1994). The discomfort and risk-taking needed by

teachers necessitates non-judgmental observations of their teaching, particularly in initial attempts (Hills, 2007).

- *Creating a classroom culture of inquiry.* Teachers need guidance to engender a culture of inquiry in their classroom. Extended time is needed to develop effective student collaborative relationships and to learn to engage students in meaningful discussions (Crawford, Krajcik, & Marx, 1998).
- *Content knowledge.* Disciplinary knowledge is central to teachers' ability to cope with the unexpected issues endemic to inquiry (Arnold, 2008b). Teacher educators should embed opportunities to develop teachers' content knowledge in investigations to model how teachers should develop students' content knowledge.
- *Accountability.* Support from teacher educators coupled with informal pressure or accountability is important for continuing improvement (Guskey, 2002).

5   Experiences in Assisting Teachers to Adopt Statistical Investigations

Teaching teachers to incorporate statistical investigations into their classrooms is an ongoing challenge. Workshops or short-term professional development programmes are not sufficient to sustain innovation. In these next two sections, we describe two projects aimed at developing teachers' practices over time which address many of the challenges we highlighted above. Both projects engage long-term scaffolds that assist teachers in adopting the curriculum and pedagogies associated with statistical investigations. Additionally, they focus on shifting teachers' epistemological beliefs about statistics from a set of methods and calculations, towards statistics as an investigative data-rich process of understanding the world. The first project, based on collaboration among researchers in the European Union, focuses on developing an online community of practice to support teachers in embracing statistical investigations (see also Meletiou-Mavrotheris & Serrado, this book). The second project, based on a longitudinal study in Australia, works with teachers both individually within their classrooms and collectively through regular professional development over several years to support their move to inquiry-based teaching in statistics. The diverse approaches of these projects are used to suggest ways to develop teachers' proficiency with teaching statistical investigations.

5.1   The EarlyStatistics (ES) Project

The aim of the project was to support teachers' knowledge of statistical investigations through online interactions and six modules that guided them through readings, teaching activities, and reflections. The online modules addressed key phases of a statistical investigation (problem-posing, data collection, data analysis, and interpretation), initial experiences in teaching statistical investigations, and

Table 33.1 Iterative stages in EarlyStatistics project

Stage	Aim
Initial moment. Preparation before classroom intervention	Analysis of: Statistical and probabilistic content; Official curriculum; Students' ideas; Models of intervention
Experimental moment. During classroom intervention	Activity: Plan their own scenario; Implement and report on classroom intervention
Reflection and Assessment moment. After classroom intervention	Reflect and assess: Statistical and probabilistic content developed; Students' learning outcomes; Classroom dynamics

reflections on their learning (EarlyStatistics Consortium, 2008; Meletiou-Mavrotheris, Paparistodemou, Mavrotheris, & Stav, 2008b).

The EarlyStatistics project design (Meletiou-Mavrotheris et al., 2008a) embraced characteristics of effective learning environments (National Research Council, 2000): Learning-centred, knowledge-centred, assessment-centred, and community-centred. It developed key ideas of statistical problem-solving using the GAISE framework (Franklin et al., 2005) through increasingly sophisticated levels of problem-posing, data collection, data analysis, and interpretation, with a focus throughout on variability concepts. Statistical tools such as *Fathom* (Finzer, 2005) and *TinkerPlots* were central to the work. To support teachers towards expertise with statistical investigations, project modules provided teachers with exemplars, classroom activities, and reflections (Table 33.1). These activities occurred through interactions with an online community of practice.

With an emphasis on reflection and teacher community embedded in experiences that build teachers' understanding and teaching repertoire of key ideas in statistics, the EarlyStatistics project highlighted the diverse ways that teachers adapt to teaching statistical investigations. For example, in their reflections, teachers focused on the significance of problem-solving in statistics, the difficulties of developing well-chosen statistical activities, the importance of creating a classroom environment that engaged students, or the challenges of envisioning and implementing statistical investigations (Azcárate, Serrado, Cardeñoso, Meletiou-Mavrotheris, & Paparistodemou, 2008).

5.2 Developing Expertise in Teaching Statistical Inquiry

The aim of this study was to understand development of primary teachers' confidence, commitment, and expertise as they gained experience with teaching inquiry in a supportive environment. Throughout the study, teachers designed or modified published units and taught three to four inquiry units per year to their students. Outcomes of the study produced a model (Makar, 2008) to describe the diversity of teachers' evolving experiences in teaching statistical inquiry over time, with four phases to describe common patterns in the teachers' classroom focus (Fig. 33.1) that are analysed below.

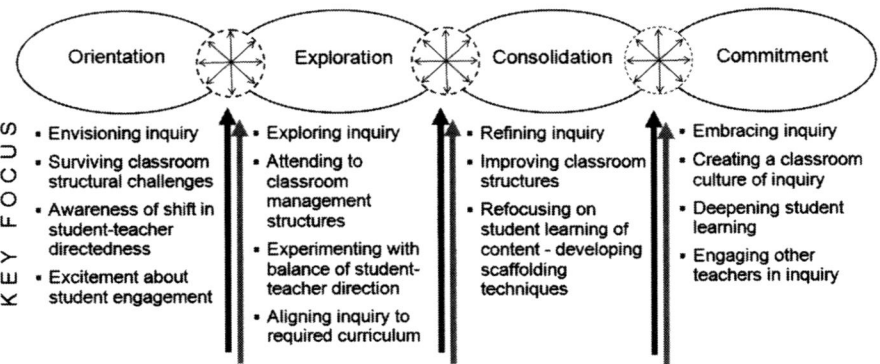

Fig. 33.1 A model of learning to teach statistical inquiry

Orientation cycle. The *orientation cycle* represented teachers' initial experience in teaching statistical inquiry. Being able to envision the inquiry process in a classroom setting was by far the most challenging hurdle for the teachers in this cycle (Makar, 2010). Teachers typically found their first unit quite difficult as they wrestled with unexpected learning outcomes that surfaced. They often blamed themselves for not anticipating outcomes rather than seeing this as the nature of inquiry. During their initial experiences, it was apparent that the teachers' main concerns were in envisioning what statistical inquiry is, coming up with an interesting problem, and engaging with structural aspects of their classroom (e.g., group work, eliciting and supporting student independence). Addressing these challenges became a focus of their teaching in the next cycle.

Exploration cycle. After the teachers experienced what a statistical inquiry looked like in their classrooms, they reacted to problems that had emerged. For example, they could see a range of potential directions in different phases of the PPDAC investigative cycle and responded to changing classroom management issues that arose in each of these cycles. The teachers continued to find logistical aspects challenging, like organising and coordinating group work, and helping students develop independence. Their growing experience helped them modify their teaching styles to begin to address these issues.

Consolidation cycle. By the next stage, the teachers had developed a "big picture" of what was involved in teaching a statistical investigation and worried less about management issues (e.g., classroom behaviours, logistical issues). They found it easier to design and locate rich driving questions to initiate the inquiry process, and in many cases, a new interest was emerging to deepen students' understandings of content by better structuring teaching of more subtle aspects of the inquiry process. Teachers in this phase felt more comfortable now negotiating the balance between student decision-making and providing scaffolding to help their inquiry stay focused. There was improved interest in supporting student learning, such as helping students make connections between the question being posed, the data they collected, and the conclusions being drawn (Hancock, Kaput, & Goldsmith, 1992; Fielding-Wells, 2010). The teachers needed to experience this issue firsthand in their own classrooms to better envision their roles in scaffolding students in this process. The *consolidation cycle* occurred after the teachers had taught with inquiry for a year or more and again emphasises the non-trivial nature of learning to teach statistical investigations.

Commitment cycle. After 2 years, some teachers were clearly committed to including statistical inquiry as a regular part of their teaching, as well as working to help other teachers develop and improve their teaching of inquiry.

6   Implications for Teaching Teachers to Teach Statistical Investigations

The process of learning to teach statistical investigations is complex. Research has been clear in the need to develop teachers' confidence with teaching statistical inquiry, but few opportunities exist for them to gain this critical experience. The two projects presented above were diverse in their approaches, but common elements suggest the following key characteristics are needed in teaching teachers to teach statistical investigations:

1. *Statistical content knowledge.* Although the EarlyStatistics (ES) project was more explicit in developing teachers' statistical content knowledge, both projects provided opportunities for teachers to deepen their understandings of the "big ideas" in statistics – variation, average, sampling, chance, and inference (Watson, 2006, see also Burrill & Biehler, in this book). The ES project developed these understandings through professional reading of statistics education literature while the Developing Expertise in Teaching Statistical Inquiry project developed statistical understandings by addressing concepts as they emerged within the statistical investigations teachers' learned and taught over time.

2. *Engaging in statistical investigations as learners.* Opportunities to experience statistical investigations as learners provide teachers with deeper understandings of complex statistical processes, such as the interrogating of data, modes of thinking, dispositions required, uncertainties and ambiguities encountered, and multiple interpretations and decisions made in each phase of a statistical investigation. These are often new experiences for teachers who are accustomed to mathematical structures and procedures that are more deterministic and predictable in their outcomes (see Gattuso & Ottaviani, in this book).

3. *Learning embedded in teachers' classrooms.* A key success of these projects was their ability to situate teachers' learning to teach statistical investigations within their own classrooms. The EarlyStatistics project supported teachers' classroom experiences remotely, yet engaged their classroom experiences as central to their learning to teach statistical investigations. They did not just *read* about statistical investigations but also *implemented* a statistical investigation and *reflected*

on their teaching. The Developing Expertise in Teaching Statistical Inquiry project partnered teachers and researchers within the teachers' classrooms over a number of years. Researchers played the role of a peer mentor in supporting teachers' development. These projects both connected teachers' learning to their own schools and maximised opportunities for teachers to transfer their learning to their classroom practices.

4. *Collaboration.* Teacher communities were key contributors to the success of both projects. By engaging teachers in collaboration with their peers and university researchers, the projects supported teacher professionalism and explicitly valued teachers' classroom expertise. The validation, collegiality, sharing of resources and experiences, and accountability as part of a learning community supported teachers in addressing challenges they encountered, particularly in their initial experiences teaching statistical investigations.

5. *Reflection.* Both projects provided teachers with time and opportunities for reflection on their learning to teach statistical investigations. Reflection is a powerful yet under-utilised tool for deepening learners' knowledge and understandings. In the case of statistical investigations, these reflections – both individual and communal – allowed teachers to recognise and attend to key contributors to their learning and improved the potential that they would apply these understandings to their students' learning.

6. *Long-term support and resources.* Finally, these projects both highlighted the importance of providing ongoing support and exemplary resources as teachers develop proficiency in teaching statistical investigations. This requires a shift from more traditional modes of teacher learning through workshops or coursework.

Although there were also significant differences in the way that these two projects were conducted, these common elements point to the need to be more conscious of the complexities in teachers' learning to teach statistical investigations. If we acknowledge and come to recognise the changing needs of teachers as they develop their expertise in teaching statistical investigations over time (Makar, 2008), new windows of opportunity for improving research and practice in this area will be provided.

Acknowledgement This research and writing was funded by the Australian Council for Research (LP0776703), Education Queensland, and The University of Queensland.

References

Allmond, S., & Makar, K. (2010). Developing primary students' ability to pose questions in statistical investigations. In C. Reading (Ed.), *Proceedings of the Eighth International Conference on Teaching Statistics.* Ljubljana, Slovenia: International Statistical Institute and International Association for Statistical Education. Online: www.stat.auckland.ac.nz/~iase/publications

Anderson, J. (2005). Implementing problem solving in mathematics classrooms: What support do teachers want? In A. D. P. Clarkson, D. Gronn, M. Horne, A. McDonough, R. Pierce, & A. Roche (Eds.), *Proceedings of Annual Conference of the Mathematics Education Research Group of Australasia. Building Connections: Theory, Research and Practice* (Vol. 1, pp. 89–96). Melbourne, Australia: Mathematics Education Research Group of Australasia.

Anderson, R. D. (2002). Reforming science teaching: What research says about inquiry. *Journal of Science Teacher Education, 13*(1), 1–12.

Arnold, P. (2008a, July). *What about the P in the PPDAC cycle? An initial look at posing questions for statistical investigation.* Paper presented at the Eleventh International Congress of Mathematics Education (ICME-11), Monterrey, Mexico.

Arnold, P. (2008b). Developing new statistical content knowledge with secondary school mathematics teachers. In C. Batanero, G. Burrill, C. Reading, & A. Rossman (2008).

Australian Curriculum, Assessment and Reporting Authority. (2009). *The shape of the Australian curriculum: Mathematics.* Barton, ACT: Commonwealth of Australia.

Azcárate, P., Serrado, A., Cardeñoso, J., Meletiou-Mavrotheris, M., & Paparistodemou, E. (2008). An online professional environment to improve the teaching of statistics. In C. Batanero, G. Burill, C. Reading, & A. Rossman (2008).

Batanero, C., Burrill, G., Reading, C., & Rossman, A. (Eds.) (2008). *Joint ICMI/IASE Study: Teaching Statistics in School Mathematics. Challenges for Teaching and Teacher Education. Proceedings of the ICMI Study 18 and 2008 IASE Round Table Conference.* Monterrey, Mexico: International Commission on Mathematical Instruction and International Association for Statistical Education. Online: www.stat.auckland.ac.nz/~iase/publications

Borasi, R. (1992). *Learning mathematics through inquiry.* Portsmouth, NH: Heinemann.

Cobb, P. (1999). Individual and collective mathematical development: The case of statistical data analysis. *Mathematical Thinking and Learning, 1*(1), 5–43.

Crawford, B. A. (2000). Embracing the essence of inquiry: New roles for science teachers. *Journal of Research in Science Teaching, 37*(9), 916–937.

Crawford, B. A., Krajcik, J. S., & Marx, R. W. (1998). Elements of a community of learners in a middle school science classroom. *Science Education, 83,* 701–723.

Davies, N. (2007). Developments in promoting the improvement of statistical education. *Proceedings of the International Statistical Institute 56th Session.* Lisbon, Portugal: International Statistical Institute. Online: www.stat.auckland.ac.nz/~iase/publications

EarlyStatistics Consortium. (2008). *EarlyStatistics: A teacher professional development course in statistics education.* Thessaloniki: Author.

Fielding-Wells, J. (2010). Linking problems, conclusions and evidence: Primary students' early experiences of planning statistical investigations. In C. Reading (Ed.), *Proceedings of the Eighth International Conference on Teaching Statistics.* Ljubljana, Slovenia: International Statistical Institute and International Association for Statistical Education. Online: www.stat. auckland.ac.nz/~iase/publications

Finzer, W. (2005). *Fathom dynamic data software.* Emeryville, CA: Key Curriculum Press.

Franklin, C., Kader, G., Mewborn, D., Moreno, J., Peck, R., Perry, R., & Scheaffer, R. (2005). *Guidelines for assessment and instruction in statistics education (GAISE) report: A pre K-12 curriculum framework.* Alexandria, VA: American Statistical Association. Online: www. amstat.org/Education/gaise/

Gal, I. (2002). Adults' statistical literacy: Meanings, components, responsibilities. *International Statistical Review, 70*(1), 1–51.

Groves, S., & Doig, B. (2004). Progressive discourse in mathematics classes: The task of the teacher. In M. J. Hoines & A. B. Fluglestad (Eds.), *Proceedings of the 28th Conference of the International Group for the Psychology of Mathematics Education* (Vol. II, pp. 495–502). Bergen, Norway: Bergen University College.

Guskey, T. R. (2002). Professional development and teacher change. *Teachers and Teaching: Theory and Practice, 8*(3/4), 381–391.

Hancock, C., Kaput, J., & Goldsmith, L. (1992). Authentic inquiry into data: Critical barriers to classroom implementation. *Educational Psychologist, 27*(3), 337–364.

Hills, T. (2007). Is constructivism risky? Social anxiety, classroom participation, competitive game play and constructivist preferences in teacher development. *Teacher Development, 11*(3), 335–352.

King, P. M., & Kitchener, K. S. (1994). *Developing reflective judgment: Understanding and promoting intellectual growth and critical thinking in adolescents and adults.* San Francisco: Jossey-Bass.

Konold, C., & Miller, C. D. (2005). *TinkerPlots: Dynamic data exploration.* Emeryville, CA: Key Curriculum Press.

Krajcik, J., Blumenfeld, P. C., Marx, R. W., Bass, K. M., Fredricks, J., & Soloway, E. (1998). Inquiry in project-based science classrooms: Initial attempts by middle school students. *Journal of the Learning Sciences, 7*(3/4), 313–350.

Lavigne, N. C., & Lajoie, S. P. (2007). Statistical reasoning of middle school children engaged in survey inquiry. *Contemporary Educational Psychology, 32*(4), 630–666.

Makar, K. (2008). *A model of learning to teach statistical inquiry.* In C. Batanero, G. Burrill, C. Reading, & A. Rossman (2008).

Makar, K. (2010). Teaching primary teachers to teach statistical inquiry: The uniqueness of initial experiences. In C. Reading (Ed.), *Proceedings of the Eighth International Conference on Teaching Statistics.* Ljubljana, Slovenia: International Statistical Institute and International Association for Statistical Education. Online: www.stat.auckland.ac.nz/~iase/publications

Makar, K., & Confrey, J. (2002). Comparing two distributions: Investigating secondary teachers' statistical thinking. In B. Phillips (Ed.), *Proceedings of the Sixth International Conference on Teaching Statistics.* Cape Town, South Africa: International Statistical Institute and International Association for Statistics Education. Online: www.stat.auckland.ac.nz/~iase/publications

Marx, R. W., Blumenfeld, P. C., Krajcik, J. S., Blunk, M., Crawford, B., Kelly, B., et al. (1994). Enacting project-based science: Experiences of four middle grade teachers. *The Elementary School Journal, 94*(5), 517–538.

Meletiou-Mavrotheris, M., Paparistodemou, E., Mavrotheris, E., Azcárate, P., Serrado, A., & Cardeñoso, J. (2008a). Teachers' professional development in statistics: The EarlyStatistics European project. In C. Batanero, G. Burill, C. Reading, & A. Rossman (2008).

Meletiou-Mavrotheris, M., Paparistodemou, E., Mavrotheris, E., & Stav, J. B. (2008b). *EarlyStatistics: Pedagogical framework.* Thessaloniki: EarlyStatistics Consortium.

Moore, D. (1997). New pedagogy and new content: The case of statistics. *International Statistical Review, 65*(2), 123–165.

National Council of Teachers of Mathematics. (2000). *Principles and standards for school mathematics.* Reston, VA: Author.

National Research Council. (2000). *How people learn: Brain, mind, experience, and school.* Washington, DC: National Academy Press.

New Zealand Ministry of Education. (2007). *The New Zealand curriculum.* Wellington: Learning Media.

Sorto, M. A. (2006). Identifying content knowledge for teaching statistics. In A. Rossman & B. Chance (Eds.), *Proceedings of the Seventh International Conference on Teaching Statistics.* Salvador, Brazil. International Statistical Institute and International Association for Statistical Education. Online: www.stat.auckland.ac.nz/~iase/publications

Walker, A., & Leary, H. (2008). A problem based learning meta analysis: Differences across problem types, implementation types, disciplines, and assessment levels. *Interdisciplinary Journal of Problem-based Learning, 3*(1). Online: docs.lib.purdue.edu/ijpbl/

Watson, J. (2006). *Statistical literacy at school: Growth and goals.* Mahwah, NJ: Lawrence Erlbaum.

Wild, C. J. (1994). Embracing the 'wider view' of statistics. *American Statistician, 48*(2), 163–171.

Wild, C. J., & Pfannkuch, M. (1999). Statistical thinking in empirical enquiry. *International Statistical Review, 67*(3), 223–265.

Chapter 34
Characterising and Developing Teachers' Knowledge for Teaching Statistics with Technology

Hollylynne S. Lee and Karen F. Hollebrands

Abstract Developing the pedagogical expertise needed to effectively engage students in learning statistics with technology requires teachers to have a depth of knowledge about statistics, technological tools for exploring statistical ideas, and of pedagogical issues related to teaching and learning statistics with technology. In this chapter, a framework for a specialised knowledge that is called technological pedagogical statistical knowledge (TPSK) is presented and examples of how aspects of this type of knowledge may assist a teacher are provided. Implications for training teachers are described.

1 Introduction

Many international organisations and curricula promote the use of technology in teaching and learning statistics. The GAISE project (Franklin et al., 2005) and the 2008 Joint ICMI/IASE Study Conference call for teachers to have a deeper understanding of statistics and an ability to use technology tools. Although technologies are becoming more prevalent in classrooms, teachers' abilities to use these tools effectively in lessons depends on many factors, including their: (a) statistical knowledge, (b) understanding of how to use technology to explore statistical ideas, and (c) understanding of pedagogical issues related to teaching statistics. These factors impact teachers' decisions, and will ultimately affect whether the use of technology will enhance or hinder students' learning of statistics. This chapter provides a framework that integrates these three factors, a discussion of issues to consider in developing knowledge for teachers of statistics, and examples of teacher education efforts that appear promising.

H.S. Lee (✉) and K.F. Hollebrands
North Carolina State University, 502D Poe Hall Box 7801, Raleigh, NC, USA
e-mail: hollylynne@ncsu.edu; Karen_hollebrands@ncsu.edu

C. Batanero, G. Burrill, and C. Reading (eds.), *Teaching Statistics in School*
Mathematics-Challenges for Teaching and Teacher Education: A Joint ICMI/IASE Study,
DOI 10.1007/978-94-007-1131-0_34, © Springer Science+Business Media B.V. 2011

1.1   Teachers' Pedagogical Content Knowledge

Teacher education and research has been greatly influenced by Shulman's (1986) pedagogical content knowledge (PCK) as an integration of teachers' content understandings with knowledge needed in teaching. For example, Simon (1995) described important components of a mathematics teaching cycle that include a teacher's knowledge of content, activities and representations, students' learning of particular content, and a teacher's hypotheses about students' knowledge.

Recently, Hill, Ball, and Schilling (2008) have extended PCK to describe mathematical knowledge needed for teaching, which includes constructs such as common content knowledge (content considered to be commonly used by many) and specialised content knowledge (content knowledge needed in the practice of teaching). Groth (2007) used these two constructs to hypothesise what statistical knowledge might be needed for teaching. In particular, he drew upon differences between mathematics and statistics as fields of study and ways of thinking (delMas, 2004; Rossman, Chance, & Medina, 2006; see also Gattuso & Ottaviani, in this book) and gave examples of specialised knowledge needed that was mathematical and non-mathematical. For example, identifying difficulties students may have in constructing algorithms for generating random data is mathematical in nature, but deciding if data collection should include random sampling or random assignment is a non-mathematical task (see Godino et al., and Callingham & Watson, in this book for other analyses of PCK components).

1.2   Teachers' Technological Pedagogical Content Knowledge

The teaching and learning of mathematics and statistics has been greatly influenced by technology (e.g., Ben-Zvi, 2000; Chance, Ben-Zvi, Garfield, & Medina, 2007; Heid & Blume, 2008; Pratt, Connor, & Hunt, this book). Others have also pondered how technology influences teaching and learning and have described technological pedagogical content knowledge (TPCK, see Fig. 34.1) as a type of knowledge needed to effectively use technology to teach specific content (Koehler & Mishra, 2005; Niess, 2005, 2006; American Association of Colleges for Teacher Education Committee on Innovation and Technology, 2008; Mishra & Koehler, 2008).

Niess (2005, 2006) describes four aspects that comprise teachers' TPCK that include a focus on understanding: (a) how to teach a subject with technology, (b) instructional strategies and representations, (c) students' thinking with technology, and (d) curriculum materials that integrate technology. Niess et al. (2009) have since recast these four aspects as being specific to mathematics and have proposed standards and indicators for mathematics teachers' TPCK. In what follows, the notion of TPCK specifically for teachers of statistics is described.

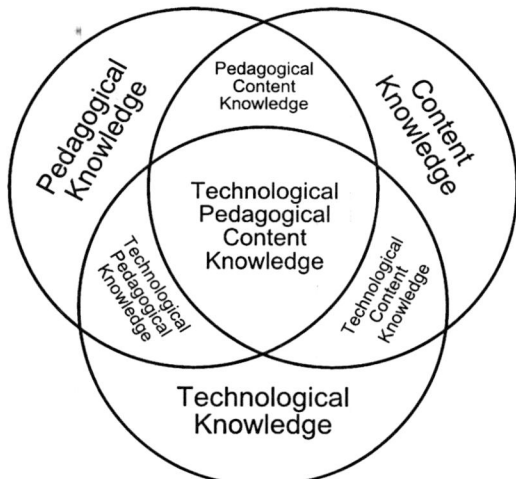

Fig. 34.1 Components of technological pedagogical content knowledge

2   A TPCK Framework for Teachers of Statistics

Rather than describing each of the seven distinct types of knowledge emphasised in the TPCK framework (Fig. 34.1), the framework described here focuses on three important types of knowledge for specifying TPCK for the teaching of statistics. Every aspect of this framework is focused on knowledge needed in the practice of teaching, and thus some pedagogical component is blended in each aspect of the framework, albeit not always an explicit focus. The development of teachers' technological pedagogical statistical knowledge (TPSK) is conceptualised as three layered circles with a foundation focused on teachers' statistical knowledge (Fig. 34.2).

Thus, a teacher's statistical knowledge needed to engage in statistical thinking is the largest of our "sets." This illustrates that a teacher's statistical knowledge and thinking abilities are paramount for their knowledge of anything related to pedagogy or the use of technology in teaching statistics. The inner-most layer represents elements of TPSK and is a subset of the sets in the outer two circles, meaning TPSK is founded on and developed with teachers' knowledge in the outer two sets of technological statistical knowledge (TSK) and statistical knowledge (SK). In addition, developing TSK and SK is essential to, but not sufficient for, teachers having the specialised TPSK.

2.1   Statistical Knowledge and Thinking as Foundational

For many teachers, engaging in statistical thinking is a different process than typically used in teaching and learning mathematics (delMas, 2004; Pfannkuch & Ben-Zvi, this book; Rossman et al., 2006). Thus, it is important to engage teachers as active

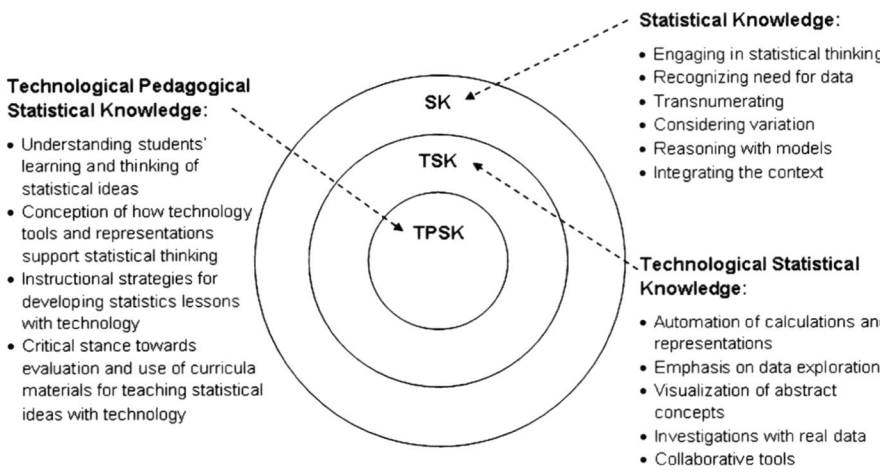

Fig. 34.2 Framework for teachers' technological pedagogical statistical knowledge

learners and doers of statistical practices and explicitly develop an understanding of and disposition towards statistical thinking as "an understanding of why and how statistical investigations are conducted and the 'big ideas' that underlie statistical investigations" (Ben-Zvi & Garfield, 2004, p. 7). The remaining five aspects are adapted from Wild and Pfannkuch (1999) and build off Pfannkuch's (2008) description of how to engage teachers in the "game of statistics" (p. 1).

To engage in statistical thinking, teachers should recognise the need for properly collected data to examine situations and make decisions, rather than relying on personal experiences or anecdotal evidence. Teachers should then be able to engage in transnumeration (Wild & Pfannkuch, 1999) as a process of transforming a representation between a real system and a statistical system with the intention of engendering understanding (Pfannkuch & Wild, 2004). Thus, teachers should be able to collect measures, represent them meaningfully with graphs and computed statistical measures, and translate their interpretations back to the context.

Statistics is founded on the fact that variations exist in phenomena and that one must use non-deterministic models and explanations to describe such phenomena with attention to variation (e.g., Moore, 1997; Wild & Pfannkuch, 1999; Ben-Zvi & Garfield, 2004; Shaughnessy, 2006). As statistical thinkers, teachers should notice variation in contexts and use strategies to reduce or eliminate sources of variation in data collection, where possible, and use models that take into account other sources of variation when making predictions or explanations (Pfannkuch, 2008).

With a focus on statistical models, teachers should be able to focus more on aggregate-based reasoning, rather than data as individuals (Konold & Higgins, 2003). Considering data in the aggregate can allow one to characterise group propensities that can include attention to centres, spread, outliers, clusters, intervals, or residuals. In accord with the notion of active graphing used by Ainley, Nardi, and Pratt (2000), teachers should not consider a statistical model such as a graph or a measure of centre as merely indicating a result of a statistical analysis, but rather as a means to reason with to tell a bigger story of the phenomenon under study.

It is fundamental to integrate a teacher's knowledge of a context in a statistical investigation. The context of data, and the reason for undertaking a statistical investigation, should always influence a teacher's thinking, choice of strategies and methods, representations, statistical summaries used, and interpretations made.

2.2 Technological Statistical Knowledge

Tools such as graphing calculators, spreadsheets, and statistical packages such as *SPSS*, *SAS*, or *Minitab*, have become commonplace in many tertiary and some secondary contexts. Newer educational technologies such as *Fathom, TinkerPlots*, and *Probability Explorer* are available and allow for dynamic control over data – meaning that as data changes, representations of that data dynamically update. For example, in *Probability Explorer* and *Fathom*, as data is randomly generated, graphs can be simultaneously "building" so that variability in a distribution can be analysed as sample size increases. Further, several tools allow users to drag data points within a graph and notice the effect on tabular representations and measures.

The availability of technologies for today's work of doing and teaching statistics calls for attention to what specialised knowledge teachers need about technology that is particular to statistics (see Pratt, Connor, & Hunt, in this book). Building from the work of Pea (1987) and Ben-Zvi (2000) provides a useful lens on ways to amplify or reorganise one's work with technology. According to Pea, technology tools are typically used in two different ways: to amplify our abilities to solve problems or reorganise the way we think about problems and their solutions. The notion of amplifier and reorganiser is used as a lens to consider the five aspects of TSK adapted from Chance et al. (2007).

The idea of an *amplifier* is that the tool expedites a process that could be completed without its use. For example, technology tools can be used to automate many activities such as, quickly organising data, generating lists of pseudorandom numbers, computing measures, and generating graphs. By automating the tasks of computing or generating graphs, technology affords an opportunity to focus on conceptual understanding and more time to engage in exploratory data analysis (Konold & Higgins, 2003).

Automation in technology also facilitates a person's capability to visualise abstract concepts and serve as a *reorganiser*, such as taking advantage of dynamic dragging capabilities to illustrate the effect of an outlier on a measure of centre in a univariate distribution. Through dynamic features of dragging, linking of multiple representations, and overlaying measures on graphs, technology can be used in ways that extends what we may be able to do without technology to help reorganise and change a student's or teacher's statistical conceptions. For example, overlaying statistical measures such as means and regression line on a graphical representation can help change the way teachers and students conceptualise these measures in relation to a bivariate distribution, particularly since the statistical measures update as data is changed by dragging points in the graph. This visualisation is not possible without technology and can provide a way of reorganising one's conceptions of bivariate distributions.

Technology can be used to view and design simulations that can enhance the study of random processes and statistical concepts such as sampling distributions (Chance et al., 2007). The flexibility of many simulation tools allow for: (a) algorithms and models to be used to input the properties of a theoretical distribution that would control the pseudorandom number generation, (b) controlling parameters such as sample size, and (c) displaying graphs generated in real time. A teacher who uses technology in their own statistical investigations will have first-hand knowledge of the power of using simulations as a pedagogical tool.

Technology also facilitates the use of large messy data sets gathered and accessible through the Internet (see Hall, in this book). Longstanding projects such as Census at School and newer projects such as Experiments at School (Connor, Davies, & Holmes, 2006) demonstrate an advantage of the Internet to gather and access data of interest to students. Knowing how to gather real data from the Internet and how to transform it into usable data in a particular piece of software or downloading it into a graphing calculator is a useful skill. In addition, teachers should develop an ethical disposition concerning the use of public data, citing sources, and being wary of data that has already been transformed by others with particular agendas. At the same time, teachers also need to consider characteristics of data sets that can be used to bring different statistical ideas to the fore. For example, data sets with a skewed tendency are typically good for investigating the usefulness of different measures of centre.

Chance et al. (2007) support the use of technologies such as discussion forums, Wikis, interactive whiteboards, and self-assessment software to promote collaboration and student involvement. While these tools can help in course management and engage students in learning and assessment, we also support a focus on collaborative tools for data collection, analysis, and visualisation. Examples of collaborative tools to collect or simulate data include networked graphing calculators (e.g., TI-Navigator™ systems), Experiments at School (www.experimentsatschool.ntu.ac.uk) GoogleDocs spreadsheets (docs.google.com), and networked computing and simulations tools such as HubNet, which uses *Netlogo* (www.ccl.northwestern.edu/netlogo/). Such technologies can be used to have individuals contribute data from local simulations to be aggregated as a class. Collaborative tools can promote a community approach to generating and analysing data that can foster both small-group investigations at a local machine and whole-group discussions to consider the phenomena in an aggregate.

2.3   Technological Pedagogical Statistical Knowledge

The ultimate goal in the preparation of teachers of statistics is to develop a specialised subset of knowledge for teachers representing TPSK (Lee & Hollebrands, 2008a). This knowledge encompasses TSK and SK (see Fig. 34.2). While pedagogical issues and implications may be implicit in aspects of teachers' SK or their TSK, pedagogy comes to the fore when considering the particular subset of TPSK.

TPSK can allow teachers to consider how students think and reason about statistics with and without technology. This implies that they have the specialised content knowledge that Groth (2007) hypothesised as particular to statistics teachers, and that they are familiar with common ways that students may approach statistical tasks. For example, teachers should know that students often initially consider data representing a characteristic of an individual (e.g., Sally's height), and have difficulty viewing data as an aggregate where they consider the entire distribution (Konold & Higgins, 2003). They should also know how technology can promote different reasoning that may facilitate a transition to aggregate-based thinking. For example, when examining distributions graphically, students can characterise the data with such ideas as "bins" (intervals) (Rubin & Hammerman, 2006) and a "modal clump", that is, a small range of data that contains many data points within a data distribution (Konold & Higgins, 2003), rather than initially focusing on computing statistical measures.

Teachers can also use TPSK to consider how technology can facilitate and support students' statistical thinking, and in essence become designers of a conceptual space for students to learn powerful ideas (Pratt, 2008). For example, teachers would know how to use *a dynamic statistical tool* to highlight a region of data in a distribution, compute the number and proportion of data within the region, and use this process to support students' natural tendencies to describe a distribution's centre using a "modal clump", to complement a formal measure of centre. This aspect of TPSK necessarily encompasses components of teachers' TSK (e.g., automation of graphs, data exploration, visualising concepts) and SK (e.g., transnumeration, consideration of variation).

Teachers are continually planning lessons for students and evaluating and choosing curricula materials for use in their classroom. In these contexts, teachers are again working as designers (Pratt, 2008) and using TPSK in their daily work. They should be able to draw upon elements of their TSK and SK that facilitates an appropriate use of technology that can positively affect students' learning of statistics. For example, consider a teacher in the midst of a lesson on least squares regression where she ascertains that her students are having difficulty understanding the concept of a residual. She uses the dynamic dragging capabilities in *Fathom* to provide a demonstration of the residual plot and how it relates to the position of a moveable line overlaid on a scatter plot (Fig. 34.3). The teacher uses vertical translations of the moveable line to illustrate how the residual plot responds if she places the moveable line entirely above all data points. She anticipates this may help students understand why the corresponding residuals would have a negative numeric value.

3   Developing Teachers' TPSK

There is an increasing trend for teacher preparation programmes to include a focus on the use of technology for teaching mathematics (e.g., Powers & Blubaugh, 2005) and research that suggests that mathematics teachers may struggle in learning how to use technology in their teaching (Zbeik & Hollebrands, 2008; Niess et al., 2009). Both Niess et al., and Zbiek and Hollebrands propose a model of how mathematics teachers may develop as technology-using teachers, with early stages including

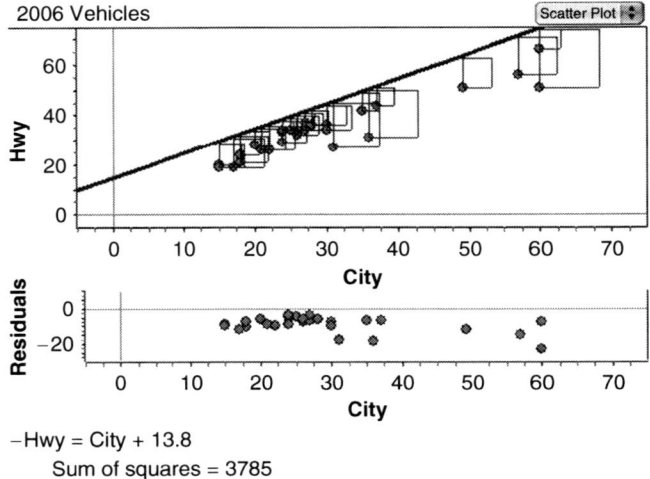

−Hwy = City + 13.8
Sum of squares = 3785

Fig. 34.3 Visualising residuals in *Fathom*

playing with the technology and using it for their personal use for doing mathematics. Latter stages include integration of tools and a focus on teachers' assessment and evaluation of how technology may be impacting students' learning.

The responsibility for developing teachers' TPSK does not only lie with teacher educators. Mathematicians and statisticians who teach statistical content courses are very influential, because this is where teachers can develop their SK and TSK as learners of statistics. Thus, courses and textbooks that include the use of educational software such as *Fathom*, rather than only statistical packages such as *MiniTab*, are needed (e.g., Rossman, Chance, & Lock, 2001). Materials for use in pedagogy-focused courses can also be helpful.

To help develop teachers' TPSK, materials should engage teachers in statistical thinking as doers of statistics with a variety of technology tools and then provide opportunities for teachers' to reflect on their own learning and to consider pedagogical issues. The work of Graham (2006), Ben-Zvi (2008), Lee and Hollebrands (2008b), Pratt (2008), and Lee, Hollebrands, and Wilson (2010) have similar aims and focus on the use of technology and pedagogical considerations in teaching and learning statistics.

For example, Table 34.1 provides two tasks that could be used to develop teachers' TPSK. Both tasks are posed to teachers after they have explored a probability context by designing and using simulations in their own learning.

Because many teachers struggle with focusing on students' thinking, the use of artefacts of practice, such as video cases and students' work, can be helpful to engage teachers in discussions about pedagogical decisions that may affect students' understanding (e.g., Groth, 2008; Wilson, 2008). The use of these artefacts can help teachers make sense of students' work and develop an understanding of how students reason about statistical ideas (Hollebrands, Wilson, & Lee, 2007; Groth, 2008; Wilson, Lee, & Hollebrands, 2011).

Table 34.1 Tasks to develop TPSK

Example 1. (Graham, 2006, p. 183)

a. How could you use the simulation possibilities of a graphics calculator to help learners test the statement "a six is hard to get?" Hint: You need to generate a number of dice rolls and see whether the outcome "6" is under-represented.

b. Some people might claim that the calculator provides a more satisfactory microworld than a spreadsheet for investigations like these. See what you think after trying this task.

Example 2. (Lee, Hollebrands, & Wilson, 2010, p. 114)

Suppose a university gives away token spirit gifts to all incoming freshmen. They get to randomly choose one of three cards. Each card displays a different gift: keychain, window decal for a car, t-shirt with college logo.

a. Explain how you would help students use the graphing calculator to simulate this context. Explicitly describe what the commands represent and how the students should interpret the results.

b. Suppose that after conducting the above simulation for 30 freshmen choosing a card, a student gets a result that a t-shirt was chosen 20% of the time. The student is surprised and claims that this proportion is too low from the expected 33.3%. What might this student be misunderstanding? What are some questions you could pose and further simulations you could suggest that might help this student?

4 Conclusion

Preparing teachers of statistics to use technology appropriately in their classrooms is a difficult and important task. While statistical knowledge is foundational in the TPSK framework, it is also important to engage teachers in opportunities to develop technological statistical knowledge and TPSK. This implies that the community of those involved in educating teachers of statistics (e.g., mathematics educators, statistics educators, mathematicians, statisticians) should also have opportunities to develop their own TPSK and join together in efforts to create our next generation of teachers of statistics.

Acknowledgements The preparation of this manuscript was supported by the National Science Foundation (DUE 04-42319 and DUE 08-17253; Online: ptmt.fi.ncsu.edu).

References

Ainley, J., Nardi, E., & Pratt, D. (2000). Towards the construction of meanings for trend in active graphing. *The International Journal of Computers for Mathematical Learning, 5*(2), 85–114.

American Association of Colleges for Teacher Education Committee on Innovation and Technology. (2008). *Handbook of technological pedagogical content knowledge (TPCK) for educators*. New York: Routledge.

Batanero, C., Burrill, G., Reading, C., & Rossman, A. (Eds.). (2008). *Joint ICMI/IASE Study: Teaching Statistics in School Mathematics. Challenges for Teaching and Teacher Education. Proceedings of the ICMI Study 18 and 2008 IASE Round Table Conference.* Monterrey, Mexico: International Commission on Mathematical Instruction and International Association for Statistical Education. Online: www.stat.auckland.ac.nz/~iase/publications

Ben-Zvi, D. (2000). Toward understanding the role of technological tools in statistical learning. *Mathematical Thinking and Learning, 2*(1 & 2), 127–155.

Ben-Zvi, D. (2008). Partners in innovation: Helping teachers to integrate technology in the teaching of statistics. In C. Batanero, G. Burrill, C. Reading, & A. Rossman (2008).

Ben-Zvi, D., & Garfield, J. (2004). Statistical literacy, reasoning, and thinking: Goals, definitions and challenges. In D. Ben-Zvi & J. Garfield (Eds.), *The challenge of developing statistical literacy, reasoning, and thinking* (pp. 79–95). Dordrecht, The Netherlands: Kluwer.

Chance, B., Ben-Zvi, D., Garfield, J., & Medina, E. (2007). The role of technology in improving student learning of statistics. *Technology Innovations in Statistics Education 1*(1). Online: repositories.cdlib.org/uclastat/cts/tise/

Connor, D., Davies, N., & Holmes, P. (2006). Using real data and technology to develop statistical thinking. In G. Burrill (Ed.), *Thinking and reasoning with data and chance: 68th NCTM Yearbook (2006)* (pp. 185–194). Reston, VA: National Council of Teachers of Mathematics.

delMas, R. C. (2004). A comparison of mathematical and statistical reasoning. In D. Ben-Zvi & J. Garfield (Eds.), *The challenge of developing statistical literacy, reasoning, and thinking* (pp. 79–95). Dordrecht, The Netherlands: Kluwer.

Franklin, C., Kader, G., Mewborn, D., Moreno, J., Peck, R., Perry, M., & Scheaffer, R. (2005). *Guidelines for assessment and instruction in statistics education (GAISE) report: A pre-k-12 curriculum framework.* Alexandria, VA: American Statistical Association. Online: www.amstat. org/Education/gaise/

Graham, A. (2006). *Developing thinking in statistics.* Thousand Oaks, CA: Sage Publications.

Groth, R. E. (2007). Toward a conceptualization of statistical knowledge for teaching. *Journal for Research in Mathematics Education, 38*(5), 427–437.

Groth, R. E. (2008). Navigating layers of uncertainty in teaching statistics through case discussion. In C. Batanero, G. Burrill, C. Reading, & A. Rossman (2008).

Heid, M. K., & Blume, G. (Eds.). (2008). *Research on technology and the teaching and learning of mathematics* (Vol. 1). Charlotte, NC: Information Age Publishers.

Hill, H., Ball, D., & Schilling, S. (2008). Unpacking pedagogical content knowledge: Conceptualizing and measuring teachers' topic-specific knowledge of students. *Journal for Research in Mathematics Education, 39,* 372–400.

Hollebrands, K. F, Wilson, P. H., & Lee, H. S. (2007). Prospective teachers use of a videocase to examine students' work when solving mathematical tasks using technology. In T. Lamberg & L. R. Wiest, (Eds.), *Proceedings of the 29th Annual Meeting of the North American Chapter of the International Group for the Psychology of Mathematics Education* (pp. 200–206). Stateline (Lake Tahoe), Reno, NV: University of Nevada.

Koehler, M. J., & Mishra, P. (2005). What happens when teachers design educational technology? The development of technological pedagogical content knowledge. *Journal of Educational Computing Research, 32*(2), 131–152.

Konold, C., & Higgins, T. (2003). Reasoning about data. In J. Kilpatrick, W. G. Martin, & D. Schifter (Eds.), *A research companion to principles and standards for school mathematics* (pp. 196–215). Reston, VA: National Council of Teachers of Mathematics.

Lee, H. S., & Hollebrands, K. F. (2008a). Preparing to teach data analysis and probability with technology. In C. Batanero, G. Burrill, C. Reading, & A. Rossman (2008).

Lee, H. S., & Hollebrands, K. F. (2008b). Preparing to teach mathematics with technology: An integrated approach to developing technological pedagogical content knowledge. *Contemporary Issues in Technology and Teacher Education 8*(4). Online: www.citejournal.org

Lee, H. S., Hollebrands, K. F., & Wilson, P. H. (2010). *Preparing to teach mathematics with technology: An integrated approach to data analysis and probability* (1st ed.). Dubuque, IA: Kendall Hunt.

Mishra, P., & Koehler, M. J. (2008, March). *Introducing technological pedagogical content knowledge.* Paper presented the Annual Meeting of the American Educational Research Association, New York.

Moore, D. S. (1997). New pedagogies and new content: The case of statistics. *International Statistical Review, 65*, 123–165.

Niess, M. L. (2005). Preparing teachers to teach science and mathematics with technology: Developing a technology pedagogical content knowledge. *Teaching and Teacher Education, 21*, 509–523.

Niess, M. L. (2006). Guest editorial: Preparing teachers to teach mathematics with technology. *Contemporary Issues in Technology and Teacher Education, 6*(2). Online: www.citejournal.org

Niess, M. L., Ronau, R. N., Shafer, K. G., Driskell, S. O., Harper S. R., Johnston, C., Browning, C., Özgün-Koca, S. A., & Kersaint, G. (2009). Mathematics teacher TPACK standards and development model. *Contemporary Issues in Technology and Teacher Education 9*(1). Online: www.citejournal.org

Pea, R. D. (1987). Cognitive technologies for mathematics education. In A. Schoenfeld (Ed.), *Cognitive science and mathematics education* (pp. 89–122). Hillsdale, NJ: Erlbaum.

Pfannkuch, M. (2008). Training teachers to develop statistical thinking. In C. Batanero, G. Burrill, C. Reading, & A. Rossman (2008).

Pfannkuch, M., & Wild, C. (2004). Towards an understanding of statistical thinking. In D. Ben-Zvi & J. Garfield (Eds.), *The challenge of developing statistical literacy, reasoning, and thinking* (pp. 79–95). Dordrecht, The Netherlands: Kluwer Academic Publishers.

Powers, R., & Blubaugh, W. (2005). Technology in mathematics education: Preparing teachers for the future. *Contemporary Issues in Technology and Teacher Education 5*(3). Online: www.citejournal.org

Pratt, D. (2008). Statistics teachers as designers of conceptual space. In C. Batanero, G. Burrill, C. Reading, & A. Rossman (2008).

Rossman, A., Chance, B., & Lock, R. (2001). *Workshop statistics: Discovery with data and Fathom*. Emeryville, CA: Key College Publishing.

Rossman, A., Chance, B., & Medina, E. (2006). Some important comparisons between statistics and mathematics, and why teachers should care. In G. Burrill (Ed.), *Thinking and reasoning with data and chance: 68th NCTM Yearbook (2006)* (pp. 323–334). Reston, VA: National Council of Teachers of Mathematics.

Rubin, A., & Hammerman, J. K. (2006). Understanding data through new software representations. In G. Burrill (Ed.), *Thinking and reasoning with data and chance: 68th NCTM Yearbook (2006)* (pp. 241–256). Reston, VA: National Council of Teachers of Mathematics.

Shaughnessy, J. M. (2006). Research on students' understanding of some big concepts in statistics. In G. Burrill (Ed.), *Thinking and reasoning with data and chance: 68th NCTM Yearbook (2006)* (pp. 77–98). Reston, VA: National Council of Teachers of Mathematics.

Shulman, L. (1986). Those who understand: Knowledge growth in teaching. *Educational Researcher, 15*(2), 4–14.

Simon, M. A. (1995). Reconstructing mathematics pedagogy from a constructivist perspective. *Journal for Research in Mathematics Education, 26*(2), 114–145.

Wild, C., & Pfannkuch, M. (1999). Statistical thinking in empirical enquiry (with discussion). *International Statistical Review, 67*(3), 223–265.

Wilson, P. H., Lee, H. S., & Hollebrands, K. F. (2011). Understanding prospective mathematics teachers' processes for making sense of students' work with technology. *Journal for Research in Mathematics Education, 42*(1), 42–67.

Wilson, P. S. (2008). Teacher education: A conduit to the classroom. In M. K. Heid & G. W. Blume (Eds.), *Research on technology and the teaching and learning of mathematics* (Cases and perspectives, Vol. 2, pp. 415–426). Charlotte, NC: Information Age Publishing.

Zbeik, R. M., & Hollebrands, K. F. (2008). A research-informed view of the process of incorporating mathematics technology into classroom practice by in-service and prospective teachers. In M. K. Heid & G. W. Blume (Eds.), *Research on technology and the teaching and learning of mathematics* (Research syntheses, Vol. 1, pp. 287–344). Charlotte, NC: Information Age Publishing.

Chapter 35
Preparing Teachers Through Case Analyses

Randall E. Groth and Shihong Xu

Abstract The aim of this chapter is to discuss the usefulness of case analysis as a resource for statistics teacher education. In order to ground the discussion, two illustrations of teacher preparation through case analysis are introduced, one of which describes a case discussion among a group of prospective secondary mathematics teachers in the Mid-Atlantic United States of America and the other a teaching research activity among a group of practising junior high school teachers in Guangzhou, China. The dynamics of case discussions from these two different environments indicated that cases can help participants consider factors in general pedagogical knowledge, content-specific pedagogical knowledge, and content knowledge. Some recommendations for further research and use of case analysis in teacher education are also included.

1 Introduction

The analysis of practice-based cases is a well-established teaching method in mathematics teacher education and can be potentially useful for statistics teacher education. Through reading and discussing cases, professionals can acquire knowledge of general principles of their fields while also developing reasoning skills necessary for navigating complex and ambiguous situations (Markovits & Smith, 2008). Because teaching is a profession fraught with uncertainty, the case approach to teacher education has gained momentum during the past two decades (Grossman, 2005).

R.E. Groth (✉)
Department of Education Specialities, Salisbury University, Salisbury,
MD 21801, USA
e-mail: regroth@salisbury.edu

S. Xu
Teaching Research Office of Guangzhou Education Bureau, Guangzhou, China
e-mail: xshhgz@126.com

C. Batanero, G. Burrill, and C. Reading (eds.), *Teaching Statistics in School*   371
Mathematics-Challenges for Teaching and Teacher Education: A Joint ICMI/IASE Study,
DOI 10.1007/978-94-007-1131-0_35, © Springer Science+Business Media B.V. 2011

Teacher educators have embraced the approach in hopes that discourse about specific cases of practice will help prepare teachers to make sound pedagogical decisions in complex classroom environments.

Cases that teachers study sometimes come directly from their own classrooms or schools, and other times come from more unfamiliar settings. For instance, the increasingly popular lesson study approach to teacher education, which involves planning, acting, observing, and reflecting, relies heavily upon engaging teachers in conversations about lessons from their classroom or school settings (Lewis, 2002; Han, 2007; Masami, 2007). One advantage to such an approach is that teachers' familiarity with the context of the school can facilitate the process of forming viable solutions to pedagogical dilemmas. Proposed solutions can then be directly tested with the students who were studied as part of the case. Nonetheless, cases that come from more unfamiliar settings also have potential value. Such cases serve the function of exposing prospective and practising teachers to contexts that would otherwise be inaccessible (Merseth, 1991).

In this chapter the types of discourse that can be catalysed by using cases of statistics lessons drawn from unfamiliar settings as well as those drawn from teachers' own schools are illustrated. In the first extended illustration a case discussion among a group of prospective secondary mathematics teachers in the Mid-Atlantic United States is presented (Groth, 2008), and in the second a case discussion among a group of practising Junior High School teachers in Guangzhou, China (Xu, Yongdong, Bangquan, & Reisheng, 2008) is described. By exploring the dynamics of case discussions from these two different environments, the types of discourse statistics teacher educators can aim to catalyse as they facilitate case discussions are being mapped and some conclusions for teachers' education and research in statistics teacher education are also drawn.

2   Illustration 1: Prospective Teachers Discussing a Written Teaching Episode

2.1   Background

The first illustration describes a discussion that occurred among 15 prospective secondary mathematics teachers about a written case entitled, "It's time for a tail" (Merseth, 2003) during the second semester of a two-semester teaching methods course sequence. While the course ordinarily met face-to-face, the case discussion described below took place in an asynchronous discussion board setting as part of an outside-of-class assignment that participants had 7 days to complete. There were no face-to-face class meetings held during the time period participants completed the assignment. Guidelines for the assignment specified posting at least four different messages on 4 different days, and that messages were to be directed towards analysing the case and conjecturing about its implications for teaching practice. The group had completed similar online discussion assignments in the past (Groth, 2007), so they were already familiar with how to use the technology provided to host the case discussion. As the case discussion occurred, all messages were captured online and retained for analysis.

2.2 Description of the Case Under Discussion

The case described events transpiring in a classroom where a teacher had given students two tasks involving the statistical ideas of randomness, sampling, and independence. In the first of the two tasks, Ms. Brady (a pseudonym for the teacher in the case) told her students that a coin was tossed and came up "heads" five consecutive times. Students were asked to predict the likelihood of obtaining "heads" on the next toss. In the second task, Ms. Brady told her students that a decision needed to be made about which of two basketball players, Dennis or Michael, should be chosen to take the last shot to try to win a game. Dennis made five of every ten shots, on average, but had made every one of his last five shots. Michael, on the other hand, made seven out of every ten shots, but had missed his last three shots. Ms. Brady considered the two tasks to be parallel in terms of statistical content, and was using the second task as an assessment of students' understanding of the statistical ideas embedded in the first.

It should be noted that the question of whether or not basketball shots constitute independent binomial trials, like coin flips, is a point that has been debated in the literature. Gilovich, Vallone, and Tversky (1985) analysed basketball game data and found no evidence of positive correlation between the outcomes of successive shots. They attributed the common belief that one successful shot makes the next shot more likely to be successful to the widespread misconception that even short random sequences should be representative of their generating process i.e., belief in the "law of small numbers" (Tversky & Kahneman, 1971). Larkey, Smith, and Kadane (1989), on the other hand, presented data indicating that some players do, in fact, have improbably long strings of success in shooting the ball. They argued that the analysis of Gilovich et al. was overly simplistic in that it analysed isolated strings of outcomes apart from the context of the games in which they occurred. Therefore, the case had a built-in opportunity for prospective teachers to take up a statistical debate that has occurred among scholars in the field.

The events of the case were portrayed through student work samples, excerpts from classroom discourse, and Ms. Brady's personal reflections on the lesson. These artefacts illuminated her students' reasoning about both the tasks they had been given. For the first task, Ms. Brady's students struggled with the notion that flips of a coin were independent of one another. Some, for example, expressed the opinion that a coin flip was more likely to produce the result "tail" than "head" after a sequence of five heads had occurred. In reasoning about the basketball task, many students believed that players' shots were not independent of one another. Some thought that the player who had missed his previous attempts was more likely to finally make a basket, and others thought that the other player was more likely to make a basket because he was having a good game. Still others felt each player had the same probability of making the basket, reasoning that the two possible outcomes (making the basket or missing it) were equally likely.

In addition to providing details about students' reasoning, the case provided a variety of background information about the school setting, students' behaviour, and the teacher's strategies for organising instruction. At the conclusion of the case, several questions were given to spark discussion among prospective teachers, such as one item asking what the teacher should do next with the class, and another asking for a judgment about the independence of a player's shots in basketball. The prospective teachers studying the case were asked to take these items into account during the case discussion. The resultant discourse centred upon statistical content, content-specific pedagogy, and content-independent pedagogy, as described below.

2.3   Statistical Discourse Catalysed by the Case

There was a substantial amount of discussion among the prospective teachers about whether or not the content of the two tasks presented in the case were statistically parallel. As in the scholarly literature, some believed that shots of a basketball constituted independent trials, like flips of a coin, and others did not. Hence, the case succeeded in generating opposing viewpoints about a controversial statistical matter.

Although the positions taken by the prospective teachers aligned with those in the literature, the justifications for the positions at times departed from normative ones, as some of the justifications involved statistical misconceptions, similar to those found in other studies involving prospective teachers (e.g., Batanero, Biehler, Maxara, Engel, & Vogel, 2005). Elaine (a pseudonym, as are the rest of the prospective teachers' names), for example, argued that Michael was "due" to make a basket since he had not made one recently. Her argument indicated a belief in the law of small numbers, since she expected a short string of outcomes to be more representative of its generating process. Other prospective teachers argued from positions grounded in the equiprobability bias (Lecoutre, 1992), believing that all random events are equally likely to occur. Harold, for example, said that it did not matter which player was chosen to take the winning shot because each player could produce two different possible outcomes: success or failure. He assigned the same probability to each outcome, even though the case provided different information about each player's long-term success percentage. Essentially, some of the same misconceptions exhibited by students in the case were voiced by the prospective teachers discussing it.

Beliefs about context also played a role in the prospective teachers' discussions. Several who were familiar with the game of basketball argued that shooting a basketball was not the same as flipping a coin because one's confidence is a factor in the former but not the latter. Elaine, who admitted to being unfamiliar with basketball, was one of the two prospective teachers willing to entertain the notion that coin flips and basketball shots might be somewhat analogous. She stated,

> Just as in a coin flip, you may flip a series of tails, but later on you may also flip a series of heads. In the long run, a player's average shot rate will not necessarily be affected by streak shooting.

Stephanie affirmed the possible validity of the analogy Elaine proposed by quoting articles from the Internet. However, the position that basketball shots might behave like coin flips gained no further support. The predominant focus remained upon the idea that basketball shots could not be independent because of a presumed confidence factor. This element of presumed knowledge about the context of the situation blocked consideration of opposing arguments.

2.4 Content-Specific Pedagogical Discourse Catalysed by the Case

Prospective teachers discussing the written case also evaluated Ms. Brady's pedagogical practices for teaching stochastics. Several conjectured that her basketball and coin flipping tasks were too lengthy for consideration in a single class period. Ivan, for example, stated that the tasks were good for understanding probability and sampling, but that they were too complex to consider together in a single class period. The conjecture that too many complex tasks were used during the class was not disputed by any of the participants. There were, however, disagreements about the overall appropriateness of the basketball task for younger students. Kelly felt that when first learning statistical ideas, students should be given examples that are easier to understand. Kelly's conjecture about the appropriateness of the task was challenged by others in the conversation. Claire, for example, challenged Kelly's idea by stating, "I don't think that there is anything wrong with her examples ... If you can only use probability with coins but not with dice or lotteries, etc. then you aren't really learning the concepts." Just as in the discussion of the solutions to the basketball task, there were competing opinions on the task's pedagogical appropriateness.

2.5 General Pedagogical Discourse Catalysed by the Case

A considerable amount of discussion among the prospective teachers was devoted to commenting on Ms. Brady's ability to manage the classroom. Several participants conjectured that she needed to change aspects of her classroom management practices. For example, at one point in the case, Ms. Brady allowed students to select their own group partners and some prospective teachers felt that assigning students to cooperative groups rather than letting them choose their own would minimise disruptions. Others conjectured that the teacher should have set clearer expectations for behaviour, although the validity of that conjecture was disputed. The occurrence of several comments about classroom management was somewhat surprising, since the prospective teachers were not specifically asked to comment on that aspect of the case. The extensive focus on classroom management was likely intensified by the fact that the discussion participants would soon be responsible for their own classrooms. At such a point in one's academic career, concerns about establishing and enforcing classroom rules and becoming authority figures can become paramount (Jones & Vesilind, 1995).

3   Illustration 2: Practising Teachers Discussing an Observed Demonstration Lesson

3.1   Background

The second illustration describes a "teaching research activity" that involved 30 junior high school mathematics teachers in Guangzhou, China, as they discussed a lesson entitled "the meaning of mean" (a more detailed description of this activity was presented in Shihong et al. 2008). The teaching research activity is a popular method of in-service teacher professional development in China. Such activities support teachers' growth by encouraging the discussion of some demonstration lessons they observe (Huang, Peng, Wang, & Li, 2010).

In the beginning of 2007, the demonstration lesson was identified and a "teaching preparing group" was established, whose members included five key teachers in the district who created the first draft of a lesson plan and then assigned Mr. Li to conduct a pilot trial of the lesson. After several revisions of the lesson, the teaching research activity was held at Mr. Li's junior high school with 36 participants, including the teaching preparing group, 29 mathematics teachers of the district, and some researchers. The demonstration lesson was taught in a class that consisted of 40 students and lasted for 49 min. Each teaching research activity participant had a copy of the final written lesson plan. The teaching research activity included a class observation and post-class reflection and discussion among the participants. Below we discuss the discourse catalysed by participants' observations of the events that transpired during the demonstration lesson.

3.2   Selection of Tasks and Content for the Lesson

Chinese teachers put great emphasis on textbooks and teaching reference books. Most teaching design originates from textbooks, so making good use of textbooks usually becomes the key topic in discussion. In this case, the textbook gave examples and exercises that tried to focus the teaching on the meaning of arithmetic mean (the concept is first introduced in primary school), and also on reading data from a variety of diagrams. However, the teaching preparing group made some adjustments to the text, such as omitting and modifying exercises and adding new questions.

The utilisation of problem contexts familiar to students was one teaching strategy used in Mr. Li's lesson. Because the teaching research activity was held during children's festival, when children younger than 14-years old can enjoy the day off, Mr. Li asked his students to collect data of students' ages and review the method of calculating the average age. In addition, the school held basketball competitions between classes the previous week, and Mr. Li's class obtained good results, so he used the students' scores in the basketball competition in an exercise that involved construction of a bar graph to show the average score of each player in the competition. The problems set in familiar contexts captured the students' attention and were approved by the observers.

In China, it is common for teachers to modify the examples and exercises in a textbook, and to stratify exercises according to students' learning ability. Mr. Li used these strategies in his teaching design and practice. For instance, the contexts of some examples in the textbook were replaced with others relating to the students in Mr. Li's class, the school, and Guangzhou local information. After new material had been taught, students were given several exercises that were divided into three categories. Each category was designed to be accessible to students at different levels of understanding. These methods also were well received by the observers.

In order to finish every part of the lesson plan, Mr. Li went 4 min beyond the typical duration of class to explain the solution to an exercise, check a group of problems, and synthesise the lesson. Most observers disagreed with this decision, arguing that 4 min was not an adequate amount of time because a number of concepts in the lesson still needed clarification and students appeared not to concentrate due to the fast explanation. Additionally, a few teachers suggested that numerical calculations could be reduced in some tasks to give students more time for interpretive activities.

The possibility of including more tasks involving the process of statistical investigation was discussed as well. Because most schools in the district use only half of the time provided to teach statistics, and mathematics teachers who teach statistics lack statistical background, few teachers address the design of statistical investigations with their students. Although some teachers proposed that Mr. Li's lesson should have included more activities like data collection, data presentation, and so on, most observers felt that it was difficult to present a complete statistical investigation process in the lesson. Nonetheless, the researchers attending the activity suggested that some examples, such as collecting data of students' age or scores of basketball competition, could be redesigned to involve statistical investigations. A few teachers argued that using class time to address another mathematics topic would be more desirable than using time for statistical investigation. These teachers felt that focusing on the mean alone was not enough and suggested introducing the weighted mean as well, believing that students could learn both concepts in one lesson. This view, however, was not shared by all the participants.

3.3 Teaching Strategies

Teaching strategies beyond those involving selection of tasks and content were also considered and debated by participants. In the demonstration lesson, many effective strategies were noticed, such as the efficient organisation of the teaching process, a well-designed worksheet, use of group discussion, and the teacher's explanations and concluding remarks after giving the students some time for independent thinking. After discussion, the observers reached agreement on four general principles: (a) providing an appropriate worksheet can help students concentrate upon their work as they solve tasks and contribute to the efficiency of a lesson, (b) articulating a clear lesson structure can promote students' systematic thinking, (c) posing problems before providing explanations can prompt students to think more deeply about the problems they are given, and (d) using a projector to show students' solutions to the whole class can be conducive to helping them understand one others' thoughts and promote further communication and exploration. The effective use of the blackboard, projector, and computer in this lesson was also noticed by the observers who agreed that writing on the blackboard helped students to understand the teacher's thinking, displaying students' worksheets on a projector was conducive to mutual learning, and using Excel and PowerPoint reflected the value of technology.

Despite the perceived strengths of the lesson, it was noted that no students asked their own questions in class. The researcher attending the teaching research activity concluded that this passivity or lack of curiosity and the failure to measure or address it was one of the weaknesses in the lesson. A cause of this weakness in the lesson may have been that more time was spent on providing step-by-step details than on encouraging students' creative thinking. Teachers believed that because the students were so busy finishing the tasks given to them by the teacher, they lost the initiative to question and had too little time to think and act.

4 Implications for Future Research and for Teacher Education in Statistics

The two case discussions described in this chapter provided venues for both pre-service and in-service teachers to deal with problems arising in statistics and teaching. The type of critical dialogue that occurred, where peers challenged one another's interpretations and recommendations, is essential to fostering reflective teaching practice (Lampert & Ball, 1998).

The two illustrations illuminate three important areas about which cases can help foster critical dialogue: statistical content, content-specific pedagogy, and general pedagogy (Table 35.1). These three areas are believed to be foundational to the enterprise of teacher education (Shulman, 1987). Some concluding thoughts on using cases to foster the study of each of the three areas are offered below, along with some suggestions for future research.

Table 35.1 Focal categories and examples of topics from case discussions

	Illustration 1	Illustration 2
Content	Randomness	Arithmetic mean
	Sampling	
	Independence	
	Overcoming probability misconceptions	
	Problem context	
	Statistical thinking	
Content-specific pedagogy	Problem selection	Modifying textbook problems
	Amount of time spent on a problem	Teaching the process of statistical investigation
		Teaching calculation versus teaching concepts
		Teaching the weighted mean along with the mean
General pedagogy	Classroom management	Lesson organisation
		Classroom discourse
		Students' engagement

4.1   Statistical Content

Since cases of statistics classrooms, by nature, involve statistics problems, they provide opportunities to develop teachers' content knowledge. In illustration 1, the embedded problems involved randomness, sampling, and independence. In illustration 2, the problems involved the arithmetic mean. The problems posed in illustration 1 elicited common misconceptions that occur in reasoning about stochastic processes. They also prompted the prospective teachers to apply beliefs about problem context to their arguments. Taking context into account is an essential component of statistical thinking, but strong contextual beliefs can also inhibit individuals from considering alternative points of view, as occurred in the prospective teachers' discussion of the basketball task. In any event, it is clear that cases can be used as mechanisms to elicit entrenched misconceptions and to cause contextual assumptions to come to the surface. As these reasoning elements are revealed, they can be challenged by the case moderator and other participants in the case discussion.

Given the potential of the case method for raising ill-formed ideas to be challenged, an important task for future research is to find ways to develop cases so they elicit prevalent misconceptions and contextual beliefs. Illustration 1 merely scratches the surface of common reasoning patterns about stochastic processes. Many more such reasoning patterns have been described by researchers (Shaughnessy, 1992, 2007; Jones, Langrall, & Mooney, 2007). Problems involving these reasoning patterns can be embedded in new cases, and then field-tested with prospective teachers. Field-testing of cases by researchers can lead to refinements that spark richer content-oriented conversations. As rich cases are developed, researchers can work towards the goal of developing a comprehensive case-based curriculum that systematically addresses prevalent known patterns of problematic statistical reasoning. The successful development of such a curriculum would be very relevant to the field of statistics teacher education, since it would effectively integrate the study of content with the study of pedagogy rather than artificially separating the two.

4.2   Content-Specific Pedagogy

In order for the case method to achieve its full potential, discussions of content should be accompanied by discussions of pedagogy necessary to address the specific statistical content under consideration. In both illustrations, case discussion participants considered the appropriate scope of problems that should be selected to address the content objectives of the lessons. In illustration 2, teachers considered additional issues, such as addressing the process of statistical investigation, focusing on concepts instead of calculations, and possibly teaching the weighted mean along with the mean. Teachers did not arrive at consensus on all of these content-specific pedagogical issues, illustrating that consensus need not be a goal when discussing content-specific pedagogy. Rather, just as when reasoning through a statistical investigation, teachers should focus on producing viable and defensible arguments rather than absolute and general solutions to classroom problems. Case facilitators can help teachers converge on a range of reasonable solutions to classroom dilemmas that involve portraying content to students, but they should not expect to obtain complete uniformity. Variability among teachers' opinions is desirable insofar as it is plausibly based on the variability in students' learning needs that one may observe across classrooms.

4.3   General Pedagogy

A striking feature in comparing the two illustrations is that in the area of general pedagogy, the prospective teachers in illustration 1 expressed more classroom-management concerns than the practising teachers in illustration 2. As with other elements of case discussion, conversations about classroom management practices have possible positive and negative aspects. One positive aspect is that prospective teachers have opportunities to vicariously experience problematic situations that may occur in their own classrooms in the future. This provides the unique opportunity to anticipate and strategise solutions to potentially serious classroom problems in a relatively relaxed, supportive environment.

As strategies are formed, however, it needs to be emphasised that there are often multiple causes for students' misbehaviour and/or disengagement from a lesson. The prospective teachers discussing Ms. Brady's classroom tended to posit generic, content-independent solutions for classroom management problems.

Hence, they needed to be reminded that students' understanding of content (or lack thereof) can also strongly influence their willingness to become meaningfully engaged in a lesson. In contrast, the experienced teachers in illustration 2 focused on the organisation of the lesson, likely understanding that students' behaviour is heavily influenced by the pedagogy employed in a lesson.

The prospective teachers' focus on classroom management issues raises an important question for case writers and researchers: What level of detail should be used in producing cases for the purpose of statistics teacher education? Some advocate for showing an extensive amount of detail about each case by video recording lessons for case analyses (e.g., McGraw, Lynch, Koc, Budak, & Brown, 2007) or having teachers observe actual classrooms (e.g., Lewis, 2002).

The strategy of exposing teachers to a great amount of detail about a statistics lesson is supported by illustration 2, since practising teachers were able to attend to multiple, simultaneously occurring aspects of the lesson. However, it seems plausible that the inclusion of extensive detail about classroom management in the written case described in illustration 1 may have diverted prospective teachers' attention from making further progress on the discussion of content and content-specific pedagogy. The discussions they had about general pedagogy were likely not as detailed as those of the practising teachers because they did not yet have as much experience dealing with complex classroom environments. Jones and Vesilind (1995) found that prospective teachers began to make connections among the variables contributing to classroom management approximately midway through their teaching experiences. Hence, actual teaching experience may be an important pre-requisite for learning from cases that contain a great amount of detail. At minimum, it seems necessary to reject a simplistic maxim of "the more detail, the better" when preparing cases for prospective teachers to discuss. Future research can focus on the optimal level of detail for fostering various learning objectives related to statistics teacher education.

References

Batanero, C., Biehler, R., Maxara, C., Engel, J., & Vogel, M. (2005, May). *Using simulation to bridge teachers' content and pedagogical knowledge in probability*. Paper presented at the ICMI Study 15 Conference: The Professional Education and Development of Teachers of Mathematics, Aguas de Lindoia, Brazil.

Gilovich, T. R., Vallone, R., & Tversky, A. (1985). The hot hand in basketball: On the misperception of random sequences. *Cognitive Psychology, 17*, 295–314.

Grossman, P. (2005). Research on pedagogical approaches in teacher education. In M. Cochran-Smith & K. M. Zeichner (Eds.), *Studying teacher education: The report of the AERA Panel on Research and Teacher Education* (pp. 425–476). Mahwah, NJ: Erlbaum.

Groth, R. E. (2007). Analysis of an online case discussion about teaching stochastics. *Mathematics Teacher Education and Development, 7*, 53–71.

Groth, R. E. (2008). Navigating layers of uncertainty in teaching statistics through case discussion. In C. Batanero, G. Burrill, C. Reading, & A. Rossman (2008).

Han, X. (2007, November). *Learning with colleagues through public lessons*. Paper presented at The World Association of Lesson Studies International Conference (WALSIC) 2007, Hong Kong. Online: www.ied.edu.hk/wals/website/

Huang, R., Peng, S., Wang, L., & Li, Y. (2010). Secondary mathematics teacher professional development in China. In F. K. S. Leung & Y. Li (Eds.), *Reforms and issues in school mathematics in East Asia* (pp. 129–152). Rotterdam: Sense Publishers.

Jones, G., Langrall, C., & Mooney, E. (2007). Research in probability: Responding to classroom realities. In F. Lester (Ed.), *Second handbook of research on mathematics teaching and learning* (Vol. 2, pp. 909–955). Charlotte, NC: Information Age Publishing and National Council of Teachers of Mathematics.

Jones, M. G., & Vesilind, E. (1995). Preservice teachers' frameworks for class management. *Teaching and Teacher Education, 11*, 313–330.

Lampert, M., & Ball, D. L. (1998). *Teaching, multimedia, and mathematics: Investigations of real practice*. New York: Teachers College Press.

Larkey, P., Smith, R., & Kadane, J. (1989). It's okay to believe in the "hot hand". *Chance, 2*(4), 22–30.

Lecoutre, M. P. (1992). Cognitive models and problem spaces in "purely random" situations. *Educational Studies in Mathematics, 23*, 557–568.

Lewis, C. C. (2002). *Lesson study: A handbook of teacher-led instructional change*. Philadelphia: Research for Better Schools.

Markovits, Z., & Smith, M. (2008). Cases as tools in mathematics teacher education. In D. Tirosh & T. Wood (Eds.), *The international handbook of mathematics teacher education* (Vol. 2, pp. 39–64). Rotterdam: Sense Publishers.

Masami, M. (2007, November). *A review of lesson study in Japan*. Paper presented at The World Association of Lesson Studies International Conference (WALSIC) 2007, Hong Kong. Online: www.ied.edu.hk/wals/website/ Retrieved on: May18, 2011

McGraw, R., Lynch, K., Koc, Y., Budak, A., & Brown, C. A. (2007). The multimedia case as a tool for professional development: An analysis of online and face-to-face interaction among mathematics pre-service teachers, in-service teachers, mathematicians, and mathematics teacher educators. *Journal of Mathematics Teacher Education, 10*, 95–121.

Merseth, K. K. (1991). *The case for cases in teacher education*. Washington, DC: American Association of Colleges for Teacher Education.

Merseth, K. K. (2003). *Windows on teaching math: Cases of middle and secondary classrooms*. New York: Teachers College Press.

Shaughnessy, J. M. (1992). Research in probability and statistics: Reflections and directions. In D. A. Grouws (Ed.), *Handbook of research on mathematics teaching and learning* (pp. 465–494). New York: Macmillan and National Council of Teachers of Mathematics.

Shaughnessy, M. (2007). Research on statistics learning and reasoning. In F. Lester (Ed.), *Second handbook of research on the teaching and learning of mathematics* (Vol. 2, pp. 957–1009). Charlotte, NC: Information Age Publishing and National Council of Teachers of Mathematics.

Shulman, L. S. (1987). Knowledge and teaching: Foundations of the new reform. *Harvard Educational Review, 57*(1), 1–22.

Tversky, A., & Kahneman, D. (1971). Belief in the law of small numbers. *Psychological Bulletin, 76*(2), 105–110.

Xu, S., Yi, L., Yongdong, L., Bangquan, T., & Reisheng, P. (2008). A teacher training case in junior high schools of Guangzhou. In C. Batanero, G. Burrill, C. Reading, & A. Rossman (2008).

Chapter 36
Distance Education of Statistics Teachers

Maria Meletiou-Mavrotheris and Ana Serradó

Abstract The affordances offered by modern Internet technologies provide new opportunities for the pre-service and in-service education of teachers, making it possible to overcome the restrictions of shrinking resources and geographical locations and to offer, in a cost-effective and non-disruptive way, high quality learning experiences to geographically dispersed teachers. The focus of this chapter is the question of how information and communication tools made available online could be exploited effectively to help improve the quality and efficiency of teacher training in statistics education.

1 Introduction

Statistics has been established as a vital part of school mathematics in many countries and has been introduced into mainstream mathematics curricula without adequate attention paid to teacher professional development. There is some evidence of poor understanding and insufficient preparation to teach statistical concepts among both pre-service and in-service teachers (e.g., Watson, 2001; Chick & Pierce, 2008). Some teachers tend to have weak knowledge of the statistical concepts and to focus their instruction on the procedural aspects of statistics, and not on conceptual understanding (Watson, 2001).

M. Meletiou-Mavrotheris (✉)
Department of Education Sciences, European University Cyprus, 6 Diogenous St.,
1516 Nicosia, Cyprus
e-mail: m.mavrotheris@euc.ac.cy

A. Serradó
Colegio La Salle-Buen Consejo, Teresa de Calcuta, 70, 11510 Puerto Real Cadiz, Spain
e-mail: aserradob@lasalleandalucia.net

C. Batanero, G. Burrill, and C. Reading (eds.), *Teaching Statistics in School*
Mathematics-Challenges for Teaching and Teacher Education: A Joint ICMI/IASE Study,
DOI 10.1007/978-94-007-1131-0_36, © Springer Science+Business Media B.V. 2011

This chapter analyses the possibilities of information and communication tools made available by modern Internet technologies to improve the quality and efficiency of teacher pre-service and in-service training in statistics education. First, the main pedagogical issues and challenges underlying distance education in general, and online teacher training in particular are discussed. Then some examples of programmes that have utilised distance education to offer at-distance teacher training in statistics education are provided. The chapter concludes with some implications for distance training of statistics teachers.

2 Distance Education in Statistics: Main Pedagogical Perspectives and Challenges

Distance education is very broad and encompasses several methods of delivery (e.g., regular mail, radio, television, Internet). It is not new, either in general education or in the field of statistics, but has its roots in correspondence courses, which can be traced back to late nineteenth century. The advent of the Internet had a profound impact on distance education, which went through a process of transformation and adaptation to emerge as a new method of e-learning, depending heavily on Information and Communication Technologies (ICT). Distance education now encompasses a variety of technologies, which support both synchronous and asynchronous communication. In recent years, we have witnessed a rapid expansion of distance education worldwide as educational institutions at all levels are becoming increasingly involved in distance education initiatives. Online course delivery has become common in a wide variety of disciplines, including statistics (Philips, 2003), and this expansion is likely to continue, given the expanding access to the Internet and the greater emphasis given to lifelong learning.

Several advantages associated with distance education have been identified in the research literature. Distance education offers flexibility and convenience, allowing learners to determine their own place, pace, time, and content of study. Further, the distance option may allow students the opportunity to take courses from prominent experts in their field of study. From the viewpoint of statistics education, network-based training creates some unique opportunities for enhancing statistics instruction (Philips, 2003) and several examples of successful programmes of teaching statistics via distance have been documented in the literature (e.g., Saporta & Bourdeau, 2003; Evans et al., 2007; Everson & Garfield, 2008; Dale, 2010).

The Internet offers a vast array of tools and resources that can be used for better understanding of statistical methods and concepts. Interactive Java-applets and virtual statistical laboratory experiments, for example, allow for visualisation of statistical concepts and hands-on simulations (Vermeire, Carbonez, Darius, & Fresen, 2002). Several statistics instructors mention using technological tools and resources in their online courses. For example, Utts, Sommer, Acredolo, Maher, and Matthews (2003) used the online textbook, *CyberStats*, which contains interactive applications and practice problems. In another example, Everson and Garfield (2008) used Fathom or SPSS in their undergraduate and graduate online introductory statistics courses and also introduced students to statistical applets.

Despite the undisputed benefits and proliferation in recent years of online professional development programmes, concerns remain about their quality, as research suggests that the effectiveness of distance education is variable (Evans et al., 2007). While most of the conducted studies indicate that students taking courses with an online component have similar achievement and satisfaction levels compared to students in traditional, face-to-face classrooms (Tallent-Runnels et al., 2006; Mathieson, 2010), there is growing evidence of many web-based distance learning courses failing to meet the expectations raised.

Early attempts at Internet-based instruction assumed that setting up an attractive website with interesting online and multimedia applications was adequate for learning to take place. It is now recognised that the level of success of a distance learning course is determined by multiple factors, such as underlying theory, technologies, teaching strategies, and support for learners. Elements in the design of a web-based course such as the content and structure of the course, the presentation of the online materials, and the amount of interaction between instructors and learners as well as among learners are important factors affecting students' learning and attitudes (Tudor, 2006). Another important criterion for the level of success of network-based statistical training is the extent to which instruction allows learners to experience the practice of statistics, and to apply statistical tools in order to tackle real-life problems (Vermeire et al., 2002).

In addition to the general issues and considerations regarding distance education in statistics, the training of statistics teachers at a distance poses special challenges that ought to be taken into account when designing an online professional development programme. These challenges are discussed briefly in the next section.

3   Distance Education of Teachers

In recent years, it has been recognised that teacher training is more effective in producing real changes in classroom practices, when promoting continuous, professional development opportunities that are cumulative and sustained over the career of a teacher (Joubert & Surtherland, 2009). The financial and logistic difficulties of engaging teachers in face-to-face professional development opportunities, as well as the need for professional development which can fit with teachers' busy schedules and can draw on powerful resources often not available locally, have encouraged the creation of online teacher professional development programmes.

The expansion in the modes of communication enabled by recent advances in communications and information technologies is revolutionising distance education, leading to the development of new forms of online professional development settings, in accord with socio-constructivist views of learning. There is increased interest in online communities of practice as vehicles which can promote teacher learning and development, by enabling geographically dispersed teachers to exchange ideas with other teachers and acquire support as they try new strategies in their classrooms. Online communities of practice are constantly evolving into many forms and styles as they embrace new and evolving technologies (Gray, 2004).

However, while online communities proliferate in cyberspace, little is known about best practices for their effective design and implementation, and our understanding of the reasons underlying their success or failure is still at the initial stage. Existing research highlights several difficulties in building and maintaining online communities involving shared professional learning (Rourke & Kanuka, 2007). Consequently, despite the early enthusiasm and encouragement of participants, some online communities of practice fail to progress. For example, after examining 28 studies, Zhao and Rop (2001) reported that there was little conclusive evidence to demonstrate the effective use of reflective online communities of practice. Davies and Graff (2005) raised several issues that consistently create challenges for community sustainability, including barriers around usability and sociability, and lack of time to spend in online discussions.

While in statistics education research it is well-documented that the incorporation of discussion and active learning in the statistics classroom can help learners to think and reason about statistical concepts, bringing these important learning approaches to an online course has been challenging (Everson & Garfield, 2008). The design of cognitive tools to support an online community of practice involves many inter-related considerations (e.g., moderator involvement, reliability, and stability of the technology), most of which are not yet well understood (Stahl, 2006). More research is needed to shed light into how to best support the development of healthy and sustaining online communities of practice. Below some experiences related to educating statistics teachers at-distance are analysed.

4   Experiences of Teacher Education at a Distance

This section provides a brief description of four professional development programmes that have utilised distance education to offer teacher training in statistics education, and of the main experiences gained from implementing these programmes.

4.1   LUDDITE Project

LUDDITE (Learning the Unlikely at Distance Delivered as an Information Technology Enterprise), a 3-year project run by the Australian Association of Mathematics Teachers during the period 1994–1997 was one of the first programmes using distance education for teacher professional development in statistics education. LUDDITE, which was funded by the Australian government to explore various technologies for the delivery of professional development in probability and statistics to teachers of grades 5–9 spread across Australia, entailed three stages of development. During the first stage, four live satellite television narrowcast programmes highlighting issues associated with statistics instruction were produced for schools with satellite technology. The second stage of LUDDITE employed videoconferencing facilities in five Australian states to introduce participants to a multimedia professional development package intended to provide comprehensive coverage of all aspects of teaching chance and data in the middle school. The final phase of the programme featured a consolidation of the produced material and the creation of a CDROM to allow for access to more print-type material as well as digitised video material relevant to teaching chance and data.

"The Chance and Data Professional Development" CDROM, which was the major development of the LUDDITE project, was organised in five different sections (data collection and sampling, data representation, chance and basic probability, data reduction, inference). Each section included an introductory overview with TV extracts, curriculum documents and materials for teachers, newspaper articles, video clips from television broadcasts, video clips of students discussing chance and data concepts, cross-curricular links, and software for probability simulation and data handling.

Watson (1998) reported on the main findings from the evaluation of LUDDITE. Responses of 19 teachers to the total professional development package were very positive, with teachers noting the well-rounded package with background as well as classroom materials, and appreciating the linking mechanism across media. Less positive comments were related to availability of, and familiarity with the technology that caused frustration for a few teachers, who found it difficult to use. Teachers' reactions to the at-distance presentation of the package were generally positive, but several noted that they missed personal contact with others and suggested the introduction of some face-to-face meetings, such as half-day workshops at the beginning, or clusters formed in local areas. After weighing up all the possibilities for distance professional development trialled in Australia through LUDDITE, Watson (1998) concluded that the package appeared to offer more value if introduced in person and/or as part of a directed professional development programme in a school system. She acknowledged the danger of a package becoming outdated if created in a fixed medium such as a CDROM, but explained that the option of providing the whole package on the Internet was unrealistic at the time of writing, due to lack of accessibility for many teachers, and low speed of transfer of information, particularly digitised videos.

4.2   INSPIRE Project

INSPIRE (Insight into Statistical Practice, Instruction, and Reasoning) was a more recent project sponsored by the National Science Foundation (NSF) and the American Statistical Association (ASA) to improve high school statistics teaching.

The objective of the project was to craft a professional development experience for practising high school mathematics teachers, which would prepare them to: (a) teach an introductory statistics course, (b) learn and understand the concepts and methods of introductory statistics, (c) use real data, active learning techniques, and technology to teach statistics, (d) understand statistics as a comprehensive approach to data analysis, and (e) become familiar with a variety of resources for teaching introductory statistics. Additionally, the project aimed at developing a long-term online community of learners who would advise and support each other about classroom practices, pedagogy, and statistical concepts (Gould & Peck, 2005).

INSPIRE brought together university faculty, secondary teachers, and ASA statisticians who developed and piloted a course for in-service high school mathematics teachers who were novices to statistics teaching. The first component was a week-long face-to-face workshop that initiated the course, and which was intended to help participants recognise the differences between mathematics and statistics, to model effective pedagogy, and to prepare teachers for the second, online component of the course. The year-long online component was an introductory statistics course, enhanced with special attention to pedagogical issues. Materials were delivered primarily online in a structured curriculum involving group work, self-study, concept-exploration, periodic milestone assessments, and small projects. To facilitate online community formation participants were assigned to small groups and were provided with group discussion questions, which addressed content as well as pedagogical concerns. The final component of the course was a year-long "practicum" that helped teachers to create deeper understanding of statistics by working with non-academic statisticians from business, industry, or government on projects dealing with real-world problems.

Gould and Peck (2005) described their experiences from the first offering of the INSPIRE course to 32 participants. The workshop component of the course had enormous success with all of the participants being uniformly enthusiastic about it; the online component, however, was less successful. The lower than anticipated level of teacher-to-teacher interaction was one of the greatest disappointments of the course. The online course also had a high attrition rate, with 11 of the 32 participants (34%) dropping out before its completion. When contacted by the project evaluator, teachers suggested difficulties with technology, (e.g., network problems, problems in file sharing), time constraints, and frustration over being required to participate in online discussions, as the biggest factors in their decision to drop out.

4.3 EarlyStatistics Project

EarlyStatistics was a 3-year project funded by the European Union under the Socrates–Comenius Programme (2005–2008). A consortium of five higher education institutions from four countries, developed and pilot-tested an intercultural online professional development course in statistics education targeting European elementary and middle school mathematics teachers. The course aimed at helping teachers improve their pedagogical and content knowledge of statistics through exposure to innovative web-based educational tools and resources, and cross-cultural exchange of experiences and ideas.

The course is made up of six modules (Azcárate et al., 2008). In modules 1–3, the focus is on enriching the participants' statistical content and pedagogical knowledge by exposing them to learning situations, technologies, and curricula similar to those they should employ in their own classrooms. Statistics is presented as an investigative process that involves four components: (a) clarifying the problem at hand and formulating questions that can be answered with data, (b) designing and employing a plan to collect appropriate data, (c) selecting appropriate graphical or numerical methods to analyse the data, and (d) interpreting the results (Franklin et al., 2005). Through participation in authentic educational activities such as projects, experiments, computer explorations with data, small group work, and whole class discussions, participating teachers learn where and how the "big ideas" of statistics apply and develop a variety of methodologies and resources for effective instruction.

In modules 4–6, the focus shifts to classroom implementation issues. Teachers customise and expand upon provided materials, and apply them in their own classrooms with the support of the design team. Once the teaching experiment is completed, they report on their experiences to other participants in the course, and also provide video-taped teaching episodes and samples of their students' work that are used for group reflection and evaluation.

The *EarlyStatistics* course has a hybrid format. At the beginning, teachers gather together to attend a 1-week intensive seminar, where they are introduced to the objectives and pedagogical framework underlying *EarlyStatistics* and are familiarised with the facilities offered by the e-learning environment. More importantly, this initial face-to-face meeting acts as an effective launch point for the distance learning part of the course, by giving teachers the opportunity to meet and interact with each other and with the course instructors. The remainder of the course is delivered online, through technology-rich interaction and problem-solving activities including text, illustrations, animations, and audio/video. To offer teachers flexibility and to accommodate different time zones, the largest portion of the course is conducted asynchronously through online discussion and e-mail groups. There is also some synchronous communication through use of technologies such as chat rooms and audio/video streaming.

A pilot delivery of the *EarlyStatistics* course and follow-up classroom experimentation took place during the final year of the project in three of the partner countries (Cyprus, Greece, Spain). Fourteen teachers, who differed considerably in their mathematical and statistical knowledge and in their confidence and experience in teaching statistics, participated in the *EarlyStatistics* course pilot delivery. The overall feedback from the participating teachers, as well as from external experts regarding the course content, services, and didactical approaches was generally very positive. Key conclusions from the analysis of the user feedback were that *EarlyStatistics* proved quite successful in helping teachers to improve their pedagogical and content knowledge of statistics (Chadjipadelis, Meletiou-Mavrotheris, & Paparistodemou, 2010). Moreover, data obtained from the teaching interventions in the course participants' classrooms suggest positive gains in student learning and attitudes towards statistics. Nonetheless, the consortium had limited success in establishing a functional community of practice, which was a main objective of the project. Similar results were reported by Gould and Peck (2005). In both cases, there was a much lower than anticipated level of teacher-to-teacher interaction (for more details see Meletiou-Mavrotheris et al., 2008).

4.4   "Becoming a Teacher of Statistics" Online Course

"Becoming a Teacher of Statistics" is a graduate-level course offered by the University of Minnesota, which prepares teachers of introductory statistics at the college and high school levels. Although originally delivered in a face-to-face setting, the course has been recently converted to an online course to make it accessible to a wider variety of pre-service and in-service teachers. The course strives to help teachers develop into knowledgeable, reflective, and effective teachers of statistics (Garfield & Everson, 2009). It is organised around the six recommendations set by the "Guidelines for Assessment and Instruction in Statistics Education" (GAISE) Report (Franklin et al., 2005): (a) emphasise statistical literacy and develop statistical thinking, (b) use real data, (c) stress conceptual understanding rather than mere knowledge of procedures, (d) foster active learning in the classroom, (e) use technology for developing conceptual understanding and analysing data, and (f) use assessments to improve and evaluate student learning.

In the face-to-face version of "Becoming a Teacher of Statistics", each class meeting focuses on a different topic and teachers are expected to prepare for class by having read the assigned readings. Teachers discuss the readings, watch demonstrations, participate in various hands-on and technology-based activities, and make assigned presentations. They also bring in materials, articles, and readings to share with the class. In designing the e-learning course, efforts were made to preserve the format and content of its face-to-face version, through the creation of an online learning environment that provides learners with many opportunities for small-group or whole-class discussions of the course material, and for collaboration. Teachers post reflections about the weekly readings, share and critique ideas for classroom activities, discuss ways to implement the GAISE recommendations in the statistics classroom, and evaluate different types of assessment tools and statistical software packages. The assignments teachers are asked to complete in the online course also mirror the types of assignments completed in the classroom-based course. Of course, the way in which the assignments are submitted or shared with classmates had to be altered in some cases. For example, instead of asking teachers to give short oral presentations, special discussion forums are set up for them to post and discuss the web resources and data sets they have found.

The first online version of the course was offered during the spring of 2008. Evaluation of the course indicated that it was successful, and provided teachers with experiences parallel to those provided in the face-to-face version of the course (Garfield & Everson, 2009). Teachers' engagement in online discussions was high; they appeared to be taking the necessary time to reflect critically on the assigned readings and on their roles as teachers of statistics. Performance on assignments and assessments was comparable to what was observed when teaching the course in a classroom setting, and teacher feedback was equally positive. The only element of the online course that was suggested to be improved was the organisation of the discussion groups. Specifically, some participants did not like the fact that the discussion groups were permanent, and would have preferred to change groups during the semester.

5 Implications for Future Training of Teachers

In order to make statistical thinking accessible to all students, there ought to be fundamental changes in the instructional methods and tools employed to teach statistical and probabilistic concepts in the mathematics classroom. Thus, it is critical for mathematics teachers to have rich teaching and learning experiences in statistics and its pedagogy. The need for training in statistics education of large numbers of teachers makes distance learning an attractive option.

Despite the potential benefits of implementing an online teacher training course in statistics education, there are a number of possible risks that could adversely affect its quality. Teachers participating in such courses are likely to be characterised by diversity in a number of parameters, including pedagogical and content knowledge of statistics and mathematics, educational level and grade they teach, cultural and/or professional backgrounds, and comfort with technology and with distance learning. While online teacher professional development courses share many features with face-to-face programmes, the review of the research literature and the experience gained from the four professional learning programmes presented in this chapter, suggest that they also present some unique challenges. Several pedagogical and technical issues should be taken into account in the course design in order to provide an effective online learning environment that motivates and supports teachers:

- Choice of accessible media
- User-friendly interface and navigation services
- User-friendly content addressing teachers' workplace educational needs
- Multimedia (e.g., audio, video, text, images) presentation of content to ensure effective knowledge transfer

- Activities and resources (e.g., simulations, animations, video clips) that stimulate and engage teachers participating in professional development, and address a variety of teaching and learning styles
- Access to multiple distance collaboration tools that promote interaction with peers and with course facilitators
- Regular assessments that can be used by instructors to monitor progress and provide feedback, and by participants to monitor their own progress
- Careful scheduling of course activities to offer teachers flexibility, and to accommodate different time zones
- Setting of realistic work expectations so as not to overburden teachers

A particularly important issue for online teacher professional development is in ensuring the successful building of an online community of practice. As the first experiences with the *EarlyStatistics* and INSPIRE teacher professional programmes indicate, building online communities is quite challenging. Although in both experiences, the course team employed several strategies to promote teacher dialogue and collaboration, they were not completely successful in establishing a functional online community. As Gould and Peck (2005) pointed out, leading a discussion of substance on a "discussion board" is more challenging than in a real classroom. Merely forming a discussion group and providing the technology does not automatically lead to the establishment of relations and group cohesion.

Online instruction is similar to, yet different from face-to-face learning, and requires new teaching skills and strategies; thus online instructors' new role as course facilitators turns them into both guides and learners (Heuer & King, 2004). Teacher educators must be trained in this new mode of instruction, to facilitate teacher success and develop online participation, as they develop in the art of becoming online guides. The success of the Garfield and Everson (2009) at-distance teacher training course in achieving learner participation and collaboration is explained by the fact that their online courses have been going through a continuous cycle of evaluation and improvement. Each time an online course is taught, changes are made in the way in which discussion assignments are structured and used, based on feedback received from participants and on careful study of the patterns of interaction occurring within different discussion groups.

Another extremely important consideration is whether teachers feel the professional development project is useful and supportive of their efforts to improve their teaching practice (see also Ponte, in this book). Historically, professional development efforts have largely been ineffective in producing reform-based classroom change. They often failed to transfer to the learners' "real-work" situations, because they were too distant from their real-work needs or organisational realities. A possible solution is the incorporation, whenever possible, of follow-up procedures – such as the teaching intervention undertaken by participants in the *EarlyStatistics* course – to help teachers apply what they learn in the course to a real classroom setting. Helping teachers meet their individual workplace goals and needs is very challenging, but necessary if they are to make the difficult leap from professional development to classroom practice.

References

Azcárate, P., Serradó, A., Cardeñoso, J. M., Meletiou-Mavrotheris, M., Paparistodemou, E., & Mavrotheris, E. (2008). An on-line professional environment to improve the teaching of statistics. In C. Batanero, G. Burrill, C. Reading, & A. Rossman (2008).

Batanero, C., Burrill, G., Reading, C., & Rossman, A. (Eds.). (2008). *Joint ICMI/IASE Study: Teaching Statistics in School Mathematics. Challenges for teaching and teacher education. Proceedings of the ICMI Study 18 and 2008 IASE Round Table Conference.* Monterrey, Mexico: International Commission on Mathematical Instruction and International Association for Statistical Education. Online: www.stat.auckland.ac.nz/~iase/publications

Chadjipadelis, T., Meletiou-Mavrotheris, M., & Paparistodemou, E. (2010). Statistics teacher of the new era: Another specialized mathematician or a totally different person? In C. Reading (Ed.), *Proceedings of the Eighth International Conference on Teaching Statistics.* Voorburg, The Netherlands: International Statistics Institute and International Association for Statistical Education. Online: www.stat.auckland.ac.nz/~iase/publications

Chick, H. L., & Pierce, R. U. (2008). Teaching statistics at the primary school level: Beliefs, affordances, and pedagogical content knowledge. In C. Batanero, G. Burrill, C. Reading, & A. Rossman (2008).

Dale, I. (2010). E-learning of statistics in Africa. In C. Reading (Ed.), *Proceedings of the Eighth International Conference on Teaching Statistics.* Voorburg, The Netherlands: International Statistics Institute and International Association for Statistical Education. Online: www.stat.auckland.ac.nz/~iase/publications

Davies, J., & Graff, M. (2005). Performance in e-learning: Online participation and students grades. *British Journal of Educational Technology, 36*(4), 657–663.

Evans, S. R., Wang, R., Haija, R., Zhang, J., Rajicic, N., Xanthakis, V., et al. (2007). Evaluation of distance learning in an introduction to biostatistics course. *Statistical Education Research Journal, 6*(2), 59–77. Online: www.stat.auckland.ac.nz/serj/

Everson, M. G., & Garfield, J. (2008). An innovative approach to teaching online statistics courses. *Technology Innovations in Statistics Education, 2*(1). Online: repositories.cdlib.org/uclastat/cts/tise/

Franklin, C. A., Kader, G., Mewborn, D., Moreno, J., Peck, R., Perry, M., & Scheaffer, R. (2005). *Guidelines for assessment and instruction in statistics education (GAISE) report: A pre-K–12 curriculum framework.* Alexandria, VA: American Statistical Association. Online: www.amstat.org/Education/gaise/

Garfield, J., & Everson, M. (2009). Preparing teachers of statistics: A graduate course for future teachers. *Journal of Statistics Education, 17*(2). Online: www.amstat.org/publications/jse/

Gould, R., & Peck, R. (2005). Inspiring secondary statistics. *MSOR Connections, 5*(3). Online: mathstore.gla.ac.uk/

Gray, B. (2004). Informal learning in an online community of practice. *Journal of Distance Education, 19*(1), 20–35. Online: www.stat.auckland.ac.nz/serj/

Heuer, B. P., & King, K. P. (2004). Leading the band: The role of the instructor in online learning for educators. *The Journal of Interactive Online Learning, 3*(1). Online: www.ncolr.org/

Joubert, M., & Surtherland, R. (2009). *A perspective on the literature: CPD for teachers of mathematics.* Bristol, UK: National Centre of Excellence in the Teaching of Mathematics.

Mathieson, K. (2010). Comparing outcomes between online and face-to-face statistics courses: A systematic review. In C. Reading (Ed.), *Proceedings of the Eighth International Conference on Teaching Statistics.* Voorburg, The Netherlands: International Statistics Institute and International Association for Statistical Education. Online: www.stat.auckland.ac.nz/~iase/publications

Meletiou-Mavrotheris, M., Paparistodemou, E., Mavrotheris, E. Azcárate, P., Serradó, A., & Cardeñoso, J. M. (2008). Teachers' professional development in statistics: The EarlyStatistics European project. In C. Batanero, G. Burrill, C. Reading & A. Rossman (2008).

Philips, B. (2003). Overview of online teaching and Internet resources for statistics education. In J. Engel (Ed.), *Statistics Education and the Internet: Proceedings of the IASE/ISI Satellite Conference on Statistical Education*. Voorburg, The Netherlands: International Statistical Institute and International Association for Statistical Education. Online: www.stat.auckland. ac.nz/~iase/publications

Rourke, L., & Kanuka, H. (2007). Barriers to online critical discourse. *International Journal of Computer Supported Collaborative Learning, 2*(1), 105–126. Online: http://ijcscl.org/

Saporta, G., & Bourdeau, M. (2003). St@tNet: An assessment and new developments. In J. Engel (Ed.), *Statistics Education and the Internet: Proceedings of the IASE/ISI Satellite Conference on Statistical Education*. Voorburg, The Netherlands: International Statistical Institute and International Association for Statistical Education. Online: www.stat.auckland.ac.nz/~iase/ publications

Stahl, G. (2006). *Group cognition: Computer support for collaborative knowledge building*. Cambridge, MA: MIT Press.

Tallent-Runnels, M. K., Thomas, J. A., Lan, W. Y., Cooper, S., Ahern, T. C., Shaw, S. M., et al. (2006). Teaching courses online: A review of the research. *Review of Educational Research, 76*(1), 93–135.

Tudor, G. (2006). Teaching introductory statistics online: Satisfying the students. *Journal of Statistics Education, 14*(3). Online: www.amstat.org/publications/jse/

Utts, J., Sommer, B., Acredolo, C., Maher, M. W., & Matthews, H. R. (2003). A study comparing traditional and hybrid Internet-based instruction in introductory statistics classes. *Journal of Statistics Education, 11*(3). Online: www.amstat.org/publications/jse/

Vermeire, L., Carbonez, A., Darius, P., & Fresen, J. (2002). Just-in-time network based statistical learning: Tools development and implementation. In B. Phillips (Ed.), *Proceedings of the Sixth International Conference on Teaching Statistics*. Cape Town, South Africa: International Statistical Institute and International Association for Statistics Education. Online: www.stat. auckland.ac.nz/~iase/publications

Watson, J. M. (1998). Professional development for teachers of probability and statistics: Into an era of technology. *International Statistical Review, 66*(3), 271–289.

Watson, J. M. (2001). Profiling teachers' competence and confidence to teach particular mathematics topics: The case of chance and data. *Journal of Mathematics Teacher Education, 4*(4), 305–337.

Zhao, Y., & Rop, S. (2001). *A critical review of the literature on electronic networks as reflective discourse communities for inservice teachers*. Online: www.ciera.org/library/reports/ inquiry-3/3-014/3-014.pdf

Chapter 37
The Role of Statistical Offices and Associations in Supporting the Teaching of Statistics at School Level

Delia North and Jackie Scheiber

Abstract In many countries mathematics curricula for primary and secondary schools have been reformed to include statistics. At the same time, national statistics offices have recognised that statistics, if taught meaningfully at school-level, would promote statistical literacy and lead to a better understanding of national statistics office activities, such as census. A number of national statistics offices and statistical associations have thus embarked on projects that develop materials for use in the classroom and/or assist school teachers to engage more meaningfully with the statistics content of the school syllabus. This chapter gives a few specific examples of the roles that national statistics offices and statistical associations around the world are playing in supporting the teaching of statistics at school-level.

1 Introduction

Over the past 30 years it has been well-documented that statistical concepts should ideally be introduced into the school curriculum as early as possible (Holmes, 1980; Wild & Pfannkuch, 1999; Gal, 2002; Franklin et al., 2005), and be built upon from year to year thereafter (Scheaffer, 1998; Shaughnessy, 2006). In many countries statistics content has consequently been included in the full spectrum of mathematics curricula, resulting in mathematics teachers at all levels being tasked with the teaching of statistical concepts. The strong call of the 1990s to develop a

D. North
University of KwaZulu-Natal, Province of KwaZulu-Natal, South Africa
e-mail: northd@ukzn.ac.za

J. Scheiber (✉)
RADMASTE Centre, University of the Witwatersrand,
Private Bag 3, Wits 2050, South Africa
e-mail: jackie.scheiber@wits.ac.za

C. Batanero, G. Burrill, and C. Reading (eds.), *Teaching Statistics in School Mathematics-Challenges for Teaching and Teacher Education: A Joint ICMI/IASE Study*, DOI 10.1007/978-94-007-1131-0_37, © Springer Science+Business Media B.V. 2011

data-orientated teaching of statistics necessitated recognition by the teacher that statistics is a discipline in its own right. Statistics should thus not be taught as only a branch of mathematics, but rather as a discipline that has its own independent intellectual method (Cobb & Moore, 1997; Ben-Zvi, Garfield & Zieffler, 2006; Scheaffer, 2006; Fields, 2008; Pfannkuch, 2008), with an emphasis on inductive reasoning rather than deductive reasoning (Gattuso, 2008; Gattuso & Ottaviani, this book).

It has been widely documented all over the world that some teachers have difficulty in teaching statistics topics, particularly where some analysis of data is called for (Coutinho, 2008). Mathematics teachers are generally more comfortable with the formula-driven approach where there is one correct answer, rather than having to deal with a statistical approach where problems are solved with more than one possible solution (Gattuso & Pannone, 2002). These teachers naturally gravitate towards emphasising the solution phase of a problem, which is typically the least cognitively demanding part of problem-solving and the easiest to teach (Makar, 2008).

Though topics in mathematics and statistics differ from a logical and pedagogical point of view, it is vital to bring home to mathematics teachers that both mathematics and statistics are important and re-enforce each other. It is particularly important for teachers to recognise that the study of statistics in the classroom is an excellent opportunity to use different parts of mathematics (arithmetic, geometry, measurement, algebra) as well as calculators, computers, and the Internet to process data sets, thus demonstrating the importance of the integration of technological tools in mathematics. Teachers should appreciate the fact that statistics brings meaningful contexts and a creative approach to the mathematics lessons, and has the potential to stimulate children's interest in "numbers" and mathematics (Gattuso, 2008). Moreover, several authors suggest that many mathematics teachers, in particular at primary education level, lack specific training in statistics education (Batanero, Godino, & Roa, 2004), resulting in an acute need for them to be trained in order to acquire the skill to design and implement activities geared towards developing statistical literacy (Reston & Bersales, 2008).

National statistical offices around the world encourage statistical literacy amongst their citizens so that the general public has confidence in, and can engage with their output. They have long recognised that the promotion of statistical literacy at school-level is a major step towards creating a statistically literate society (Ridgway, Nicholson, & McCusker, this book). Thus, in order for statistics to be taught more meaningfully in the classroom, it is not unusual for statistics offices to embark on projects that promote access, understanding, and the greater use of their data in the school sector by training school teachers and developing material for their use (Lehohla, 2002; Kong & Harradine, 2006).

This chapter is devoted to describing specific examples of projects run by national statistics offices that aim to improve the teaching of statistics in the classroom. Though most of these projects are specific to a particular country, some, such as CensusAtSchool and the International Statistical Literacy Project (ISLP), have been adopted and adapted by a number of countries, becoming true international projects. Though the space limitation does not allow a comprehensive list of such projects, the projects discussed are those that the authors feel have made a meaningful contribution towards the promotion of statistical literacy at school-level. More information about projects run by statistics offices may be found in Sanchez (2008).

2   National Projects Aimed at Developing Classroom Material

Some national statistical offices, such as those in New Zealand, Canada, and Portugal, have assisted with the development of statistics material whose purpose is to encourage school teachers and students to appreciate statistics, and to ultimately understand the output of the national statistics office. In this section we summarise some of these projects.

2.1   Statistics New Zealand

Under the New Zealand Statistics Act (Statistics New Zealand, 1975) there is an obligation, under the Treaty of Waitangi (Te Triti O Waitangi), to meet the statistical needs of the Maori. In 1990, an achievement initiative for schools was announced by the New Zealand Government. This led to the development of the New Zealand Curriculum Framework and supporting national curriculum statements in the two official languages, English and Maori. Curriculum statements were subsequently developed in Maori, providing for teaching and learning in Maori through Maori knowledge and experiences.

A consultation process was subsequently undertaken between Statistics New Zealand, the education sector, teacher training providers, and teachers to determine the content and the most appropriate medium for the development and delivery of a cross-curricular statistics resource for schools. This resource was developed to meet the outcomes from the essential learning areas of Mathematics, Social Studies, English, Technology, and Geography. The resource consisted of an English booklet, a Maori resource booklet (Te kete Tatauranga), and a map, with supporting material on the Internet. The resource provides a cross-curricula learning programme about the census, strongly emphasising the importance of understanding and interpreting the significance of information, for both English and Maori teachers.

The resource was made freely available to schools and well publicised in the media, and a series of workshops were held throughout New Zealand to raise awareness of the resource. Eighty-nine percent of all primary and secondary schools in the country registered to receive the resource, and subsequent evaluation has shown that teachers found the resource to be user friendly. They particularly appreciated the good links with the national curriculum (Hooper, 2002). In the mid-1990s, the Schools' Corner was introduced as a dedicated part of the Statistics New Zealand website (www.stats.govt.nz/methods_and_services/schools_corner.aspx). This website targets teachers and contains background information to official statistics, tables, links, and activities directly related to the curriculum that can be printed and used in the classroom. The site is currently visited approximately 5,000 times per month and is viewed as a key way to reach the school audience and to achieve greater statistical literacy (Forbes, 2008).

2.2   Statistics Canada

Statistics Canada launched the *Census results teachers' kits*, a set of lessons for teachers to use in the classroom that deal with the results of the 2001 Census. These activities are appropriate for English, Mathematics, Theater Arts, Social Sciences, Geography, History, Family Studies, and Informatics, and suggested grade levels are indicated on each activity. Similar census results teachers' kits were produced based on the 2006 Census. Each activity follows the critical challenge approach and presents an engaging question or task. Learners are taught to become competent in reaching reasoned judgments as they locate, use, interpret, and assess information.

Two census results websites (www12.statcan.ca/english/census01/teacher's_kit/ and www12.statcan.ca/english/census06/teacherskit/) have been developed so that teachers could access the activities for use in their classroom.

Statistics Canada's interactive website (www.statcan.ca/english/edu) offers free access to more curriculum-relevant information, learning tools and resources specifically designed for teachers and students. The education community accounts for about 40% of the access to the Statistics Canada website, up from 19% in 1997. Close to 20,000 users a day log onto the website looking for information to help with homework or classroom assignments (Townsend, 2008).

2.3   Statistics Portugal

ALEA (Local Action of Applied Statistics, www.alea.pt) is a project based in Portugal and is aimed at providing teaching materials for the study of statistics to both teachers and pupils via the Internet. It is a joint project between Statistics Portugal and Tomaz Pelayo Secondary School together with the Portuguese Ministry of Education through its Northern Regional Education Department (DREN) (Campos, 2008).

The website was launched in 1999 and aims at providing effective tools for the understanding and teaching of statistics. Assorted content is provided; including introductory courses to statistics and probability; teaching materials, important names and dates in statistics, and educational games. The ALEA project also offers regular e-learning courses for both basic and secondary education teachers. Since its launch, the site has received around half a million visitors, and was, in 2007, the first winner of the Best Cooperative Project Award conferred by the International Association for Statistical Education (IASE).

3   International Projects Aimed at Developing Classroom Material

Several statistics offices and associations have launched projects with an international focus, where the emphasis is on introducing the school children to statistics outputs on a global scale. This section focuses on two such projects: CensusAtSchool and the International Statistical Literacy Project (ISLP), each of which is discussed extensively.

3.1   CensusAtSchool

A paper on a census of school children in New Zealand, written by Forbes (1996), was reviewed in Induzioni (Forbes, 1997) and subsequently caught the attention of Italy's National Institute for Statistics (ISTAT). The potential of a census of school children as a vehicle for creating awareness of the national Italian census was immediately recognised by ISTAT. Subsequently the Italian Society for Statistics, the Ministry for Public Education, and the Italian Mathematics Society for Statistics joined forces to plan the CensusAtSchool activities in Italy. ISTAT was tasked with project management of the CensusAtSchool initiative, and supplied the schools with all the necessary material and offered technical advice to teachers. This national project ran from early 2000 until 2001 and a total of 190,000 school children in primary and secondary school participated (Conti & Lombardo, 2002).

The United Kingdom census of April 2001 was seen as an ideal opportunity to take the message of census into the classroom. The Royal Statistical Society Centre for Statistical Education (RSSCSE) approached the United Kingdom Office of National Statistics (ONS) in 1999 for funding to launch the CensusAtSchool project in the United Kingdom. This project involves children between the ages of 7 and 16 gathering information about themselves. This information forms the basis of a national database which school children can use in the data handling part of various subject areas in the school curriculum. Though the idea of involving children at school in a census activity originated in New Zealand, the United Kingdom made a very significant contribution to this initiative by launching a dynamic website (www.censusatschool.com). This website contains a huge database of 60,000 children's responses. The database is anonymised and a random selector is used to access raw records of around 200 candidates which are then sent to schools on request. A variety of curriculum tasks, in the form of worksheets, are posted on the web, along with the results from CensusAtSchool projects from other countries including Australia (Queensland), Canada, New Zealand, and South Africa. The children are thus able to get an international perspective on how their class data compares with the United Kingdom as a whole, and also with other countries (Connor, 2002; Knights, 2006).

South Africa launched a CensusAtSchool project to coincide with its 2001 Census. In the video which was made to promote the initiative, and that is summarised by Connor (2002), Pali Lehohla, Statistician General of South Africa, stated that CensusAtSchool brings the data home to the people and the teachers so that they can relate to it.

Statistics South Africa called on the services of the RSSCSE in the United Kingdom to act as consultants to plan the initial stages of the CensusAtSchool project. The project targeted both primary and secondary school children, gathering information which was then used to devise worksheets and other materials to enhance the teaching of relevant data-handling topics by using fun activities. A total of 277 schools and 43,500 children spread around all provinces in South Africa took part in the pilot programme. Though the CensusAtSchool project in South Africa had the advantage of drawing on the successes of the projects launched in other countries, the constraint of inadequate access to technology in many South African schools had a dampening effect on the potential of offering a rich source of materials on the website, as had been done in the United Kingdom. The South African CensusAtSchool initiative was thus not as "high tech" as the United Kingdom, focusing rather on a resource pack (hard copy) being delivered to participating schools.

In 2003, Statistics Canada joined the international trend of having a CensusAtSchool project. Their CensusAtSchool project took the form of an online survey where class data was subsequently analysed and compared to national and international data. CensusAtSchool was recognised as being a wonderful vehicle for the new math and technology curricula in Canada as it was a vehicle for obtaining rich data bases about student lifestyles. The Canadian CensusAtSchool project had a very strong level of teacher involvement. Statistics Canada decided to set up a Teacher Advisory Board made up of teachers from across Canada to assist with the setting up of the fundamental infrastructure, as well as helping to formulate questions for the online survey. Teachers were thus heavily involved from the outset and had a sense of ownership of the project, leading them to help build a community of over 900 participating teachers who received e-mail communications through Statistics Canada's bi-monthly learning resources listserv (Townsend, 2006).

The CensusAtSchool project was also adopted by the Australian Bureau of Statistics (ABS) and commenced in October 2005. The project was aimed at school children between ages of 10 and 17. The main drivers were the promotion of statistical literacy amongst students, the encouragement of effective and practical use of ICT in teaching and learning, the creation of a stimulating student-centred learning environment, and the education of future users of ABS publications. In Australia the CensusAtSchool project took the form of an online data collection project designed for primary and secondary school children, very much as was the case in the United Kingdom. The ABS obtained permission from the RSS in United Kingdom for the use of the CensusAtSchool branding. Further details about the experiences, the processes followed, and the successes/challenges of the CensusAtSchool initiative in Australia may be found in Kong and Harradine (2006).

3.2 The International Statistical Literacy Project (ISLP)

The International Statistical Literacy Project (ISLP) of the International Statistical Institute has as its main objective the contribution to the statistical literacy of the young and of adults in all walks of life across the world, and falls under the umbrella of the International Association for Statistical Education (IASE), a section of the International Statistical Institute (ISI). The project provides an online repository of resources that are useful for acquiring and developing statistical literacy at all levels from primary/elementary school through to adult learners. There are also web pages for official statisticians and for journalists and the mass media. Further, there are web pages devoted to statistical literacy projects, websites, etc that have been developed by national statistics offices, national statistics societies, and other non-profit organisations. These are found at www.stat.auckland. ac.nz/~iase/islp/.

The International Statistical Literacy Competition is run by the ISLP with the threefold aim of increasing the awareness of statistics among students and teachers throughout the world, of promoting statistical literacy resources, and of bringing together parties interested in statistical literacy in each country. Registration for the 2009 competition opened in September 2007 and was aimed at children in the 10–18 year age group. Participation was free and students registered through their school teachers. This project had three phases. In phase one, teachers received questions which the students answered within their own school. The completed tests were sent off to ISLP for grading. From these results, winners in each school were identified. In phase two, winners from the various schools competed under the auspices of the national statistics offices. The process ran until a national winner was identified for each participating country. In phase three, the winners from each participating country travelled to Durban, South Africa, for the final stage of the competition held during the 57th session of the ISI.

The 2011 ISLP competition takes the form of a poster competition, and is divided into two age categories: 12–15 year-old students and 16–18 year-old students. In phase one, the national competition, students will have to submit their posters to the ISLP coordinator for their country. In phase two, the international competition, the winning posters from each country compete against each other. Prizes will be awarded to the international winners at the 58th session of the ISI in Dublin, Republic of Ireland in 2011.

4 Projects Aimed at Training Teachers

Some statistics offices like those in Canada and South Africa have projects whose aim is to assist school teachers to teach statistics more meaningfully. Below is a short summary.

4.1   Statistics Canada

A new grade 12 course on the mathematics of data management was introduced in Ontario, Canada in 2003. Teachers were expected to teach how to apply sampling, modelling, and statistical analysis techniques, with 20% of the students' final mark based on an analytical data project.

Statistics Canada has a long history of being a trusted resource in the education community and immediately helped by providing large quantities of relevant data for student projects, and educational representatives at Statistics Canada assisted teachers and school children with manipulation of data. Their network of regional representatives help educators integrate Canadian statistics into teaching, learning, and research activities. They offer workshops at professional development days and conferences. These regional representatives distribute communication materials to 40,000 in-service teachers each year, as well as to more than 20,000 pre-service teachers at 60 faculties of education. The effectiveness of a project such as this is difficult to quantify. Statistics Canada looks at this as a long-term project, but has been able to demonstrate remarkable success in the short-term, and is continuing with their educational outreach programme.

4.2   Statistics South Africa

A new school curriculum became policy in South Africa in 2002. Due to the recognition of the cross-curricular need for statistics, statistical topics were included at all levels (North & Scheiber, 2008). This was in direct contrast to what had been the case previously when statistics had virtually been totally absent from the school syllabus (North & Zewotir, 2006). South Africa thus introduced statistics into the school syllabus long after more developed countries had done so, offering the distinct advantage that teachers did not have to be "untrained" in bad habits, but could rather benefit from research and experiences in statistics education from around the world (North & Zewotir, 2006).

Training teachers in South Africa to deliver the new statistics content meaningfully is a challenge, as is the case all over the world. However, South Africa is a very complex society with social, economic, and cultural diversity, confounding the problem of training teachers to produce statistically-literate school leavers. It would thus take a huge commitment and careful planning to put initiatives in place to upgrade statistical knowledge of teachers (North & Scheiber, 2008).

Following the emergence from a political legacy that strove to create a dysfunctional society, the South African government pledged to address the task of improving levels of basic literacy and numeracy in particular, striving for the economic literacy of the nation. Statistical literacy, being the basic ingredient for economic literacy, was thus given specific prominence in post-apartheid restructuring, resulting in an impressive budget allocation to assist with the teaching of statistics at school-level (Lehohla, 2002). Accordingly, in order to address the problem of providing statistics training to roughly 10,000 mathematics educators (grades 10–12) in South Africa, Statistics South Africa, launched the *maths4stats* campaign (North & Scheiber, 2008).

Teachers were selected via a competition advertised in the National Press and were taken to the Seventh International Conference on Teaching of Statistics (ICOTS-7) in Brazil as a means of kick-starting the project. They then attended courses over weekends and during vacations where they were taught the new statistics content. They had to obtain at least 75% of the maximum score in a test based on the content contained in the new curriculum and had to also pass an oral presentation which was judged by the project facilitators. Once fully endorsed *maths4stats* trainers, they were then eligible to run further workshops in their districts for teachers using the material prepared by the two project facilitators.

A national coordinator was appointed to oversee and drive the project and coordinators were appointed in each of the nine provinces. These provincial coordinators were fully trained and were in a position to meaningfully engage with the statistics content of the school syllabus. They were expected to run workshops in their province, and had a group of *maths4stats* trainers to call on, and oversee. Both the national coordinator and provincial coordinators were fully employed by Statistics South Africa.

In April 2008 a group of 172 Department of Education (DoE) subject advisors, from all nine provinces in South Africa, were trained by the *maths4stats* project facilitators. These subject advisors form a link between the Statistics South Africa initiative and the DoE, a link that is crucially necessary to ensure that the roll-out plan of the *maths4stats* campaign will be taken seriously by mathematics teachers. The *maths4stats* campaign has managed to excite teachers about the new statistics content of the school curriculum, and teachers are becoming more and more enthusiastic about the relevance of statistics in teaching "thinking maths", seeing the statistics lessons as an opportunity to make the mathematics class more fun. The *maths4stats* project won the award for the Best Cooperative Project in Statistics Literacy during the International Statistical Institute conference in Durban, South Africa in August 2009.

5   Implications for Training the Teachers

All over the world the inclusion of statistics into the mathematics curriculum for primary and secondary school has challenged teachers to engage more meaningfully with statistics topics, more specifically with the call for data-orientated teaching of statistics. Many statistics offices and associations around the world have recognised the potential for promoting statistical literacy at school level, and have accordingly launched various projects to support the statistics training of teachers and to provide them with material which they can use in their classrooms. The challenge is to get every statistics office and association to do the same.

Teachers want resources that are relevant to their curriculum and are statistically correct. In the projects described, teachers with access to the Internet could obtain resources from their statistics office's website. The challenge is to provide resources like this to teachers in the rest of the countries around the world, especially to those teachers who do not have access to the Internet. At the same time, as more and more countries are ensuring that their schools are connected to the Internet, both pre-service and in-service teachers will need to attend courses related to the pedagogical knowledge needed to use computers in the classroom (see Lee & Hollebrands, this book), otherwise these valuable resources will be lost to them. The challenge is to encourage the teacher trainers and/or statistics offices and associations to provide these courses.

References

Batanero, C., Burrill, G., Reading, C., & Rossman, A. (Eds.). (2008). *Joint ICMI/IASE Study: Teaching Statistics in School Mathematics. Challenges for Teaching and Teacher Education. Proceedings of the ICMI Study 18 and 2008 IASE Round Table Conference.* Monterrey, Mexico: International Commission on Mathematical Instruction and International Association for Statistical Education. Online: www.stat.auckland.ac.nz/~iase/publications

Batanero, C., Godino, J., & Roa, R. (2004). Training teachers to teach probability. *Journal of Statistics Education, 12*(1). Online: www.amstat.org/publications/jse/

Ben-Zvi, D., Garfield, J. B., & Zieffler, A. (2006). Research in the statistics classroom: Learning from teaching experiments. In G. Burrill (Ed.), *Thinking and reasoning with data and chance: 68th NCTM Yearbook (2006)* (pp. 467–481). Reston, VA: National Council of Teachers of Mathematics.

Campos, P. (2008). Thinking with data: The role of ALEA in promoting statistical literacy in Portugal. In J. Sanchez (Coord.), *Government statistical offices and statistical literacy.* Los Angeles, CA: International Statistical Literacy Project. Online: www.stat.auckland.ac.nz/~iase/islp/

Cobb, G. W., & Moore, D. S. (1997). Mathematics, statistics and teaching. *American Mathematical Monthly, 104*(9), 801–823.

Connor, D. (2002). CensusAtSchool 2000: Creation to collation to classroom. In B. Phillips (Ed.), *Proceedings of the Sixth International Conference on Teaching Statistics.* Cape Town, South Africa: International Statistical Institute and International Association for Statistics Education. Online: www.stat.auckland.ac.nz/~iase/publications

Conti, C., & Lombardo, E. (2002). The Italian CensusAtSchool. In B. Phillips (Ed.), *Proceedings of the Sixth International Conference on Teaching Statistics.* Cape Town, South Africa: International Statistical Institute and International Association for Statistics Education. Online: www.stat.auckland.ac.nz/~iase/publications

Coutinho, S. (2008). Teaching statistics in elementary and high school and teacher training. In C. Batanero, G. Burrill, C. Reading, & A. Rossman (2008).

Fields, P. J. (2008). A case study in collaboration preparing secondary education teachers. In C. Batanero, G. Burrill, C. Reading, & A. Rossman (2008).

Forbes, D. S. (1996). Raising statistical awareness. *Teaching Statistics, 18*(3), 66–69.

Forbes, D. S. (1997). Il "Censimento dei bambini" in Nuova Zelanda: Ovvero come elevare la consapevolezza statistica (The children's census in New Zealand: How to raise an awareness of statistics). *Induzioni, 14,* 105–113.

Forbes D. S. (2008). Raising statistical capability: Statistics New Zealand's contribution. In J. Sanchez (Coord.), *Government statistical offices and statistical literacy.* Los Angeles, CA: International Statistical Literacy Project. Online: www.stat.auckland.ac.nz/~iase/islp/

Franklin, C., Kader, G., Mewborn, D. S., Moreno, J., Peck, R., Perry, M., & Scheaffer, R. (2005). *A curriculum framework for a pre-K-12 statistics education. GAISE report.* American Statistical Association. Online: www.amstat.org/education/gaise/

Gal, I. (2002). Adults' statistical literacy. Meanings, components, responsibilities. *International Statistical Review, 70*(1), 1–25.

Gattuso, L. (2008). Mathematics in a statistical context? In C. Batanero, G. Burrill, C. Reading, & A. Rossman (2008).

Gattuso, L., & Pannone, M. A. (2002). Teacher's training in a statistics teaching experiment. In B. Phillips (Ed.), *Proceedings of the Sixth International Conference on Teaching Statistics.* Cape Town, South Africa: International Statistical Institute and International Association for Statistics Education. Online: www.stat.auckland.ac.nz/~iase/publications

Holmes, P. (1980). *Teaching statistics 11–16.* Slough, UK: Foulsham Educational.

Hooper, L. (2002). Making census count in the classroom. In B. Phillips (Ed.), *Proceedings of the Sixth International Conference on Teaching Statistics.* Cape Town, South Africa: International Statistical Institute and International Association for Statistics Education. Online: www.stat.auckland.ac.nz/~iase/publications

Knights, E. (2006). Data handling from a classroom perspective in the United Kingdom. In A. Rossman & B. Chance (Eds.), *Proceedings of the Seventh International Conference on Teaching Statistics.* Salvador, Bahia, Brazil: International Statistical Institute and International Association for Statistical Education. Online: www.stat.auckland.ac.nz/~iase/publications

Kong, S., & Harradine A. (2006). CensusAtSchool in Australia. In A. Rossman & B. Chance (Eds.), *Proceedings of the Seventh International Conference on Teaching Statistics.* Salvador, Bahia, Brazil: International Statistical Institute and International Association for Statistical Education. Online: www.stat.auckland.ac.nz/~iase/publications

Lehohla, P. (2002). Promoting statistical literacy: A South African perspective. In B. Phillips (Ed.), *Proceedings of the Sixth International Conference on Teaching Statistics.* Cape Town, South Africa: International Statistical Institute and International Association for Statistics Education. Online: www.stat.auckland.ac.nz/~iase/publications

Makar, K. (2008). A model of learning to teach statistical inquiry. In C. Batanero, G. Burrill, C. Reading, & A. Rossman (2008).

North, D., & Scheiber, J. (2008). Introducing statistics at school level in South Africa. The crucial role played by the National Statistics Office in training in-service teachers. In C. Batanero, G. Burrill, C. Reading, & A. Rossman (2008).

North, D., & Zewotir, T. (2006). Introducing statistics at school level in South Africa. In A. Rossman & B. Chance (Eds.), *Proceedings of the Seventh International Conference on Teaching Statistics.* Salvador, Bahia, Brazil: International Statistical Institute and International Association for Statistical Education. Online: www.stat.auckland.ac.nz/~iase/publications

Pfannkuch, M. (2008). Training teachers to develop statistical thinking. In C. Batanero, G. Burrill, C. Reading, & A. Rossman (2008).

Reston, E., & Bersales, L. G. (2008). Reform efforts in training statistics teachers in the Philippines: Challenges and prospects. In C. Batanero, G. Burrill, C. Reading, & A. Rossman (2008).

Sanchez, J. (Coord.). (2008). *Government statistical offices and statistical literacy.* Los Angeles: International Statistical Literacy Project. Online: www.stat.auckland.ac.nz/~iase/islp/

Scheaffer, R. L. (1998). Statistics education: Bridging the gaps among school, college and the workplace. In L. Pereira-Mendoza (Ed.), *Proceedings of the Fifth International Conference on Teaching Statistics.* Singapore: International Statistical Institute and International Association for Statistical Education. Online: www.stat.auckland.ac.nz/~iase/publications

Scheaffer, R. L. (2006). Statistics and mathematics: On making a happy marriage. In G. Burrill (Ed.), *Thinking and reasoning with data and chance: 68th NCTM Yearbook (2006)* (pp. 309–321). Reston: National Council of Teachers of Mathematics.

Shaughnessy, J. M. (2006). Student work and student thinking: An invaluable source for teaching and research. In A. Rossman & B. Chance (Eds.), *Proceedings of the Seventh International Conference on Teaching Statistics.* Salvador, Bahia, Brazil: International Statistical Institute and International Association for Statistical Education. Online: www.stat.auckland.ac.nz/~iase/publications

Statistics New Zealand. (1975). New Zealand Statistics Act. Online: http://www.stats.govt.nz/

Townsend, M. (2006). Measuring success: How CensusAtSchool engages Canadian students in active learning outcome. In A. Rossman & B. Chance (Eds.), *Proceedings of the Seventh International Conference on Teaching Statistics*. Salvador, Bahia, Brazil: International Statistical Institute and International Association for Statistical Education. Online: www.stat.auckland.ac.nz/~iase/publications

Townsend, M. (2008). Statistics Canada. In J. Sanchez (Coord.), *Government statistical offices and statistical literacy*. Los Angeles, CA: International Statistical Literacy Project. Online: www.stat.auckland.ac.nz/~iase/islp/

Wild, C., & Pfannkuch, M. (1999). Statistical thinking in empirical enquiry. *International Statistics Review, 67*(3), 223–265.

Overview: Challenges for Teaching Statistics in School Mathematics and Preparing Mathematics Teachers

Carmen Batanero, Gail Burrill, and Chris Reading

1 Introduction

For five years, a group of mathematics and statistics educators worked in collaboration to reflect on the teaching of statistics in school mathematics and on the training of those teachers who are responsible for this teaching, under the auspices of the International Commission on Mathematical Instruction (ICMI) and the International Association for Statistical Education (IASE). Results from this work are reflected first in the Proceedings of the Joint ICMI/IASE Study Conference held in Monterrey in 2008 (Batanero, Burrill, Reading, & Rossman, 2008) and second in this book. These two documents have contributed to raising awareness of the need for increased statistical content at school levels to improve statistical literacy in young students around the world as well as awareness of the related challenges in training and supporting mathematics teachers who teach statistics.

For each of the initial Topics in the Joint Study, this final chapter gives a reflective summary of the discussions held at the Study Conference (part of which were first analysed in Batanero & Díaz, 2010) and the main ideas discussed throughout this book.

C. Batanero (✉)
Departamento de Didáctica de la Matemática, Universidad de Granada,
Facultad de Ciencias de la Educación, Campus de Cartuja, 18071 Granada, Spain
e-mail: batanero@ugr.es

G. Burrill
Michigan State University, 240 Erickson, East Lansing, MI 48824, USA
e-mail: burrill@msu.edu

C. Reading
SiMERR National Centre, University of New England, Education Building,
Armidale, NSW 2351, Australia
e-mail: creading@une.edu.au

C. Batanero, G. Burrill, and C. Reading (eds.), *Teaching Statistics in School*   407
Mathematics-Challenges for Teaching and Teacher Education: A Joint ICMI/IASE Study,
DOI 10.1007/978-94-007-1131-0, © Springer Science+Business Media B.V. 2011

2   The Situation of Teaching Statistics at the School Level

As suggested by Batanero and Díaz (2010), reasons for including statistics at the school level were repeatedly highlighted over the last decades of the past century (e.g., by Holmes, 1980; Hawkins, Jolliffe, & Glickman, 1991; Gal, 2002), for example, the usefulness of statistics for daily life, the important role of statistics in developing critical reasoning and the instrumental role of statistics in other disciplines and in many professions. More recently, the *Principles and Standards for School Mathematics* (National Council of Teachers of Mathematics, 2000) and the *Guidelines for Assessment and Instruction in Statistics Education* (GAISE) project (Franklin et al., 2005) were influential in further developing statistics education in the school curriculum in the United States of America and in other countries.

Papers included in the Joint Study Conference Topic 1 and in Chaps. 1–4 describe different perspectives and approaches to teaching statistics in the school curricula, depending on national policies, availability of resources within a country and the relevance given to different topics and grade level. However, a general tendency is that statistics is now taught at very early ages in many countries; in some, 6-year-old children start studying basic statistical concepts and continue to develop these concepts in all the curricular levels until secondary school, where students may study elements of statistical inference. In addition, quick innovation and globalisation in the past decade led to a new perceived complexity of reality that affected the mathematics curriculum, with a shift from content knowledge to competences (Gattuso & Ottaviani, this book) that has also been reflected in statistics.

3   Teachers' Attitudes, Conceptions and Beliefs

While the world is changing rapidly with respect to the prevalence and use of statistics, the curriculum in schools tends to be slow to respond to these changes. Although statistics as a content domain is widely accepted, typically statistics is not an independent topic in the school curriculum but is taught as part of mathematics. Consequently there is a need for a better preparation of primary and secondary school mathematics teachers, who are responsible for teaching statistics at these levels.

Teachers' statistical conceptions and beliefs deserve attention, since mathematics teachers' thinking is the key factor in any movement towards changing mathematics teaching and determines both the students' knowledge and the students' beliefs concerning mathematics and hence statistics (Batanero & Díaz, 2010). These issues were debated in the Joint Study Conference Topic 2 and in Part III of this book.

3.1   Teachers' Attitudes and Beliefs

Teacher education is usually focused on improving teachers' knowledge with relatively little attention paid to teachers' beliefs or attitudes. However, such factors can influence the way teachers teach statistics and the extent to which teachers will apply statistics outside the classroom (Gal & Ginsburg, 1994). Research presented in the Joint Study Conference (Arnold, 2008; Estrada & Batanero, 2008; Lancaster, 2008) and summarised in this book (Chick & Pierce; Estrada, Batanero & Lancaster) suggests that while teachers are willing to learn about and spend more time teaching statistics and acknowledge the practical importance of statistics, they feel their students experience greater difficulties in statistics than in other mathematical topics, and they consider themselves not well prepared to help their students face these difficulties.

Teachers also have beliefs about instructional goals and how they are linked with instructional content. For example, some teachers may have a dynamic versus a static view of mathematics and an orientation towards formal mathematics versus mathematical applications (Eichler, this book), which will in turn affect how they present topics in statistics. Thus, the implemented curricula for similar content might differ considerably depending on the teachers' objectives or beliefs.

Some researchers suggest that certain types of knowledge – including an understanding of how students learn specific statistical concepts – are best obtained in continuing professional development after the teacher has had some experience in the classroom (Ponte, this book). However, there is lack of opportunity for teachers' professional development in statistics because they do not actually teach much statistics and rarely use statistics to analyse educational data even though in general it is relevant to their work as teachers.

3.2   Teachers' Statistical Knowledge

Many activities in which teachers engage, such as "figuring out what students know; choosing and managing representations of mathematical ideas; appraising, selecting and modifying textbooks; deciding among alternative courses of action" involve mathematical reasoning and thinking (Ball, Lubienski, & Mewborn, 2001, p. 453). Consequently, teachers' statistical knowledge plays a significant role in the quality of their teaching since teachers' instructional decisions in the statistics classroom are dependent on this knowledge (Batanero & Díaz, 2010).

This is cause for concern as the research summarised in this book shows that many teachers unconsciously share a variety of difficulties and misconceptions with their students with respect to fundamental statistical ideas. Examples of teachers' difficulties with statistical concepts described in this book include: having little real understanding of the mean and median, having difficulties in creating or interpreting graphs, using only verbal reasoning with respect to variation, having little understanding of standard deviation as a measure of sample homogeneity, comparing distributions only in terms of averages, confusing correlation and causation, or viewing a statistical test as a mathematical proof of a hypothesis.

In addition to specific statistical concepts, teachers may have difficulty when implementing an experimental approach to teaching probability or teaching through statistical investigations (Stohl, 2005). Because few teachers have prior experience using statistical investigations to conduct probability experiments or simulations, they may miss opportunities to foster students' statistical reasoning when engaging students in statistical investigations or experiments. For example, their approaches to using an empirical approach to probability may rely almost exclusively on small sample sizes and fail to address the heart of the issue (Lee & Hollebrands, 2008).

3.3   Teachers' Pedagogical Content Knowledge to Teach Statistics

In addition to being proficient in mathematics, Shulman (1987) described other types of knowledge needed by teachers to be competent in the mathematics classroom. Highly relevant to several chapters in this book is Shulman's conceptualisation of pedagogical content knowledge (PCK) as a special mixture of content and pedagogy that is specific for a topic and that teachers develop as a consequence of professional practice.

Different models to describe the professional knowledge needed to teach statistics were discussed in this book. Some of them derive from frameworks taken from mathematics education and include complex components such as epistemology, instructional resources, knowledge about students' learning, capacity to implement adequate discourse and communication in the classroom, and capacity to adapt to the global school curriculum and social factors (Godino, this book).

Other authors (e.g., Burgess, this book) offer their own specific model of pedagogical content knowledge for statistics education that takes into account statistical reasoning (e.g., as described by Wild & Pfannkuch, 1999) or concerns the pedagogical expertise for effectively engaging students in learning data analysis and probability with technology (Lee, Hollebrands, & Wilson, 2010; Lee & Hollebrands, this book).

The scarce research related to PCK presented at the Joint Study Conference and summarised in this book suggests that the knowledge required for teaching is often weak. For example, in González and Pinto's (2008) research, pre-service secondary school mathematics teachers had no training in matters related to the curriculum and the processes of learning and teaching; had a scant knowledge of graphical representation; and did not perceive the different cognitive levels associated with graphs or the various components and processes linked to their interpretation. In another example, Chick and Pierce's (2008) research showed that some teachers lacked the competence to plan a lesson; they did not recognise the statistical concepts that could be developed from a particular task or data set and missed opportunities that are inherent in the task.

The last issue raised in Topic 2 was the need to prepare instruments to measure teacher's statistical knowledge (Callingham & Watson, this book). Questionnaires, with PCK items based on student survey items used in earlier studies and students' actual responses that ask teachers to predict a range of responses their students might give or how they might intervene to address inappropriate responses together with statistical analysis, can be used to obtain a measure of teacher expertise in relation to professional knowledge for teaching statistics.

4   Analysing Current Practices in the Training of Teachers, Including Developing Countries

Reports from different participants at the Joint Study Conference agreed that many of the current teacher training programmes do not yet adequately educate those who are teaching statistics for their task to prepare statistically-literate citizens. Even when many prospective secondary teachers have a major in mathematics, few of them have received specific preparation in designing sample collections or experiments, analysing data from real applications or using statistical software (Batanero & Díaz, 2010). These teachers also need education in the pedagogical knowledge related to teaching statistics as described above, given that teaching mathematics is different from teaching statistics (see Franklin et al., 2005, or Burrill & Biehler, this book).

The situation is even more challenging for primary teachers, since in many countries statistics is included in the school curriculum for children beginning in Grade 1 (6-year olds). Clearly, teaching statistics to these children needs different approaches, tasks and methods than teaching statistics in secondary or high school, so primary school teachers, in addition to their knowledge of other basic disciplines, require a profound knowledge of children's cognitive development in statistics and probability. In spite of this need, few primary school teachers have had suitable training in either theoretical or applied statistics, and traditional introductory statistics courses will not provide them with the didactical knowledge they need (Batanero, Godino, & Roa, 2004; Stohl, 2005; Franklin & Mewborn, 2006). Papers included in the Joint Study Conference Topic 3 and in Chaps. 5–9 discussed different examples of successful experiences with courses specifically directed to train teachers to teach statistics in different countries, some of them based on theoretical models prescribing how this training should be.

Topic 5, *Training teachers in developing countries*, was included in the conference to engage countries to study their specific problems. Presentations from Botswana, Central-America, China, Iran, the Philippines, South Africa and Uganda, among other countries, showed that the problems concerning the way in which teachers are specifically educated to teach statistics are similar to those described for developed countries. Because successful educational initiatives for teachers from statistical agencies or educational authorities in Central America, China, Iran, the Philippines, South Africa and Uganda were presented and discussed in other sections of this book, Training teachers in developing countries does not appear as a separate section in the book.

5   Empowering Teachers to Teach Statistics

Part IV in this book collects together suggestions and experiences in the education of teachers that were presented at the Joint Study Conference Topic 4. A consensus in the chapters in this part is the need for finding meaningful approaches for preparing teachers, as teachers do not seem to automatically gain new knowledge through participation in professional development courses (Arnold, 2008). Some suggested approaches in the training of teachers include: promoting teachers' statistical literacy and statistical reasoning; engaging teachers with real data and training teachers with project work and statistical investigations; working with technology; and connecting teacher education to their own practice. Below are comments on these approaches.

Promoting teachers' statistical literacy (Ridgway, Nicholson, & McCusker, this book) *and statistical reasoning* (Pfannkunch & Ben-Zvi, this book). In many countries, statistical offices and agencies are providing resources that can be used to support the introduction of statistical literacy in schools. However, without wide-reaching education and professional development of teachers, such resources are unlikely to have an impact on students. Moreover, in order for teachers to develop a deep and meaningful understanding of statistics that later they can use to help students develop the ability to think and reason statistically, it is important to create *a statistical reasoning learning environment* in courses they take that later they can use in their own teaching (Garfield & Ben-Zvi, 2009).

Engaging teachers with real data (Hall, this book) *and statistical investigations* (Makar & Fielding-Wells, this book). A conclusion of the conference discussion was that teachers should experience the full cycle of research with statistical projects, if the goal is to change how statistics is experienced in the classroom. Moreover, when time available for working with teachers is scarce, some papers (e.g., Godino, Batanero, Roa, & Wilhelmi, 2008; Batanero & Díaz, 2010) suggested that a formative cycle where teachers are first given a statistical project and then carry out a didactical analysis of the project can help to simultaneously increase the teachers' statistical and pedagogical knowledge.

Working with technology can be used both as amplifier and reorganiser to engage teachers in tasks that simultaneously develop their understanding of statistical ideas and allow them to experience how technology tools can be useful in fostering statistical thinking (Lee & Hollebrands, 2008). However, teachers also need adequate pedagogical knowledge about how to use technology in the statistics classroom.

Connecting teacher education to their own practice and promoting collaborative work among teachers (Ponte, this book) are essential to improving professional practice. It is through the exchange of ideas and materials among teachers who have common problems and needs that new ideas emerge for the introduction of new activities, new practices or new competencies (Arnold, 2008). In particular, *analysing collective case studies* and discussing teaching experiences and students' responses to given tasks can reveal the teachers' lack of specific knowledge of some statistical concepts and promote their statistical and pedagogical content knowledge (Groth & Xu, this book). The affordances offered by modern Internet technologies provide new distance-learning opportunities for the pre-service and in-service training of teachers, making it possible to overcome the restrictions of shrinking resources and geographical locations and to offer high-quality learning experiences to geographically dispersed teachers (Meletiou & Serrado, this book).

6 Collaboration in Teacher Education

Because of the inter-disciplinary nature of statistics, cooperation is both natural and beneficial for those involved in all aspects of statistics education. Topic 6 in the Joint Study Conference solicited the presentations of successful experiences of collaboration between countries, institutions or university departments in the training of teachers. Book chapters that describe examples of such collaborations have been spread throughout different sections.

The preparation of mathematics teachers has historically been the responsibility of mathematicians and mathematics educators, although recently statisticians have started to play a major role in teacher preparation in a few countries (Batanero & Díaz, 2010). For example, in the United States of America the GAISE framework (Franklin et al., 2005) was written, in collaboration between mathematics educators and statisticians, to provide guidance to those involved with teacher preparation.

In addition, in many countries statistical offices and associations are increasingly involved in producing materials and organising initiatives to help increase statistical literacy. Two examples are North and Scheiber (this book) who describe the data and materials provided by Statistics South African and the CensusAtSchool project and associated professional development workshops provided by Canada's National Statistical Agency (Hall, this book). Other examples include work by the Philippines Statistical System and the Philippines Statistical Association (Reston & Bersales, this book), the Iranian Statistical Society (Persian & Rejali, this book) and the institutions collaborating in Guangzhou, China in teacher training (Shihong, Yongdong, Bangquan, & Reisheng, 2008). Collaboration between countries on research projects also served to join efforts of mathematics educators to develop a professional development programme for teachers (Meletiou-Mavrotheris et al., 2008).

7 New Issues Raised at the Conference

The interest of mathematics and statistics educators towards this Joint ICMI/IASE Study is evident in the chapters of this book, many of which have been written by teams that have not previously collaborated and include people from different countries and different academic backgrounds. The Joint ICMI/IASE Study Conference Proceedings and now this book covered the topics and questions raised in the Study Conference Discussion Document, although not all of them with the same intensity. The discussions held during the Joint Study Conference showed the need to analyse the following new issues of particular relevance that would serve to orient future curricular development in statistics and that have been included in Part II of this book: fundamental statistical ideas; the role of probability in the statistics curriculum; the use of technology in teaching and learning statistics; the differences and similarities of mathematical and statistical thinking; and the value of assessment in guiding the learning process. These issues are described below.

Fundamental ideas in the school statistics curriculum. Some common agreement about which basic ideas should be included at school level in number sense, measurement or geometry seems to exist with respect to international curricula, but there is no such agreement with respect to statistics, as curricula around the world show a notable variation. An important area of work was the identification of those statistical ideas that seem to be fundamental for understanding and being able to use statistics in the workplace, in personal lives and as citizens. Burrill and Biehler (this book) use different educational perspectives in statistics to propose a list of fundamental statistical ideas that should be taught to every student.

The role of probability in teaching and learning statistics. Although the focus of the study is statistics, since statistics and probability are linked in school mathematics in many countries and within mathematics theory and practice, a reference to probability in the book was needed, as didactic problems still need to be solved in the teaching of probability (Girard & Henry, 2005). Probability is a field that can connect to the study of mathematical modelling; but while probability theory often when taught in a finite context can be very simple, its abstract model part is not direct and could require a long period of learning (Chaput, Girard, & Henry, 2008). Finally the school curriculum seems to ignore the subjective point of view of probability, which is widely used today in the applications of statistics (Carranza & Kuzniak, 2008).

Technology. Technology has changed many aspects of modern life, and this change has been reflected in statistics education. With software such as Fathom™ and Tinkerplots™ designed to support learning statistics, data analysis is no longer the exclusive domain of statisticians; students and teachers today can work on their own statistical projects and be engaged in the game of statistics, experimenting with the complete cycle of statistical reasoning (Wild & Pfannkuch, 1999). In addition to exploring data, technology now is used to explore complex statistical ideas or processes via simulation. Computer software offers the opportunity for students to learn about modelling, enabling students to build their own models to describe data and to generate simulations that can be explored. According to Pratt, Davies, and Connor (this book), by taking advantage of this kind of software students can see real world phenomenon through a mathematical model (rather than seeing the model through the data).

Teaching through project work. Projects and investigations are ideal vehicles for student engagement, for learning to solve problems in context, and for synthesising components of learning (Makar, 2010; McGilliwray & Pereira-Mendoza, this book). The emphasis should be on students posing their own questions about the data, interrogating the data and learning new information about the real world from the data (Pfannkuch & Ben-Zvi, this book). The amount of data that can today be accessed on the Internet suggests that students can choose nearly any topic of interest to them for their work in the statistics classroom, which can increase student motivation. Working with real data also helps students investigate issues that do not often appear in textbook problems, for example, recognising different types of data, managing missing or incomplete data, defining variables and categories of classification, dealing with reliability and validity issues in measurement, designing questionnaires or experiments, screening data and dealing with outliers (Hall, this book).

Mathematical and statistical thinking. An ongoing discussion in the statistics education community is how to make teachers aware of *statistical thinking* as something different from *mathematical thinking*, both of them being essential to modern society and complementing each other in ways that strengthen the overall mathematics curriculum for students (Gattuso, 2006; Scheaffer, 2006). The differences between statistics and mathematics are reflected in the philosophical, ethical, procedural and even political questions that are still being debated within statistics and its applications, a debate that does not happen often in most areas of mathematics. Statistics is much more closely related than mathematics to other sciences (from linguistics or geography to physics, engineering, agriculture or economy) where it is used as the language and method of scientific enquiry and from which many statistical methods were developed (e.g., agriculture). In this sense it is also easier in statistics than in mathematics to establish connections with other school curricular areas. In spite of these differences, teachers often teach statistics in a similar manner to the way they teach mathematics, which is not well-suited to the unique nature of statistics (Makar & Confrey, 2003).

Assessment. Assessment of student learning is an important part in every educational process as it provides information about student achievement in relation to the intended learning outcomes. Consequently, assessment has received much attention in statistics education in recent years (see, for example, Gal & Garfield, 1997). Garfield and Franklin (this book) analysed three basic components, *cognition*, *observation* and *interpretation* that underlie all assessment and that must be explicitly connected in designing a coordinated whole relative to the purpose of assessment. In addition to the classical distinction between assessment *of* learning (summative), and assessment *for* learning (formative), the authors suggest that assessment *as* learning could combine both summative and formative methods and situate the student at the centre of the process, engaging students in new learning by monitoring and adapting their own understanding via the assessment process.

8 Final Thoughts

The success of the Joint ICMI/IASE Study indicated that the time was ripe for collaboration between mathematicians and statisticians to address challenges related to the advancement of both teaching and research in statistics education and in the preparation of teachers to teach statistics. However, continuous changes and the rapid development of statistics education as part of the mathematics curriculum at the school level and the subsequent need for a better preparation of teachers imply that this collaboration is not finished with the publication of this book but should continue in the coming years.

While the chapters in the book provide directions to improve the education of teachers, it is important to expand the empirical base of studies to larger samples and different contexts to assure their validity. Thus, the hope is that the analyses, research and case studies presented and discussed in the book will provide a rich starting point for new research related to improving the teaching of statistics at the school level and the preparation of teachers to deliver that teaching. The recommendations for further research included at the end of each chapter constitute a rich research agenda and show the existence of statistics education as a research field where international collaboration is not only possible but fruitful.

Many people, across many countries, have contributed to the ICMI/IASE Joint Study and to the production of his book. Each has shown a keen interest in improving the teaching of statistics in school mathematics in one way or another, but like all large-scale implementation of change there is always more that can be done. Now that you have read the book do not just put it down and forget. Focus on your area of interest, decide how you can contribute through teaching, research or teacher training and become an active component of the changing profile of the teaching of statistics in school mathematics and/or the training of teachers to teach statistics.

References

Arnold, P. (2008). Developing new statistical content knowledge with secondary school mathematics teachers. In C. Batanero, G. Burrill, C. Reading, & A. Rossman (2008).

Ball, D. L., Lubienski, S. T., & Mewborn, D. S. (2001). Research on teaching mathematics: The unsolved problem of teachers' mathematical knowledge. In V. Richardson (Ed.), *Handbook of research on teaching* (4th ed., pp. 433–456). Washington, DC: American Educational Research Association.

Batanero, C., Burrill, G., Reading, C., & Rossman, A. (2008). *Joint ICMI/IASE Study: Teaching Statistics in School Mathematics. Challenges for Teaching and Teacher Education. Proceedings of the ICMI Study 18 and 2008 IASE Round Table Conference.* Monterrey, Mexico: International Commission on Mathematical Instruction and International Association for Statistics Education. Online: www.stat.auckland.ac.nz/~iase/publications

Batanero, C., & Díaz, C. (2010). Training teachers to teach statistics: What can we learn from research? *Statistique et enseignement, 1*(1), 5–20. Online: http://statistique-et-enseignement.fr/ojs/

Batanero, C., Godino, J. D., & Roa, R. (2004). Training teachers to teach probability. *Journal of Statistics Education, 12.* Online: www.amstat.org/publications/jse/

Carranza, P., & Kuzniak, A. (2008). Duality of probability and statistics teaching in French education. In C. Batanero, G. Burrill, C. Reading, & A. Rossman (2008).

Chaput, B., Girard, J. C., & Henry, M. (2008). Modeling and simulations in statistics education. In C. Batanero, G. Burrill, C. Reading, & A. Rossman (2008).

Chick, H. L., & Pierce, R. U. (2008). Teaching statistics at the primary school level: Beliefs, affordances, and pedagogical content knowledge. In C. Batanero, G. Burrill, C. Reading, & A. Rossman (2008).

Estrada, A., & Batanero, C. (2008). Explaining teachers' attitudes towards statistics. In C. Batanero, G. Burrill, C. Reading, & A. Rossman (2008).

Franklin, C., Kader, G., Mewborn, D. S., Moreno, J., Peck, R., Perry, M., & Scheaffer, R. (2005). *A curriculum framework for K-12 statistics education. GAISE report.* American Statistical Association. Online: www.amstat.org/education/gaise/

Franklin, C., & Mewborn, D. (2006). The statistical education of PreK-12 teachers: A shared responsibility. In G. Burrill (Ed.), *NCTM 2006 Yearbook: Thinking and reasoning with data and chance* (pp. 335–344). Reston, VA: National Council of Teachers of Mathematics.

Gal, I. (2002). Adult's statistical literacy. Meanings, components, responsibilities. *International Statistical Review, 70*(1), 1–25.

Gal, I., & Garfield, J. (Eds.). (1997). *The assessment challenge in statistics education.* Amsterdam: IOS Press.

Gal, I., & Ginsburg, L. (1994). The role of beliefs and attitudes in learning statistics: Towards an assessment framework. *Journal of Statistics Education, 2*(2). Online: www.amstat.org/publications/jse/

Garfield, J., & Ben-Zvi, D. (2009). Helping students develop statistical reasoning: Implementing a statistical reasoning learning environment. *Teaching Statistics, 31*(3), 72–77.

Gattuso, L. (2006). Statistics and mathematics: Is it possible to create fruitful links? In A. Rossman & B. Chance (Eds.), *Proceedings of the Seventh International Conference on Teaching Statistics.* Salvador, Bahia, Brazil: International Statistical Institute and International Association for Statistical Education. Online: www.stat.auckland.ac.nz/~iase/publications

Girard, J. C., & Henry, M. (2005). Pourquoi est-il si difficile d'enseigner la statistique? (Why teaching statistics is so difficult?) In B. Chaput & M. Henry (Coord.), *Statistique au lycée. V. 1. Les outils de la statistique* (pp. 13–21). Paris: Commission Inter- Irem Statistique et Probabilité.

Godino, J. D., Batanero, C., Roa, R., & Wilhelmi, M. R. (2008). Assessing and developing pedagogical content and statistical knowledge of primary school teachers through project work. In C. Batanero, G. Burrill, C. Reading, & A. Rossman (2008).

González, M. T., & Pinto, J. (2008). Conceptions of four pre-service teachers on graphical representation. In C. Batanero, G. Burrill, C. Reading, & A. Rossman (2008).

Hawkins, A., Jolliffe, F., & Glickman, L. (1991). *Teaching statistical concepts.* London: Longman.

Holmes, P. (1980). *Teaching statistics 11–16.* Sloug: Foulsham Educational.

Lancaster, S. (2008). A study of preservice teachers' attitudes toward their role as students of statistics and implications for future professional development in statistics. In C. Batanero, G. Burrill, C. Reading, & A. Rossman (2008).

Lee, H. S., & Hollebrands, K. (2008). Preparing to teach data analysis and probability with technology. In C. Batanero, G. Burrill, C. Reading, & A. Rossman (2008).

Lee, H. S., Hollebrands, K. F., & Wilson, P. H. (2010). *Preparing to teach mathematics with technology: An integrated approach to data analysis and probability.* Dubuque, IA: Kendall Hunt.

Makar, K. (2010). Teaching primary teachers to teach statistical inquiry: The uniqueness of initial experiences. In C. Reading (Ed.), *Proceedings of the Eighth International Conference on Teaching Statistics.* Ljubljana, Slovenia: International Statistical Institute and International Association for Statistical Education. Online: www.stat.auckland.ac.nz/~iase/publications

Makar, K., & Confrey, J. (2003). Clumps, chunks, and spread out: Secondary preservice teachers' reasoning about variation. In C. Lee (Ed.), *Proceedings of the Third International Research Forum on Statistical Reasoning, Thinking and Literacy [CD-ROM].* Mount Carmel, MI: Eastern Michigan University.

Meletiou-Mavrotheris, M., Paparistodemou, E., Mavrotheris, E., Azcárate, P., Serradó, A., & Cardeñoso, J. M. (2008). Teachers' professional development in statistics: The EarlyStatistics European project. In C. Batanero, G. Burrill, C. Reading, & A. Rossman (2008).

National Council of Teachers of Mathematics (2000). *Principles and standards for school mathematics.* Reston, VA: Author. Online: http://standards.nctm.org/

Scheaffer, R. L. (2006). Statistics and mathematics: On making a happy marriage. In G. Burrill (Ed.), *NCTM 2006 Yearbook: Thinking and reasoning with data and chance* (pp. 309–321). Reston, VA: National Council of Teachers of Mathematics.

Shihong, X., Yi, L., Yongdong, L., Bangquan, T., & Reisheng, P. (2008). A teacher training case in junior high schools of Guangzhou. In C. Batanero, G. Burrill, C. Reading, & A. Rossman (2008).

Shulman, L. S. (1987). Knowledge and teaching: Foundations of the new reform. *Harvard Educational Review, 57*, 1–22.

Stohl, H. (2005). Probability in teacher education and development. In G. Jones (Ed.), *Exploring probability in schools: Challenges for teaching and learning* (pp. 345–366). New York: Springer.

Wild, C., & Pfannkuch, M. (1999). Statistical thinking in empirical enquiry. *International Statistical Review, 67*(3), 221–248.

Index